Machine Learning Applications in Subsurface Energy Resource Management

The utilization of machine learning (ML) techniques to understand hidden patterns and build data-driven predictive models from complex multivariate datasets is rapidly increasing in many applied science and engineering disciplines, including geo-energy. Motivated by these developments, *Machine Learning Applications in Subsurface Energy Resource Management* presents a current snapshot of the state of the art and future outlook for ML applications to manage subsurface energy resources (e.g., oil and gas, geologic carbon sequestration, and geothermal energy).

- Covers ML applications across multiple application domains (reservoir characterization, drilling, production, reservoir modeling, and predictive maintenance)
- Offers a variety of perspectives from authors representing operating companies, universities, and research organizations
- Provides an array of case studies illustrating the latest applications of several ML techniques
- Includes a literature review and future outlook for each application domain

This book is targeted at practicing petroleum engineers or geoscientists interested in developing a broad understanding of ML applications across several subsurface domains. It is also aimed as a supplementary reading for graduate-level courses and will also appeal to professionals and researchers working with hydrogeology and nuclear waste disposal.

Machine Learning Applications in Subsurface Energy Resource Management

State of the Art and Future Prognosis

Edited by
Srikanta Mishra

CRC Press
Taylor & Francis Group
Boca Raton London New York

CRC Press is an imprint of the
Taylor & Francis Group, an **informa** business

First edition published 2023
by CRC Press
6000 Broken Sound Parkway NW, Suite 300, Boca Raton, FL 33487-2742

and by CRC Press
4 Park Square, Milton Park, Abingdon, Oxon, OX14 4RN

CRC Press is an imprint of Taylor & Francis Group, LLC

ISBN: 978-1-032-07452-8 (hbk)
ISBN: 978-1-032-07455-9 (pbk)
ISBN: 978-1-003-20700-9 (ebk)

DOI: 10.1201/9781003207009

Typeset in Times
by KnowledgeWorks Global Ltd.

Dedicated to

the engineers, scientists, and technicians of

Battelle Memorial Institute –

90+ years of history and tradition in applying

science, technology, and analytics

for the benefit of mankind.

विद्या परमं बलम्

vidyA paramaM balam

Knowledge is supreme power

Contents

SECTION VI Predictive Maintenance Applications

SECTION VII Summary and Future Outlook

Preface

The utilization of machine learning techniques to understand hidden patterns and to build data-driven predictive models from complex multivariate datasets is rapidly increasing in many applied science and engineering disciplines – including geo-energy. Motivated by these developments, this book seeks to provide a compilation of papers that collectively present a current snapshot of the state of the art and future outlook for machine learning applications in subsurface energy resource management (e.g., oil and gas, geologic carbon sequestration, and geothermal energy). To that end, this book includes a broad spectrum of case studies using various supervised and unsupervised machine learning techniques. It is organized along different sections illustrating the analysis of structured (numeric) data for domains such as reservoir characterization, drilling operations, production data analysis, reservoir modeling, and preventive maintenance.

Several textbook type publications have recently appeared in the market that describe machine learning techniques individually with their corresponding petroleum engineering and geoscience applications. In contrast, this edited volume is a unique topical and timely collection that presents example applications of multiple machine learning techniques on a domain-by-domain basis (i.e., reservoir characterization, drilling, etc.). These chapters, written by leading industry professionals, researchers, and academics, not only describe cutting edge practical applications but also discuss relevant future directions and trends. Thus, this book provides a perfect complement to the existing suite of books currently available on the topic.

This book is targeted for the practicing petroleum engineer or geoscientist interested in developing a broad understanding of state-of-the-art machine learning applications across several subsurface domains (e.g., oil and gas development, geologic carbon sequestration, and geothermal energy). It can also be used as supplementary reading for graduate-level courses. As such, the book will appeal to professionals, researchers, academics, and students. It will also be useful for professionals dealing with hydrogeology and nuclear waste disposal.

Acknowledgments

I owe a tremendous debt of gratitude to the lead authors and co-authors of various chapters in this book. You responded to my early inquiries about the book project with enthusiasm; tried to meet my ever-expanding list of deadlines (for chapter outlines, drafts, response to review comments, chapter revisions, and miscellaneous forms) despite your busy schedules; prepared your chapters diligently and also acted as reviewers of other chapters; and above all, made this compiled volume the outstanding product that it is. Thank you.

This book would not have been possible without the perseverance, patience, and positivity demonstrated by Allison Shatkin, Senior Publisher, Engineering, and my contact at CRC Press/Taylor & Francis. You have been a constant source of encouragement throughout the publication process – directing me toward the light at the end of the tunnel, even during the trying times from the early days of the pandemic. It has been a pleasure working with you on this project. Thank you.

My journey in statistical modeling and machine learning has been enriched significantly through collaboration with my Battelle colleague Jared Schuetter and interactions with attendees from my lectures, workshops, and short courses on this topic across many different countries. The depth and diversity of your perspectives, questions, and comments has constantly kept me challenged, and gradually added clarity to my own view of how data-driven modeling for subsurface resource development projects should be carried out. Thank you.

Finally, a shout-out and hat-tip to my family – my wife Snigdha, daughter Ananya, and son Anshuman. You have once again indulged this passion of mine (for sharing my knowledge) by giving me the gift of time, when needed, to sneak away from family activities and take care of my editorial chores. Thank you.

Editor

Srikanta Mishra, PhD, is Senior Research Leader and Technical Director for Geo-energy Resource Modeling and Analytics at Battelle Memorial Institute, the world's largest independent contract R&D organization. He is nationally and internationally recognized for his expertise in developing and communicating physics-based and data-driven predictive models for subsurface resource management problems.

Dr. Mishra also serves as the Technical Lead for the SMART (Science Informed Machine Learning for Accelerating Real-Time Decisions for Subsurface Applications) initiative, organized by the U.S. Department of Energy and involving multiple national laboratories and universities. He received the 2022 Society of Petroleum Engineers (SPE) International Award for Data Science and Engineering Analytics, was named as an SPE Distinguished Member in 2021 and served as an SPE Distinguished Lecturer on Big Data Analytics during 2018–2019.

Dr. Mishra is the author of approximately 200 technical publications and the book *Applied Statistical Modeling and Data Analytics: A Practical Guide for the Petroleum Geosciences*, published by Elsevier. He has taught short courses on statistical modeling, data analytics, and uncertainty quantification at various professional conferences and client locations worldwide. He has also held the position of Adjunct Professor of Petroleum and Geosystems Engineering at the University of Texas at Austin, and Honorary Faculty in the Department of Chemical Engineering at Indian Institute of Technology Guwahati. Dr. Mishra earned his PhD at Stanford University, an MS at The University of Texas at Austin, and a BTech at the Indian Institute of Technology (Indian School of Mines) Dhanbad – all in petroleum engineering.

Contributors

Bulbul Ahmmed
Los Alamos National Laboratory
Los Alamos, New Mexico

Emre Artun
Istanbul Technical University
Istanbul, Turkey

Luis Ayala
Pennsylvania State University
State College, Pennsylvania

Raj Banerjee
Schlumberger
Houston, Texas

Patrick Bangert
Samsung SDS
San Jose, California
Algorithmica Technologies Inc.
Cupertino, California

Cesar Bravo
Frontender
Houston, Texas

Dingzhou Cao
Devon Energy
Oklahoma City, Oklahoma

Hongquan Chen
Texas A&M University
College Station, Texas

Akhil Datta-Gupta
Texas A&M University
College Station, Texas

Louis Durlofsky
Stanford University
Stanford, California

David Fulford
ARM Energy
Houston, Texas

Jonathan Garcez
Pennsylvania State University
State College, Pennsylvania

Jincong He
Chevron Technical Center
Houston, Texas

Brandon Hill
Battelle Memorial Institute
Columbus, Ohio

Seyyed Hosseini
The University of Texas at Austin
Austin, Texas

Vikram Jayaram
Pioneer Natural Resources
Dallas, Texas

Su Jiang
Stanford University
Stanford, California

Satish Karra
Los Alamos National Laboratory
Los Alamos, New Mexico

Dan Kifer
Pennsylvania State University
State College, Pennsylvania

Vikas Kumar
Pennsylvania State University
State College, Pennsylvania

Richard Larson
University of Texas at Austin
Austin, Texas

Srikanta Mishra
Battelle Memorial Institute
Columbus, Ohio

Maruti Mudunuru
Pacific Northwest National Laboratory
Richland, Washington

Yusuf Nasir
Chevron Technical Center
Houston, Texas

Dan O'Malley
Los Alamos National Laboratory
Los Alamos, New Mexico

Tsubasa Onishi
Texas A&M University
College Station, Texas

Carlos Palacios
Frontender
Houston, Texas

Sumedha Prathipati
Pennsylvania State University
State College, Pennsylvania

Michael Riedl
Battelle Memorial Institute
Columbus, Ohio

Sathish Sankaran
Xecta Digital Labs
Houston, Texas

Luigi Saputelli
Frontender
Abu Dhabi, United Arab Emirates

Pallav Sarma
Tachyus
Berkeley, California

Jared Schuetter
Battelle Memorial Institute
Columbus, Ohio

Parisa Shokouhi
Pennsylvania State University
State College, Pennsylvania

Yu Sun
Our Lady of the Lake University
San Antonio, Texas

Sanjay Tamrakar
Battelle Memorial Institute
Columbus, Ohio

Shusei Tanaka
Chevron Technical Center
Houston, Texas

Vladimir Vesselinov
Los Alamos National Laboratory
Los Alamos, New Mexico

Jingshuang Xue
Pioneer Natural Resources
Dallas, Texas

Hardik Zalavadia
Xecta Digital Labs
Houston, Texas

Tao Zhao
Schlumberger
Houston, Texas

Ruizhi Zhong
University of Queensland
Brisbane, Australia

Section I

Introduction

1 Machine Learning Applications in Subsurface Energy Resource Management

State of the Art

Srikanta Mishra
Battelle Memorial Institute,
Columbus, Ohio

CONTENTS

1.1 WHY THIS BOOK?

There is increasing buzz about how machine learning (ML) and artificial intelligence (AI) are "taking over the world" – given their growing influence in diverse domains such as consumer marketing, health care, finance, design and manufacturing, politics, sports, etc. Although lagging behind a bit, there has also been an explosion in studies employing ML for subsurface energy resource problems. For example, the number of papers on ML topics in the Society of Petroleum Engineers (SPE) OnePetro database has increased exponentially since 1990 from less than 100 to more than 5,000 last year (Mishra et al., 2021).

Data-driven ML-assisted models stand in sharp contrast to physics-based models, which have been traditionally used for managing subsurface domains such as conventional and unconventional oil and gas, geologic carbon sequestration, and geothermal energy. While ML models generally do not provide insight into the

DOI: 10.1201/9781003207009-2

causal mechanisms of the underlying processes, their use is easy to justify if the relevant physics-based model is computationally intensive and/or immature, or a suitable mechanistic modeling paradigm does not exist. The utility of such "black box" models is particularly attractive when (1) the cost of a wrong answer is low relative to the value of a correct answer, (2) they produce the best results, and (3) used as tools to inspire and guide human inquiry (Holm, 2019).

The subject of ML is not yet fully integrated into petroleum engineering and geoscience curricula. This has necessitated considerable self-learning and upskilling on the part of technical staff. Multiple textbook type publications that have recently appeared in the market (i.e., Holdaway, 2014; Mishra and Datta-Gupta, 2017; Mohaghegh, 2017; Misra et al., 2019; Bhattacharya, 2021) provide some help in this regard by describing ML techniques individually alongside their corresponding subsurface applications. Papers in the SPE, American Association of Petroleum Geologists (AAPG), and Society of Exploration Geophysicists (SEG) literature also serve as additional resources, albeit with very specific and single application examples.

The goal of this book is to present a broader platform showcasing applications of multiple ML techniques on a domain-by-domain basis (i.e., reservoir characterization, drilling operations, production data analysis, reservoir modeling, predictive maintenance). The advantage of such a structure is twofold. First, the collection of cutting-edge practical applications as well as a discussion of relevant future directions and trends allows readers to take a deeper dive into their specific area of interest. Second, the chapters taken together provide an opportunity for "***analog learning,***" e.g., a production engineer reading about an application in drilling and seeing the similarities vis-à-vis problem formulation, techniques used, etc. Thus, this compiled volume – with contributions from distinguished practitioners, professors, and researchers who are both domain experts and skilled in ML – can be viewed as a useful complement to the books and research papers currently available on the topic.

It is important to note that there are many practical issues that should be properly addressed for making data-driven models more robust (i.e., accurate, efficient, understandable, and useful) while promoting a foundational understanding of ML-related technologies among petroleum engineers and geoscientists (e.g., Sankaran et al., 2020; Mishra et al., 2021). Essentially, data-driven modeling problems should not be treated as simply curve-fitting exercises with flexible and powerful algorithms easily abused for overfitting; instead, the goal should be to extract insights based on data that can be translated into actionable information for making better decisions. As the eminent computational scientist Richard Hamming opined in 1962, "*the purpose of computing is insight, not numbers.*"

To that end, this book begins with Chapter 2 titled ***Solving Problems with Data Science (Schuetter)***. This chapter provides a practical guide to the application of data science beyond the mechanics (i.e., model fitting, programming), focusing on a systematic thought process for problem-solving that is needed to ensure the extraction of meaningful insights, avoidance of erroneous conclusions, and effective communication of results. Topics addressed include the following: (1) what is data science and what is it used for, (2) a framework for its application, and (3) data science tools (data generation, exploratory data analysis, predictive modeling, inference, reporting, and visualization) and how to use them. The chapter sets the stage for the detailed domain-specific applications discussed in subsequent chapters, as summarized below.

1.2 SUMMARY OF CHAPTERS

1.2.1 Reservoir Characterization Applications

Chapter 3, "Machine Learning-Aided Characterization Using Geophysical Data Modalities (Jayaram and Zhao)," reviews several ML algorithms and workflows applied to subsurface characterization using different geophysical measurements. Beginning with a general discussion of how physics-based and data-driven approaches can be integrated for geophysical data analysis, examples are presented to demonstrate (1) ML in seismic facies interpretation, (2) ML aiding petrophysical automation, and (3) geoscience and ML-aided wellbore diagnostics. Collectively, these provide a flavor to the multi-resolution scale at which ML is applied and adapted for subsurface characterization with various geophysical data modalities.

Chapter 4, "Machine Learning to Discover, Characterize, and Produce Geothermal Energy" (Vesselinov, Mudunuru et al.)," provides details on the state-of-the-art algorithms and the data requirements to build accurate and reliable ML models suitable for geothermal applications. Next, they present some case studies and preliminary success stories of applying ML for the geothermal project exploration, development, and operations. Building on the insights from these case studies, they also identify challenge problems and high-priority research areas where ML can unlock the potential to accelerate geothermal resource discovery and energy production.

1.2.2 Drilling Operations Applications

In Chapter 5, "Real-Time Drilling and Completion Analytics: From Cloud Computing to Edge Computing and Their Machine Learning Applications (Cao, et al.)," the authors share their experience on the principles and lessons of building real-time systems for drilling and completion operations and related real-time analytics with an emphasis on shale operations. The first part is devoted to the history of real-time drilling systems, selection of stream processing technology, and transition from cloud to edge computing. The second part focuses on real-time ML applications for drilling and completion with discussion of corresponding use cases.

In Chapter 6, "Using Machine Learning to Improve Drilling of Unconventional Resources (Zhong)," the focus is on the application of ML for shale oil and gas or coal seam gas-related drilling. The chapter begins with a review of ML studies in areas such as drilling hydraulics, drilling dynamics, drilling problems, drilling target identifications, and miscellaneous drilling applications, which tend to create a huge volume of high-frequency drilling data. This is followed by two case studies illustrating ML applications: (1) a classification task to identify pay zones in coal seam gas reservoirs and (2) a regression task to estimate the height of cuttings bed.

1.2.3 Production Data Analysis Applications

Chapter 7, "Machine Learning-Assisted Production Data Filtering and Decline Curve Analysis in Unconventional Plays (Fulford)," provides an overview of best practices and current state-of-the-art ML-assisted methods for empirical forecasting with production data. The author first discusses the important step of data filtering using heuristic, unsupervised, and supervised ML methods prior to carrying out the analysis of choice. These feed into linear, nonlinear, and Bayesian regression analyses – with the chapter presenting basic principles (e.g., linearization from empirical basis models, maximum likelihood nonlinear regression with regularization, Markov chain Monte Carlo for probabilistic regression), limitations of various approaches, and case studies.

In Chapter 8, "Hybrid Data-Driven and Physics-Informed Reservoir Modeling for Unconventional Reservoirs (Sankaran and Zalavadia)," the authors propose a hybrid data-driven and physics-based workflow to characterize production performance through computation of the transient productivity index by combining diffusive time of flight and material balance concepts. Here, unsupervised learning methods can be used to identify distinct clusters within a multi-well dataset, and supervised learning methods (including time-series-based forecasting models) can be used to extrapolate productivity index, water cut, and gas-oil ratio trends. In this respect, the current chapter is similar to the previous one in that both emphasize physics-based data-driven approaches rather than purely ML-based strategies.

Chapter 9, "Role of Analytics in Extracting Data-Driven Models from Reservoir Surveillance (Banerjee)," addresses the building of analytics-based models from surveillance data and their use in workflows for optimizing oil and gas operations. First, novel enabling technologies such as the Internet of Things, big data and cloud computing, and ML algorithms are summarized. Next, their applications are discussed for representative workflows such as (1) artificial lift surveillance, (2) automated interpretation of pressure and rate data, (3) workover candidate recognition, (4) waterflood surveillance, and (5) reservoir simulation in surveillance (closed-loop reservoir management).

Chapter 10, "Machine Learning Assisted Forecasting of Reservoir Performance (Artun)," focuses on ML-based performance forecasting models that are purely based on real (field) observations. The workflow is illustrated through the following three practical case studies: (1) forecasting water cut as high (>50%) or low (<50%) in a waterflooded fractured carbonate reservoir under given static and dynamic conditions, (b) forecasting incremental oil recovery and economic value of single-well cyclic nitrogen injection where the results of detailed simulations are captured in terms of ML-based proxy models based on design parameters, and (c) forecasting cumulative gas recovery from shale wells with 1-year production data and a combination of rock and fluid properties and completion variables.

1.2.4 Reservoir Modeling Applications

Chapter 11, "An Efficient Deep Learning Based Workflow Incorporating a Reduced Physics Model for Drainage Volume Visualization in Unconventional Reservoirs (Onishi, et al.)," proposes a workflow with several novel features: (1) using the fast-marching method computed diffusive time of flight as a proxy for three-dimensional (3D) pressure and saturation maps and (2) a deep learning autoencoder-regression approach to capture the results of full-physics simulations with a lower-dimensional representation. This allows the rapid inversion of field-based pressure and flow rate observations into corresponding subsurface images without going through traditional simulation-based history matching.

Chapter 12, "Reservoir Modeling Using Fast Predictive Machine Learning Algorithms for Geological Carbon Storage (Hosseini, et al.)," provides a comparative assessment of different implementations of deep learning networks for modeling CO_2 storage for a specific Gulf Coast geological case. Numerical simulations are run under various conditions based on the same static geological reservoir model to generate datasets. Each ML algorithm is then trained from this generated data. The inputs are coordinate data, permeability, porosity, injection rate, pressure, and time. Model performance is then tested through the accuracy of target outputs such as phase saturation, pressure distribution, and production of brine with held-out testing data.

In Chapter 13, "Physics-Embedded Machine Learning for Modeling and Optimization of Mature Fields (Sarma)," the author describes a modeling approach termed "Data Physics," which is the amalgamation of state-of-the-art ML and the underlying physics present in reservoir simulators. The approach is similar to physics-informed neural networks, with the distinction that the physical equations are treated as constraints rather than adding them to the objective function. The data physics construct is described in the context of (1) traditional data assimilation (history matching) to develop a robust predictive model and (2) multi-objective optimization to predict and optimize future performance given different operational plans. An application is presented for a complex waterflood field in Argentina, where the injected water is redistributed to optimize multiple objectives.

Chapter 14, "Deep Neural Network Surrogate Flow Models for History Matching and Uncertainty Quantification (Jiang and Durlofsky)," focuses on deep neural network (DNN) surrogate models that provide both spatial distribution of temporal evolution of system states in oil and gas reservoirs. They discuss the key features of physics informed/theory-guided neural networks as well as data-driven surrogate models (which require considerably more training data). A DNN surrogate model application for a 3D channelized system to efficiently capture the response of high-fidelity simulations is presented with saturation maps and injection/production rate time histories used as key performance metrics. The surrogate model is then used for history matching.

In Chapter 15, "Generalizable Field Development Optimization Using Deep Reinforcement Learning with Field Examples (He, et al.)," generalizable field development optimization is investigated, where an ML-trained model based on a deep reinforcement learning framework is used to provide efficient solutions for different scenarios. The authors describe an improved neural network architecture that allows for a larger template model and accelerates convergence, improved location masking that allows for the incorporation of engineering knowledge, and full integration with a commercial simulator that allows for consideration of practical constraints. such as facility capacity. An example application is presented for general deepwater Gulf of Mexico reservoirs under primary depletion.

1.2.5 Predictive Maintenance Applications

Chapter 16, "Case Studies Involving Machine Learning for Predictive Maintenance in Oil and Gas Production Operations (Saputelli et al.)," provides an overview of data analytics and ML-based predictive maintenance use cases in oil and gas operations including sound data management practice. Other major topics covered include: (1) suggested techniques for identifying optimum maintenance frequency analysis, (2) detecting equipment anomalies and predicting imminent and unplanned failures as well as remaining useful life, (3) role of digital twins in predictive maintenance, and (4) predictive maintenance for well integrity management and asset life extension. It also outlines the challenges in applying such techniques in real-life applications.

In Chapter 17, "Machine Learning for Multiphase Flow Metering (Bangert)," the goal is to demonstrate how ML can realistically replace physical multiphase flow meters in industrial applications saving significant costs and efforts. Topics discussed include the virtual multiphase flow metering problem, setting up the data analysis pipeline via soft sensing, feature engineering, virtual metering, establishing the ML workflow, and judging the accuracy of the model and its usefulness to solve the original problem. A sample model is described with accuracy comparable to a physical multiphase flow meter.

1.3 ORGANIZATION OF CONTENT

The chapters are generally structured in a consistent manner (although some chapters may have a more integrated or a more cross-distributed flavor) as follows: (1) an "Introduction" section to provide a context for the subsequent sections, (2) a "Literature Review" section that summarizes pertinent prior work in the appropriate topical area, (3) a "Methodology" section that provides a brief overview of the key ML techniques utilized in the case studies presented (without going into too much detail), (4) "Case Studies" and "Lessons Learned" sections that demonstrate the application of ML techniques for the topical area of interest, and (5) a "Future Outlook" section that discusses current challenges in ML applications for that topical area and how to (potentially) address them.

The book concludes with the chapter "Machine Learning Applications in Subsurface Energy Resource Management: Future Prognosis." In this chapter, a synthesis of the "Future Outlook" section from each of the chapters is undertaken to arrive at a cross-cutting set of issues that could help advance the next generation of ML-based models for subsurface applications.

REFERENCES

Bhattacharya, S. (2021). *A Primer on Machine Learning in Subsurface Geosciences*, Springer Briefs in Petroleum Geosciences & Engineering, Cham, Switzerland.

Holdaway, K. (2014). *Harness Oil and Gas Big Data with Analytics: Optimize Exploration and Production with Data-Driven Models*, John Wiley & Sons, New York, NY.

Holm, E. (2019). "In defense of the black box," *Science*, vol. 364, pp. 26–27.

Mishra, S., and A. Datta-Gupta (2017). *Applied Statistical Modeling and Data Analytics for the Petroleum Geosciences*, Elsevier, New York, NY.

Misra, S., H. Li, and J. He (2019). *Machine Learning in Subsurface Characterization*, Gulf Professional Publishing, Houston, TX.

Mishra, S., J. Schuetter, A. Datta-Gupta, and G. Bromhal (2021). "Robust data-driven machine learning models for subsurface applications: Are we there yet?" *Journal of Petroleum Technology*. Vol. vol. 73, pp. 15–21.

Mohaghegh, S. (2017). *Data-Driven Reservoir Modeling*, Society of Petroleum Engineers, Richardson, TX.

Sankaran, S., S. Matringe, M. Sidahmed, L. Saputelli, X.-H. Wen, A. Popa, and S. Dursun (2020). *Data Analytics in Reservoir Engineering eBook*, Society of Petroleum Engineers, Richardson, TX.

2 Solving Problems with Data Science

Jared Schuetter
Battelle Memorial Institute,
Columbus, Ohio

CONTENTS

2.1 INTRODUCTION

"Data Science" is the study of methodologies for drawing insights in the face of uncertainty using evidence-based reasoning. Typically, these insights are used in decision-making processes to choose actions that will solve a problem. In my opinion, these components are necessary and sufficient to fall within the data science umbrella. Without relying on evidence (i.e., data), one is simply guessing about what actions are best to take. Without uncertainty, the best action is known and there is no reason to gather evidence. Without the goal of drawing insights, one is not doing science.

The term "data science" has been around since the 1970s and 1980s, but the first coalescence around a definition similar to its use today occurred toward the

DOI: 10.1201/9781003207009-3

late 1990s (Wu, 1997). Data science is generally accepted to contain and/or overlap with many other fields of study that are based on the scientific method and evidence-based reasoning, including statistics, information theory, mathematics and probability, computer science, machine learning, and deep learning.

The goal of this chapter is to provide a practical guide for those on the front lines "doing data science" in applied work, whatever that may be. Many resources are available for the nuts and bolts of model fitting, programming, and theory, but there is another layer around these techniques that is often not discussed in much detail, even though it is critically important. Simply put, this layer is the thought process that ensures that the data science effort will produce meaningful insights, avoid erroneous conclusions, and effectively communicate findings to stakeholders, and lead to decisions that have a real impact in the world. Without tying the techniques to the bigger picture of the problem that is being solved, you end up with a research project that doesn't change anything at all.

The remainder of the chapter will walk through this thought process layer in decreasing levels of abstraction. Section 2.2 contains a philosophical discussion about solving problems and how data science is one possible set of techniques that can be used for this purpose. It also includes an overview of a process for systematically thinking through a problem and setting yourself up for success in both your analysis and the messaging back to decision-makers. Section 2.3 can be used for reference to identify search phrases, concepts, and tools for solving different types of problems. Because the field of data science is changing rapidly, much of the detail is left out of this section in favor of general categories of techniques and tools that are likely to endure over time. Finally, Section 2.4 contains a brief summary to reiterate the points made in this introduction.

As you read through this chapter, remember that fitting models, running statistical tests, and generating plots are a small part of what data science is about. It is also about being systematic in your thinking, drawing conclusions based on evidence, and sharing what you've learned with others.

2.2 WHAT IS DATA SCIENCE AND WHAT IS IT USED FOR?

2.2.1 On Solving Problems

A search for the term "data science" across research papers, blogs, press releases, and videos will uncover a multitude of topics ranging from deep technical discussion of methodologies to whimsical use of artificial intelligence (AI) models for the latest smartphone app. Across all the complexity and breadth of the field, it is often easy to lose sight of the fact that data science, for all its terminology, methods, and theory, is only useful because it provides tools for *solving problems*. To frame the discussion in this chapter, let us start with a few definitions regarding the type of problem data science is meant to address. While the following descriptions may not apply to all problems, they do map well to "real-world" applied problems (see also Figure 2.1).

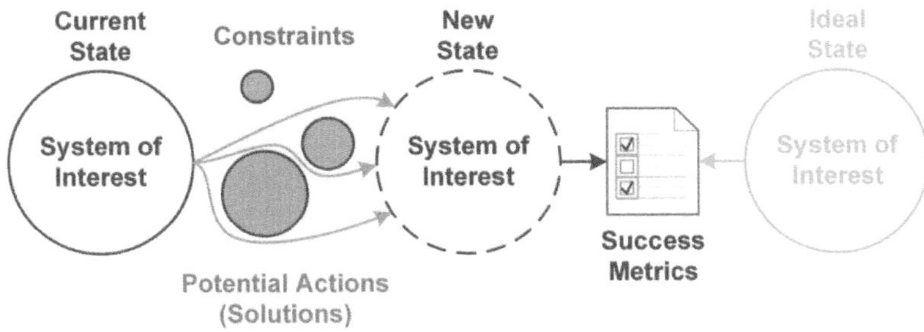

FIGURE 2.1 A general description of problem-solving.

- A **system** is a set of associated concepts tied together within a single frame of reference. It could be something physical (e.g., a drilling rig or a reservoir), a process (e.g., a drilling permit application process), or a theoretical construct (e.g., a network of customers).
- A **problem** is when a system of interest is in an undesirable state and there are constraints preventing the system from being in the desired ideal state.
- A **solution** is a set of actions taken within those constraints to move the system from the current state toward the ideal state.
- A set of **success metrics** can be used to evaluate how well the solution worked by comparing the new state with the ideal state, because usually the ideal state is not reached. In some cases (e.g., when optimizing a process or outcome of interest), that ideal state may not be explicitly known, and success metrics may instead be measured by comparing the new state to the *original state.*

Take, for example, carbon capture and storage problem. Here, the system of interest may be a saline reservoir that is currently full of brine. The ideal state is that the reservoir is filled with as much CO_2 as possible without leaking out over a 30-year post-injection period. The constraints are many (e.g., the physics of fluid flow, the structure of the subsurface, cost, time, personnel), as are the possible actions, each of which will result in some percentage of CO_2 ending up in the reservoir. The success metrics in this case may include the amount of CO_2 stored, the stability of that storage over 30 years, the total time and budget spent to inject it, and public awareness and acceptance of the project.

The difficulty in solving a problem like this is that typically the goal is to choose the best course of action (i.e., maximize the success metrics across the space of possible actions), but there are knowledge holes that make the best action unclear. There may be unknown constraints limiting possible solutions, our success metrics may be nebulous, or there is uncertainty about what the new state will be after following a series of actions. Simply put, it is hard to choose the best solution when **the system is not fully understood**. To describe this through a clichéd analogy, it is like the system is a giant puzzle with many missing pieces. To make the best decision, we need to see the full picture, but the holes in the puzzle (i.e., missing knowledge) prevent us from doing this.

For most problems, especially involving the subsurface, the system won't ever be fully understood, so to solve them we must fill the knowledge holes with approximate truths and then proceed with selecting the best course of action within our limited understanding. The success with which we can fill those holes has a direct impact on the quality of the solution, and the materials we use to fill the holes include data and inferences made using those data.

2.2.2 Using Data to Fill Knowledge Holes

Consider the carbon storage problem in a greenfield application where no wells have been drilled at a site. There is a considerable hole in our knowledge about the true structure of the reservoir in the subsurface because it is hidden from view. Someone with zero expertise in the carbon capture and storage field can, at the very least, draw from past experience (e.g., secondary school education) to surmise that there will be rocks of varying types beneath the surface organized into strata. One may claim that this deduction is made without data, but this is not true; rather, it is based on observations (i.e., data) collected during long-ago lectures in a classroom, images and descriptions seen in a nature documentary, stories told by friends, or direct observation of a cliff face while driving on a road trip. All of these observations led to an understanding of what constitutes the subsurface.

Someone with more expertise in this domain could do a better job of predicting what the reservoir looks like based on a conceptual understanding of geology and the history of the local region, again informally based on past observational data. To get a more accurate estimate of the subsurface, one would need to collect more traditional data, like core samples, seismic sensor outputs, wireline logs from pilot wells, and the like, then follow various methodologies to draw inferences from those data.

When considering problems in general, a knowledge hole is a barrier to proposing a solution. As shown in the previous example, one must use data to fill this hole, and the quality and quantity of such data falls along a continuum. When very little data are available, all we can do is fill the hole in our puzzle with a fuzzy and vague impression; however, as the data become more abundant and relevant, that hole can be filled with a sharper and more detailed image. **Data science** is the study of methodologies, processes, and procedures that can be used to guide scientists in collecting the best data and drawing the right inferences to mitigate the impact of knowledge holes as much as possible.

2.2.3 How Data Science Principles Provide Better Solutions

Humans have made decisions for tens of thousands of years, and most of those are made using observation, trial, and error. This works well for most of the decisions we make, like finding something to eat and avoiding danger, but when a situation is complex, our knowledge is not complete, or the stakes are high, it is a good idea to be more systematic. Several key barriers to making good decisions are **bias**, **uncertainty**, and **complexity** in the decision-making process. Reducing these will help steer toward faster and more impactful decisions.

One way to make decisions without data science is to rely solely on subject matter expertise, where teams of experts use knowledge gained from past successes and failures to recommend a course of action. This is common in situations where the collection of observational data is difficult, perhaps because there is not a lot of time (e.g., in emergencies or disaster response), capability (e.g., when collecting data is expensive), or opportunity (e.g., for rare events). It can also happen due to inertia, where an organization starts making decisions through experts and, even though new technologies make it possible to collect and analyze data to make better decisions, they do not invest in them.

When decisions are made with experts, it can often end up taking a lot of time and effort; there are a lot of meetings involved and manual review of material. If there is a lot of information to sift through, experts may miss something or latch onto a particular piece of evidence and end up recommending suboptimal actions. Regardless, these recommendations will be biased due to subjectivity, so to reduce that bias it may be necessary to get multiple opinions, arbitrate them, and decide what to do. It will also be difficult to update decisions in light of new information, as all of the experts will need to be reconvened to hash everything out again. This is a process that will yield **high bias** and **high uncertainty** in the sense that the recommended course of action may be completely different depending on what team is assembled to solve the problem.

To reduce bias and uncertainty, data science follows the same kind of approach that has been used in any other scientific field. It creates a structure and a formality to solving problems so that anyone who follows it can reproduce the same results. This structure is designed in a way to avoid common mistakes that could lead to bias and uses methodologies that build confidence in any inferences that are made, thereby reducing uncertainty. Some key tools in the data science toolbox are listed in the following:

- **Mathematical abstraction** is what makes data science so versatile. By abstracting problem-specific details to a common mathematical language, it is possible to share methodologies across disciplines and draw from the insights of others. Much of the work in solving problems is figuring out the best way to describe them mathematically.
- **Standardization** and **automation** are two principles that work hand in hand in data science to improve reproducibility, make it easier to collaborate with colleagues, and streamline implementation in a production environment. By using standard terminology, algorithms, software packages, and visualization tools, it is now faster and easier to assemble a data science pipeline than it has been in the past. Standardization also applies to data – primarily in the form of "data cleaning" and other quality control measures to ensure that artifacts in the data do not have unintended impacts on any downstream decision-making.

- **Statistics** and **Machine Learning** are large subfields of data science that use mathematical constructs to capture the concepts of bias and uncertainty, models to map real-world systems to those constructs, and methodologies for drawing inferences about those systems. The big difference between the two fields is that in statistics, the mathematical description of the real-world system is meant to be a valid representation of how the observed data were generated; in contrast, machine learning uses mathematical descriptions only to make accurate predictions, and there is no expectation that they reflect the true data generation mechanisms (see, e.g., Breiman's discussion in (Breiman, 2001a)). **Deep learning** is a subset of machine learning that focuses on more complex but flexible neural network models.
- **Exploratory data analysis** is a collection of techniques that can be used to understand how different concepts or variables in a system are related. Simple plots and tabular summaries can reveal the relationships between these variables to help better plan further stages of analysis, and approaches like **cluster analysis** or **dimensionality reduction** can be used to identify subgroups in the dataset, which could indicate a potential bias that was not accounted for during data collection.
- **Controlled experimentation** is a statistical technique to efficiently collect data using concepts of blocking, replication, and randomization to reduce uncertainty and improve any inferences made.
- **Estimators** are approximations calculated from collected data that are intended to measure an unknown quantity of interest (e.g., a sample mean and standard deviation). Commonly used statistical estimators have been thoroughly studied to understand how robustness and accuracy change with respect to sampling variability and different sample sizes. Machine learning tends to use statistical estimators as well but places more stock in estimators that have been shown to work well in applied work across multiple domains.
- **Predictive modeling** formalizes the relationship between a group of controllable quantities of interest (known as *input parameters, features, predictors,* or *independent variables*) and a group of less controllable quantities (known as *output parameters, responses,* or *dependent variables*). Given a set of values for the former, one can predict the values of the latter. This is where a lot of method development in data science occurs, and it is with tools like these that we fill knowledge holes.
- **Performance metrics** are used to evaluate how well a knowledge hole has been filled, or as a success metric to evaluate an outcome. They all capture some judgment of quality (e.g., time, cost, accuracy, etc.) using one or more qualitative or quantitative values. Standard sets of performance metrics tend to be used in the data science community for certain tasks, although the preferred metrics may vary depending on the application area or discipline. There are also methods of computing performance metrics that attempt to reduce bias, like **train/test/validation partitioning** or **cross-validation**.

While not part of data science per se, there are also some general science activities that can help to fill knowledge holes more accurately to improve the quality of solutions.

- **Literature review** is one of the most important activities to undertake when solving a problem. This can help to better understand the problem, identify new data sources, find new constraints or actions to explore, and make sure someone else hasn't already done a lot of the work for you. This is also a great way to find **third-party software** that allows you to skip over some implementation yourself.[1]
- **Requirements gathering**, **specification of assumptions**, and **outlining the approach** (i.e., planning what you will do ahead of time) are all good preparatory activities to make sure that nothing obvious is being missed when trying to solve a problem. This also makes it easier to **generate and test hypotheses**.

Table 2.1 describes sources of bias and uncertainty in terms of the general problem-solving framework described in Section 2.2.1 and maps them to the data science concepts described previously that address them. In general, there will tend to be two stages of analysis needed to solve a problem. The first is an exploratory stage that involves a lot of research and planning to prioritize how to spend

TABLE 2.1
Common Gaps in Knowledge, and How Data Science Bridges Them

	Knowledge Holes	Data Science Concepts to Fill those Holes	Stage of Analysis
Sources of Bias	Incomplete understanding of the system of interest	Literature review, mathematical abstraction, specification of assumptions, outlining the approach, exploratory data analysis, cluster analysis	Understand and Plan
	Incomplete list of possible actions	Literature review, automation, third-party software, exploratory data analysis	
	Poor success metrics	Performance metrics, requirements gathering, estimators, exploratory data analysis	
Sources of Uncertainty	Incomplete list of constraints	Literature review, controlled experimentation, exploratory data analysis	Learn and Evaluate
	Incomplete understanding of how actions affect the system of interest	Statistics, machine learning, deep learning, predictive modeling, train/test/validation partitioning, cross-validation, standardization, exploratory data analysis	

resources and identify candidate courses of action. The second is a learning stage where those candidate approaches are modeled and/or tested, then evaluated. The best evaluated course of action is the one most likely to solve the problem. These two stages will be described in more detail in Section 2.2.4 as part of a larger framework.

2.2.4 A Framework for Applying Data Science to Real Problems

The takeaway from the earlier philosophical discussion is that when solving a problem, data science provides you a toolbox for overcoming knowledge holes in a way that will reduce bias and uncertainty, helping to ensure that the conclusions you reach are more likely to be successful and the ultimate solution you implement will work. It can also help to have an overall decision framework for how to use those tools in a practical setting. There are many such problem-solving frameworks described in publications, blog posts, books, and presentations. Many of these are geared toward business decision-making or are broadly defined to cover an array of possible purposes. Others are more specific for certain applications, like software development or data analysis. Data science problem-solving workflows often suffer from tunnel vision on the exploration and model-building aspects but ignore much of the work taking place before and after those steps. The "UPLE" approach (Understand, Plan, Learn, Evaluate) described in this section is intended to provide a more holistic view of problem-solving that still targets data science as the method of solving the problems, but also considers the planning steps before and the evaluation steps after its application. The approach is outlined in Table 2.2, where its steps are also mapped to those from a few other popular problem-solving frameworks, i.e., define-measure-analyze-design-verify (DMADV), or guidelines for assessment and instruction in statistics education (GAISE). Each of the steps is elaborated on in the subsections below.

2.2.4.1 Understand

This step captures the common adage "look before you leap." For most data scientists, the really fun part of a project is digging into the data, making some plots, building models, and discovering something new. Unfortunately, that means there is a strong sirens' call that urges us to jump right into the deep end and start throwing data into semi-automated pipelines that generate pages of plots and tables to look at. **It is critical to avoid doing this.** Try to understand the system of interest before doing anything else. Come up with requirements and success metrics to help choose the proper course of action. Look through the literature to see if anyone else has tried to solve a similar problem, and jot down names of any methods they may have used. Search online for relevant third-party tools that others have already developed and tested. Most importantly, **talk to subject matter experts** to get their perspectives on the problem and the system of interest. Only after understanding the moving parts around the problem should you begin working on the solution.

TABLE 2.2
The UPLE Approach and How It Compares with Other Common Problem-Solving Frameworks

The UPLE Approach		Common Problem-Solving Frameworks		
Step	**Sub-Steps**	**Six Sigma DMADV (Selvi and Majumdar, 2014)**	**Waterfall Model (Ruparelia, 2010)**	**GAISE Framework (Franklin et al., 2007)**
Understand	Listen and ask questions	Define	Evaluation	Formulate Questions
	List requirements		Requirements	
	Review what has been done before	Measure	Analysis	
	Explore on your own		Design	
Plan	Identify possible actions	Analyze		
	Collect and explore new data, if needed			Collect Data
	Choose which actions to evaluate			
Learn	Build models for those actions	Design	Development	Analyze Data
	Infer how the actions impact the system			
	Select recommended actions			Interpret Results
Evaluate	Communicate back to stakeholders		Validation	
	Decide on the best approach			
	See if it works in practice	Verify	Deployment	
	Return to an earlier step if needed			

2.2.4.2 Plan

Like the "understanding" step, planning is also often skipped by data scientists who are eager to get to the modeling and plotting. If you jump straight into data analysis without planning the work out, you will inevitably run into a situation where you discover halfway through a project that what you are doing is incompatible with the larger picture and you must redo all of your analysis again from scratch. Or worse yet, sometimes without thinking through the entire analysis ahead of time, you may find partway through that you are missing data that cannot be collected, so the course of action cannot even be evaluated at all.

To avoid this, consider some high-level questions like the ones in the following list. Sketch out your approach ahead of time and break it into tasks. Identify the expected outcomes of each task, list the assumptions you are making, and ballpark the level of effort of those tasks.

- What are the high-level constraints on time, budget, and personnel?
- Are existing data available, and how easily can you collect any new data that will be needed?
- Do you have access to necessary computing resources or hardware?
- Did past projects already explore these options? If so, what was discovered?
- Do you have the technical capabilities to implement what is needed to model the outcomes of interest?
- How can you ensure that something useful results from this study given the previous constraints?

2.2.4.3 Learn

The "Learn" step of UPLE is where most data science workflows focus, and as a result there is abundant literature to be found regarding best practices. The "learning" in this step is referring to the knowledge hole you are addressing in a given task. Regardless of the approach you take, there will be some level of exploration, discovery, and error checking involved; that is, you are learning about the relationships you are modeling, the quality of the models you build or use, and the overall problem you are trying to solve. In some sense, executing this step is almost a mini-UPLE cycle in and of itself.

2.2.4.4 Evaluate

The role of the data scientist in this final step of the UPLE process is primarily to communicate findings to decision-makers and other stakeholders. On the basis of the data scientist's recommendations, the larger team can put a particular course of action in place and see if it solves the problem. If the course of action does not sufficiently solve the problem, a postmortem will be needed to identify remaining knowledge holes and return to an earlier stage of the framework to build a better understanding or consider other courses of action. Then the UPLE cycle can begin anew.

2.3 DATA SCIENCE TOOLS AND HOW TO USE THEM

Section 2.2.4 suggested a systematic process for solving a problem from start to finish, including framing the research questions, breaking the problem into tasks, tracking assumptions and expectations, producing and interpreting results, and communicating the findings to stakeholders. It is, however, intentionally scant on details about what kinds of methodologies might be used to model relationships and draw inferences. This is done for two reasons. First, there is abundant literature available that describes the newest methods, suggests what methods are best for which situations, and provides examples of how to implement them. Those sources will contain far more necessary detail than can be shared here. Second, the newest and best models and tools are always changing. Rather than providing a list of examples that will become quickly outdated, the goal of this section is to share an overview of what kinds of tools are out there and how they can be found. Often as a data scientist, it is beneficial simply to know that certain techniques exist, because you can always read up on the theory and implementation details once you know what to search for. The intention is for this section to provide some search names and phrases that may help.

Depending on your research goals, each task you identify is likely to fall under one of the categories of analysis shown in Table 2.3. Although this list is not exhaustive, it should cover most cases. Following the table are subsections devoted to each of the categories.

TABLE 2.3
Categories of Analysis, Depending on the Research Goal

Category	Description	Example Research Goals
Data Generation	A set of techniques for collecting new data for analysis	• "I want to build a dataset that I can use to analyze something of interest."
Exploratory Data Analysis	Methods for exploring relationships between variables in a dataset, prior to modeling	• "I want to describe what is happening in my dataset so I can formulate more specific research questions."
Predictive Modeling	Models for describing relationships between controllable variables (inputs) and the outcomes that are dependent on them, as well as methods for evaluating those models	• "I want to predict one or more outcomes of interest given some input data." • "I want to evaluate a model I built." • "I want to optimize some aspect of my system of interest."
Inference	Approaches for looking at a trained model and drawing insights about variables and their relationships based on the model structure or predictions	• "I want to use my model to understand the relationships between components in my system of interest." • "I want to quantify the impact or importance of those components and their relationships." • "I have too many components and not enough data, so I need to narrow down to only the important ones and focus on those."
Reporting and Visualization	Sets of tools for communicating findings with others and standardizing workflows for easier collaboration	• "I need to share my findings with others." • "I want to be able to update my results periodically on new data." • "I want my implementation to be transferrable to others."

2.3.1 Data Generation

Any solid data science effort will need to have a foundation of good data, otherwise you are in a "garbage in, garbage out" situation. However, the right type of data may vary depending on the goals of the research, including what kinds of models may be under consideration and what insights are expected to be gained. If insight into the data generation process is going to be important, you will likely end up pursuing statistical methodologies, so you should collect data that fall neatly into bins or categories of components of interest. If the components in your system can be controlled to a high degree, and the relationships between components are not likely to be more complex than a linear or quadratic relationship, **classical experimental design** approaches will be appropriate, since they will create observations at different combinations of "factor levels" (i.e., component states) and will be amenable to models that allow for statistical tests about those components. If the components in the system cannot be controlled that easily, your goal will be to ensure that your dataset reflects the full range of variability that is possible in your data-generation process. Use **blocking**, **randomization**, and **replication** to avoid headaches later when you fit models and make inferences. Any data collected from people will likely need to draw from **survey sampling** approaches.

For more complex systems, or cases where learning about the data generation process is not a concern, **space-filling designs** will be better because they will capture more nonlinear behavior in the dataset. Scattering observations across many different values in all of the input variables will also allow for better visualization. Of course, when the system is complex, more data will be needed to understand the nonlinear behavior. Note that there are **rejection sampling** techniques that can be used to avoid regions that you know are infeasible (i.e., combinations of inputs that will not work together).

In all cases, you should strive to **avoid any missing data**. If data do end up missing, try to ensure that they are not missing for a systematic reason. If, unfortunately, they are missing systematically, try to figure out the cause for the missingness at the time of data collection, as it will help with interpretation down the road.

Finally, don't be afraid to start with a small pilot study if you are early in the problem-solving process. When you invest a lot of time and effort to collect a large dataset, make sure you have a specific goal in mind and confidence in your approach before doing it.

2.3.2 Exploratory Data Analysis

If you want to explore what is happening in a dataset, you will want to select from techniques in a few different areas (see (Tukey, 1977; Cleveland, 1993) for an overview). To quantify what is happening, you will want to use **summary statistics** like the mean, variance, and 5-number summary. **Distributional tests** can be used to validate statistical assumptions about data distributions; typically, this will be a normality assumption.

To visualize what is happening, you will want to look for appropriate **univariate and bivariate plots**. When you use **multivariate plots** that involve three or more variables at once, it will be difficult to understand much of what is going on without relying on techniques like **dimension reduction** or **cluster analysis**. If you see relationships in the dataset that are nonlinear but smooth, you may consider doing a simple **variable transformation** (e.g., a logarithmic transform) to see if the relationships become linear. Linear relationship will be much easier to model, especially for statistical methods. Depending on the complexity of the system you are analyzing, you may consider moving beyond simple univariate variable transformations to **feature engineering** approaches, or **multivariate transforms**.

As you grow your experience solving problems, you begin to identify patterns that indicate a likely need for certain techniques in downstream analysis. Table 2.4 has a list of some of these common clues. You are likely to notice many of these things during exploratory data analysis, and it can help to formulate a plan of attack for predictive modeling and inference.

A typical goal at this stage of analysis is often to organize the data and prepare it for further modeling. Searching the literature for phrases like **data cleaning** or **data wrangling** can offer many suggestions for how effectively to do this.

2.3.3 Predictive Modeling

Once you have a defined set of input data and one or more outcomes of interest, you should be able to fit different types of predictive models. Bishop and Nasrabadi (2006), Hastie, Tibshirani, and Friedman (2008), and James et al. (2013) provide thorough overviews of many of these approaches, and Table 2.5

TABLE 2.4
Common Clues That Certain Approaches Will Be Needed

When I See This...	...I Should Check into This
Non-negative variables	Log transforms, log-normal or gamma distributions
Variables bounded between [0,1]	Beta distributions, binomial proportions
Any kind of periodicity	Fourier transforms
Time-varying data, sensor data	Time-series analysis, wavelet decomposition, signal processing
Spatially varying data	Gaussian process modeling, kriging
Image data	Image analysis, deep learning, signal processing
Unstructured text or speech data	Natural language processing
A need to learn or update over time, a complex process that can be broken into simpler parts, or a need to incorporate subject matter expertise	Bayesian modeling
A lot of count data or categorical data (e.g., labels)	Discrete data analysis, categorical data analysis
Multiple patterns not accounted for by a variable in the model	Cluster analysis

has a short overview of popular methods. For predicting numerical outcomes, **regression models** will be appropriate, while for categorical outcomes, **classification models** will be needed. Many model architectures can handle both cases. Selecting a good model will depend a lot on whether you want to infer anything about the process that is generating the data or not. **Statistical models** should be used if inference about the data generation process is needed. If prediction is the only goal, or the system is too complex to be modeled with a statistical model,

TABLE 2.5
Popular Predictive Models

Category	Model Type	Brief Description
Statistical Models	Linear Models	Assumes a linear relationship between inputs and outcomes, but these models are flexible. They can model polynomial, logarithmic, and other parameterized relationships.
	Generalized Linear Models (McCullagh and Nelder, 1989)	Often used when linear models are desired, but the outcome is constrained (e.g., count data or proportion data). Models the outcome as a transformed linear combination of the inputs. Logistic and Poisson regression models are examples.
	Linear/Quadratic Discriminant Analysis (Klecka, Iversen, and Klecka, 1980)	
Regularized, Hybrid, or Non-Parametric Approaches	Elastic Nets (Zou and Hastie, 2005) (LASSO (Tibshirani, 1996), Ridge (Hoerl and Kennard, 1970))	A family of linear models with an additional penalty on model complexity. LASSO can be used for down selection of inputs as well.
	Generalized Additive Model	Flexible models that assume a linear relationship between inputs and outcomes after a transform is applied. The transform is estimated in a non-parametric way. See, e.g., Rank Regression (Conover and Iman, 1981), ACE (Breiman and Friedman, 1985), and AVAS (Tibshirani, 1988).
	Naïve Bayes (Murphy, 2006)	Popular and fast benchmark classifier that assumes independence between inputs and predicts the outcome based on its prevalence in the dataset for different input combinations.
	K-Nearest Neighbor (Bhatia, 2010)	Another benchmark classifier that assigns labels for a new instance by looking at labels for similar training instances.

(Continued)

TABLE 2.5 *(Continued)*
Popular Predictive Models

Category	Model Type	Brief Description
"Traditional" Machine Learning	Classification and Regression Trees (CART) (Breiman et al., 1984)	Uses simple threshold-based rules to predict the outcome.
	Tree Ensembles (Random Forests (Breiman, 2001b), Gradient Boosting Machines (Friedman, 2001))	Builds on the CART idea to consider ensembles of trees that look at the dataset from different perspectives. Random forests use a voting scheme called bagging, while gradient boosting machines build sequential models that try to learn from the mistakes of the previous trees in the ensemble.
	Support Vector Machine (SVM) (Vapnik, 2013)	Linear models that work with highly nonlinear problems by using the "kernel trick" to transform the inputs into a different space before making predictions. SVMs were one of the most accurate and popular methods before neural networks resurged in the 2010s.
Deep Learning	Neural Networks (McCulloch and Pitts, 1943; Hopfield, 1982)	Models that relate inputs to outcomes using a chain of neuron-like nodes that compute weighted combinations of features and transform them through an "activation function." Early versions were fully-connected networks called multilayer perceptron (MLP) models. Over time, these have grown to include many variants, including the following: • Convolutional Neural Networks (CNNs) to capture local correlation in inputs (e.g., for image analysis and signal processing) (Collobert and Weston, 2008) • Recurrent Neural Networks (RNNs) and Long Short-Term Networks (LSTMs) to deal with sequential inputs (e.g., time series or text data) (Hochreiter and Schmidhuber, 1997; Tealab, 2018) • Autoencoders to do dimensionality reduction (Rumelhart, Hinton, and Williams, 1985) • Generative Adversarial Networks (GANs) to create realistic synthetic data (Goodfellow et al., 2014)

then you will want to turn to a **machine learning** or **deep learning model**. There are some types of models that fall somewhere between these approaches, like **regularized regression/classification** and **non-parametric models**. There are also some common benchmark models like ***k*-nearest neighbors**, **naïve Bayes**, and **decision trees** that are quick and easy to implement in most situations. Note that some domains will have their own commonly used models (e.g., **time-series analysis**, **survival analysis**, or **statistical process control**), so it will be good to be aware of those if they apply.

To ensure models are avoiding issues with overfitting, make sure to use a **train/test/validation** split or a **cross-validation** scheme. To evaluate models after they are trained, there are many commonly used **goodness-of-fit metrics** that are implemented in most data science software, and **residual plots** can reveal violations of assumptions you made during the modeling. You may also consider **ensemble modeling** to improve predictions through combinations of different models you have trained.

2.3.4 Inference

To generate insights into the system you are modeling, there are many statistical approaches that are specifically built to do this, and they are covered at length in most statistics-oriented books (e.g., Montgomery and Runger, 2010; Wasserman, 2010; Mishra and Datta-Gupta, 2017). **Hypothesis testing** and **analysis of variance (ANOVA)** can be used to generate insights into the system you are modeling. **Significance tests** or **confidence intervals** for coefficients can be used in statistical models to identify which inputs and relationships are important, as well. There is another category of approaches that can be used to draw inferences by probing trained models with new input data and seeing what happens to the predictions as a result. This includes techniques like **partial dependence plots** (Greenwell, Boehmke and McCarthy, 2018), **accumulated local effects (ALE) plots** (Apley and Zhu, 2020), and other **feature importance** measures. Finally, there are some approaches that use randomization or perturbation of the training data to better understand how the variables in the model are related. These would include approaches like **recursive feature elimination** (Guyon et al., 2002), **stepwise regression** (Montgomery, Peck and Geoffrey Vining, 2012), and **permutation testing** (Fisher, 1937). Random sampling methods like **bootstrap sampling** (Montgomery and Runger, 2010) and **Monte Carlo sampling** (Bishop and Nasrabadi, 2006) also often have uses in exploring the variability of outcomes that may be generated, or producing non-parametric estimates of quantities of interest, with some associated measure of confidence.

2.3.5 Reporting and Visualization

A commonly used workflow in data science today is **reproducible reporting** (e.g., Jupyter notebooks (Kluyver et al., 2016)), where scripts combine text and code to generate living documents that can be quickly updated as new data are available. These documents are also used to organize thoughts and provide a starting point for any communication of results to stakeholders. Another common solution for reporting is to build **interactive graphical user interfaces (GUIs)**. Some tools like Dash (Plotly Team, n.d.) and R-Shiny (Cheng, 2013) make it possible to generate browser-based demonstrations with little effort or computer science background needed.

To collaborate with others, ensure that all source code is stored in a **version control repository** and data are made available through a **data management** system, whether that is a fully functional database and data portal or a dataset that is stored in a shared read-only network location. To improve transferability of code between data scientists, make sure to follow **coding best practices**, including proper **code documentation**, **object-oriented programming**, **unit testing**, and breaking scripts into **functions** that consider issues of **cohesion and coupling**. Also be aware of tools for **dependency management** (e.g., Conda (Anaconda Inc., n.d.) or renv (Ushey, n.d.)) and deploying models into new environments through **containerization** (e.g., Docker (Merkel, 2014) and Singularity (Kurtzer, Sochat, and Bauer, 2017)).

2.4 SUMMARY

As shared in the introduction, data science is a term used to describe not only a set of mathematical tools, but also a rational thought process for gaining insights and making decisions based on evidence. The previous sections have hopefully provided one possible approach for ensuring that your data science efforts are in line with these objectives. In Section 2.2, we saw that in the context of solving problems, data science methodologies are one of many ways to fill holes in knowledge that arise from uncertainty. What data science provides is a systematic and rational path toward filling these knowledge holes in a way that raises the chances of extracting high quality insights. Section 2.2.4 outlined an approach called UPLE for tackling research projects that use data science. The goal of this approach is to “look before you leap” by making sure to understand the problem and plan the solution before doing any data analysis. Only after capturing assumptions and thinking about the expected outcomes does the learning start, through model building and inference. This evaluation stage is the final part of the process, where findings are communicated, and a course of action is taken. After describing this general process, Section 2.3 contained a practical discussion about commonly occurring categories of analyses and tools that may be useful as you solve problems with data science. Searching the literature for the key phrases mentioned in that section should result in additional reading material about the theory and implementations of specific approaches.

Data Science is a field that has historically run up against computational constraints in organizing, modeling, summarizing, and interpreting data. This resulted in research throughout the 20th century that focused on the theoretical considerations of applying different methodologies to a suite of problems facing growing demands in manufacturing, public health, the environment, and national security sectors. As computing power has increased, data science has grown in popularity and the development of new tools and methodologies has accelerated.

In many ways, this is an exciting time for data science, as we are seeing incredible improvements in predictive modeling accuracy, simpler workflows for model training and validation, and complex decision frameworks coming close to convincingly pass the Turing test. However, data science has also reached an inflection point where new methods can be easily applied to data without needing to understand how they work and whether they are appropriate for the problem at hand. A naïve or careless data scientist can easily draw the wrong inferences and unknowingly mislead decision-makers who are becoming increasingly comfortable with "trusting the data" even when the results are counter-intuitive.

There are a few things you can do to avoid making such costly mistakes. First, make sure you understand the methods you are implementing at a reasonable level of detail. It may not be necessary to understand all of the inner workings of the models and their theoretical properties, but you should know which types of problems the models are meant to solve, what advantages and disadvantages they have, and where potential pitfalls lie in using them. Second, make sure to keep abreast of what is happening in the field. Aside from scientific literature, there are many blogs and newsfeeds you can follow to see high-level surveys of methods, their usage on example problems, code examples, and comparisons to the state of the art. Finally, by keeping the concepts from this chapter in mind as you solve problems, you will be organized in your research approach, know what kinds of results you are expecting, and be skeptical when the data are pointing you to conclusions that deviate from those expectations. If you stay vigilant, you can make sure that you are drawing reasonable conclusions that will not only lead to good decisions but will also protect your reputation as a data scientist. Good luck!

NOTE

1. Of course, it is still recommended to thoroughly review third-party software before using it to make sure the results can be trusted!

REFERENCES

Anaconda Inc. "Conda Documentation." https://docs.conda.io/en/latest/.

Apley, Daniel W, and Jingyu Zhu. "Visualizing the Effects of Predictor Variables in Black Box Supervised Learning Models." *Journal of the Royal Statistical Society: Series B (Statistical Methodology)* 82, no. 4 (2020): 1059–86.

Bishop, Christopher M, and Nasser M Nasrabadi. *Pattern Recognition and Machine Learning*. Vol. 4: New York: Springer, 2006.

Bhatia, Nitin. "Survey of Nearest Neighbor Techniques." arXiv preprint *arXiv:1007.0085* (2010).

Breiman, Leo. "Random Forests." *Machine Learning* 45, no. 1 (2001a): 5–32.

Breiman, Leo. "Statistical Modeling: The Two Cultures (with Comments and a Rejoinder by the Author)." *Statistical Science* 16, no. 3 (2001b): 199–231.

Breiman, Leo, Jerome Friedman, Charles J Stone, and Richard A Olshen. *Classification and Regression Trees*. Boca Raton, FL: CRC press, 1984.

Breiman, Leo, and Jerome H Friedman. "Estimating Optimal Transformations for Multiple Regression and Correlation." *Journal of the American Statistical Association* 80, no. 391 (1985): 580–98.

Cheng, Joe. "Shiny: Easy Web Applications in R." Paper presented at the R User Conference: useR! 2013, University of Castilla-La Mancha, Albacete, Spain, 2013.

Cleveland, William S. *Visualizing Data.* Murray Hill, NJ: Hobart press, 1993.

Collobert, Ronan, and Jason Weston. "A Unified Architecture for Natural Language Processing: Deep Neural Networks with Multitask Learning." Paper presented at the Proceedings of the 25th International Conference on Machine Learning, 2008.

Conover, William J, and Ronald L Iman. "Rank Transformations as a Bridge between Parametric and Nonparametric Statistics." *The American Statistician* 35, no. 3 (1981): 124–29.

Fisher, Ronald Aylmer. "The Design of Experiments." *The Design of Experiments*, Second Edition. Edinburgh, UK: Oliver and Boyd, 1937.

Franklin, Christine, Gary Kader, Denise Mewborn, Jerry Moreno, Roxy Peck, Mike Perry, and Richard Scheaffer. *Guidelines for Assessment and Instruction in Statistics Education (Gaise) Report.* Alexandria, VA: American Statistical Association, 2007.

Friedman, Jerome H. "Greedy Function Approximation: A Gradient Boosting Machine." *Annals of Statistics* (2001): 1189–232.

Goodfellow, Ian, Jean Pouget-Abadie, Mehdi Mirza, Bing Xu, David Warde-Farley, Sherjil Ozair, Aaron Courville, and Yoshua Bengio. "Generative Adversarial Nets." *Advances in Neural Information Processing Systems* 27 (2014).

Greenwell, Brandon M, Bradley C Boehmke, and Andrew J McCarthy. "A Simple and Effective Model-Based Variable Importance Measure." arXiv preprint arXiv:*1805.04755* (2018).

Guyon, Isabelle, Jason Weston, Stephen Barnhill, and Vladimir Vapnik. "Gene Selection for Cancer Classification Using Support Vector Machines." *Machine Learning* 46, no. 1–3 (2002): 389–422.

Hastie, T, R Tibshirani, and J.H. Friedman. *The Elements of Statistical Learning: Data Mining, Inference, and Prediction.* Second Edition. New York: Springer, 2008.

Hochreiter, Sepp, and Jürgen Schmidhuber. "Long Short-Term Memory." *Neural Computation* 9, no. 8 (1997): 1735–80.

Hoerl, Arthur E, and Robert W Kennard. "Ridge Regression: Biased Estimation for Nonorthogonal Problems." *Technometrics* 12, no. 1 (1970): 55–67.

Hopfield, John J. "Neural Networks and Physical Systems with Emergent Collective Computational Abilities." *Proceedings of the National Academy of Sciences* 79, no. 8 (1982): 2554–58.

James, Gareth, Daniela Witten, Trevor Hastie, and Robert Tibshirani. *An Introduction to Statistical Learning.* New York: Springer, 2013.

Klecka, William R, Gudmund R Iversen, and William R Klecka. *Discriminant Analysis.* Vol. 19: Beverly Hills, CA: Sage, 1980.

Kluyver, Thomas, Benjamin Ragan-Kelley, Fernando Pérez, Brian E Granger, Matthias Bussonnier, Jonathan Frederic, Kyle Kelley, Jessica Hamrick, Jason Grout, Sylvain Corlay, Paul Ivanov, Damián Avila, Safia Abdalla, Carol Willing, and Jupyter development team. Jupyter Notebooks – A Publishing Format for Reproducible Computational Workflows. In Loizides, Fernando and Scmidt, Birgit (eds.) *Positioning and Power in Academic Publishing: Players, Agents and Agendas.* IOS Press, pp. 87–90, 2016. doi:10.3233/978-1-61499-649-1-87..

Kurtzer, Gregory M, Vanessa Sochat, and Michael W Bauer. "Singularity: Scientific Containers for Mobility of Compute." *PLoS One* 12, no. 5 (2017): e0177459.

McCullagh, Peter, and John A Nelder. *Generalized Linear Models*. Vol. 37: Boca Raton, FL: CRC press, 1989.

McCulloch, Warren S, and Walter Pitts. "A Logical Calculus of the Ideas Immanent in Nervous Activity." *The Bulletin of Mathematical Niophysics* 5, no. 4 (1943): 115–33.

Merkel, Dirk. "Docker: Lightweight Linux Containers for Consistent Development and Deployment." *Linux Journal*, no. 239 (2014): 2.

Mishra, Srikanta, and Akhil Datta-Gupta. *Applied Statistical Modeling and Data Analytics: A Practical Guide for the Petroleum Geosciences*. Cambridge, MA: Elsevier, 2017.

Montgomery, Douglas C, Elizabeth A Peck, and G Geoffrey Vining. *Introduction to Linear Regression Analysis*. Vol. 821: Hoboken, NJ: Wiley, 2012.

Montgomery, Douglas C, and George C Runger. *Applied Statistics and Probability for Engineers*. Hoboken, NJ: Wiley, 2010.

Murphy, Kevin P. "Naive Bayes Classifiers." *University of British Columbia* 18, no. 60 (2006): 1–8.

Plotly Team. "Dash Documentation & User Guide." https://dash.plotly.com/.

Rumelhart, David E, Geoffrey E Hinton, and Ronald J Williams. "Learning Internal Representations by Error Propagation." California Univ San Diego La Jolla Inst for Cognitive Science, 1985.

Ruparelia, Nayan B. "Software Development Lifecycle Models." *ACM SIGSOFT Software Engineering Notes* 35, no. 3 (2010): 8–13.

Selvi, K, and Rana Majumdar. "Six Sigma-Overview of DMAIC and DMADV." *International Journal of Innovative Science and Modern Engineering* 2, no. 5 (2014): 16–19.

Tibshirani, Robert. "Estimating Transformations for Regression via Additivity and Variance Stabilization." *Journal of the American Statistical Association* 83, no. 402 (1988): 394–405.

Tibshirani, Robert. "Regression Shrinkage and Selection via the Lasso." *Journal of the Royal Statistical Society. Series B (Methodological)* (1996): 267–88.

Tukey, John W. *Exploratory Data Analysis*. Vol. 2: Reading, MA: Addison-Wesley, 1977.

Tealab, Ahmed. "Time Series Forecasting Using Artificial Neural Networks Methodologies: A Systematic Review." *Future Computing and Informatics Journal* 3, no. 2 (2018): 334–40.

Ushey, Kevin. "Cran - Package Renv." The Comprehensive R Archive Network, https://cran.r-project.org/web/packages/renv/.

Vapnik, Vladimir. *The Nature of Statistical Learning Theory*. New York: Springer Science & Business Media, 2013.

Wasserman, Larry. *All of Statistics: A Concise Course in Statistical Inference*. New York: Springer, 2010.

Wu, CFJ. "Statistics = Data Science." In *HC Carver Professorship Lecture*, http://www2.isye.gatech.edu/~jeffwu/presentations/datascience.pdf. The University of Michigan, Ann Arbor, 1997.

Zou, Hui, and Trevor Hastie. "Regularization and Variable Selection via the Elastic Net." *Journal of the Royal Statistical Society: Series B (Statistical Methodology)* 67, no. 2 (2005): 301–20.

Section II

Reservoir Characterization Applications

3 Machine Learning-Aided Characterization Using Geophysical Data Modalities

Vikram Jayaram[1] *and Tao Zhao*[2]
[1]Pioneer Natural Resources,
Dallas, Texas
[2]Schlumberger,
Houston, Texas

CONTENTS

3.1 INTRODUCTION

Rapid advances in machine learning (ML) algorithms and increases in computational efficiency have driven modern-day decision-making through the analysis and interpretation of voluminous geophysical data modalities. The digital and connected oil fields have led to petabytes of rich, heterogeneous data from a diverse set of sources such as drilling operations, production data, completions, surveillance operations, GIS, 3D, and 4D seismic, geological interpretation, distributed acoustic sensing (DAS)/distributed temperature sensing (DTS), well data such as logs, core, thin-sections, and even scanning electron microscopic (SEM) images. To gain actionable insights as well as to uncover hidden patterns and correlations from such diverse data necessitates seamless interactions between computational scale, intelligent algorithms, databases, and web services. A comprehensive set

DOI: 10.1201/9781003207009-5

of cloud-based solutions from leading technology companies are in some ways addressing these needs. ML systems staged on cloud-based computational frameworks have now become an integral part of the operations of most oil and gas companies, allowing them to gather large volumes of information in real time and interpret these datasets into actionable insights more quickly. Data and its information content have now come to be viewed as an extremely valuable resource with a huge upside for companies with innovative, robust ML strategies. In the ongoing "new normal" low commodity price environment, saving time, reducing costs, boosting efficiencies, managing risk, and improving safety are all crucial outcomes that can be gained using ML in oil and gas operations.

However, investing in predictive technology to become more productive is challenging and easier said than done. What are the strategic considerations and challenges for adopting ML platforms? How can we seamlessly integrate ML platforms with other systems without prohibitively high costs? And how do we ensure there is a return on investment? With some of the examples in this chapter, we address the studies surrounding the heterogeneous integrations, assumptions, case studies, and applications of robust algorithms that can partially automate the interpretation and characterization of geological, geophysical, petrophysical, and engineering data within an integrated framework for decision-making.

3.2 GEOPHYSICAL DATA ANALYSIS – PHYSICS-BASED AND DATA-DRIVEN APPROACHES

Being the buzzword of many scientific communities these days, ML seems to be an easy choice when it comes to solving problems, either legacy or emerging challenges. However, we should always perform due diligence when deciding where, when, and how to apply ML to a task. Traditionally, geoscience and engineering communities have a long history of solving problems with both physics-based models and data-driven algorithms (Hampson et al., 2005; Virieux and Operto, 2009). Over the past few decades, people started using statistical approaches for geoscience and engineering problems, in addition to relying solely on physics-based models (Hegde et al., 2017; Carrassi et al., 2018). More recently, ML approaches have gained popularity rapidly among the geoscience and engineering communities (Bergen et al., 2019). In general, the advantage of ML solutions should fall under at least one of the three categories: (1) improving the efficiency over existing solutions, (2) improving the quality over existing solutions, and (3) providing solutions to unsolved problems. Over the years, we have witnessed many successful instances where ML helps advance our understanding of geoscience and engineering problems, as well as makes existing workflows more efficient. On the other hand, we have also observed examples, both from academia and industry, showing that ML does not always bring additional value over existing solutions if applied all by itself.

In many successful use cases, ML works in tandem with physics-based approaches to solve problems of a larger scale. One such example can be found in seismic-based earth model building. The objective for earth model building is to derive the velocity model (and, if possible, density model as well) from the seismic data. The current state-of-the-art solution for earth model building is seismic full waveform inversion (FWI) (Virieux et al., 2017). Interestingly, FWI itself is a good example of combining physics-based and data-driven approaches, in that people use physics-based wave equations to define the forward modeling step, whereas the inverse problem is usually solved by data-driven methods. In the past few years, various attempts have been made to directly predict the velocity model from seismic data using deep learning (Araya-Polo et al., 2018; Yang and Ma, 2019). This is an extremely challenging task, and most of the published examples on this topic have been on synthetic data with low structural and stratigraphic complexity. Alternatively, if we use deep learning in combination with FWI, we can indeed enhance the existing FWI solution. Yao et al. (2020) proposed a workflow that uses deep learning to transform elastic seismic data into their acoustic form, which greatly reduces the cost of running elastic FWI. Multiple researchers also implemented workflows to extrapolate low-frequency seismic data from the original data using deep learning models for FWI (Hu et al., 2019; Ovcharenko et al., 2019; Sun and Demanet, 2020; Zhao et al., 2020). Such low-frequency data introduce stability to the FWI algorithm by reducing the risk of cycle-skipping.

In the rest of this chapter, we would like to share some examples in which ML is used to improve the efficiency of existing workflow, as well as improving the quality of the final products by bringing additional insights to the data. We wish that these examples inspire our readers to discover more suitable use cases of ML in the geoscience and engineering communities, and eventually create more value from applications of these ML methods.

3.3 MACHINE LEARNING IN SEISMIC FACIES INTERPRETATION

Seismic facies are formally defined as mappable, 3D seismic units composed of groups of reflections whose parameters differ from those of adjacent facies units (Mitchum et al., 1977). Traditionally, seismic interpreters identify and characterize seismic facies such as channels, pinch-outs, mass transport complexes, salt bodies, and collapses, which are defined by reflection patterns that are different from the surroundings. With the development of seismic attributes (Chopra and Marfurt, 2005), interpreters have been equipped with abundant tools to inspect the seismic wiggles in different aspects and in different domains. Unfortunately, our human eyes are not trained to sense and analyze data in a space of more than three or four dimensions, and with the ever-increasing data volumes, taking full advantage of all the data accessible to us becomes increasingly challenging.

Researchers investigated computer-aided seismic interpretation utilizing multiple seismic attributes soon after the development of seismic attributes (Sonneland, 1983; Justice et al., 1985). In the early 2000s, unsupervised learning techniques such as clustering represented by *k*-means (MacQueen, 1967) and self-organizing maps (SOMs) (Kohonen, 1982) became popular in the seismic facies interpretation community (Barnes and Laughlin, 2002; Strecker and Uden, 2002; Coleou et al., 2003). Represented by fully connected and probabilistic neural networks, supervised learning approaches also gained steam around the same time, with

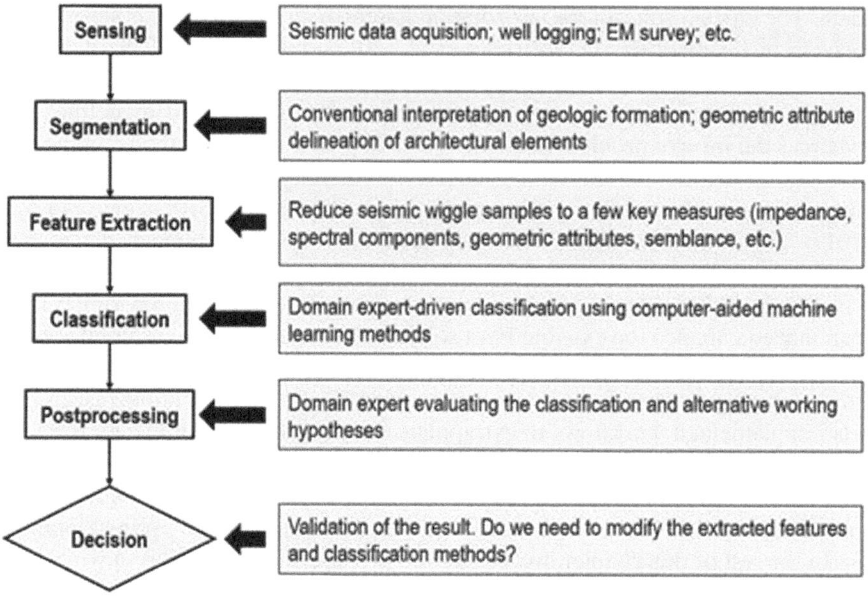

FIGURE 3.1 ML-based seismic facies interpretation workflow. Modified from Duda et al. (2001).

much broader application than seismic facies interpretation (van der Baan and Jutten, 2000; West et al., 2002). Zhao et al. (2015) provided a review of many of the then popular ML algorithms for seismic facies interpretation. In general, an ML-based seismic facies interpretation workflow resembles the steps in Figure 3.1. Here, "sensing" is the process to acquire data; "segmentation" is the practice of focusing on a certain geographical area or a group of formations of interest; "feature extraction" means obtaining data representations that best describe the features of interest; "classification" is the step in which interpreters use ML to assign each data voxel into one of a finite number of facies; "post-processing" provides interpreters a chance to evaluate the interpreted facies and interactively refine the results, if necessary; and finally, "decision" is the stage to decide whether to proceed with the results or go back to modify any of the previous steps and reanalyze the data. One of the examples described in Zhao et al. (2015) utilizing such workflow is to use an ML algorithm, called generative topographic mapping (GTM) (Bishop et al., 1998), to first obtain a smooth-varying seismic facies distribution map from multiple seismic attributes, then assign discrete facies classes interactively based on both geologic understanding and data distribution. Figure 3.2 shows the seismic facies interpretation after performing such workflow on seismic data from the Canterbury Basin, offshore New Zealand. In this example, GTM successfully projects the multi-attribute seismic data into a 2D space, in which seismic samples belonging to similar facies are assigned to the same data clusters. By interactively defining the boundary of the data clusters, discrete seismic facies are delineated.

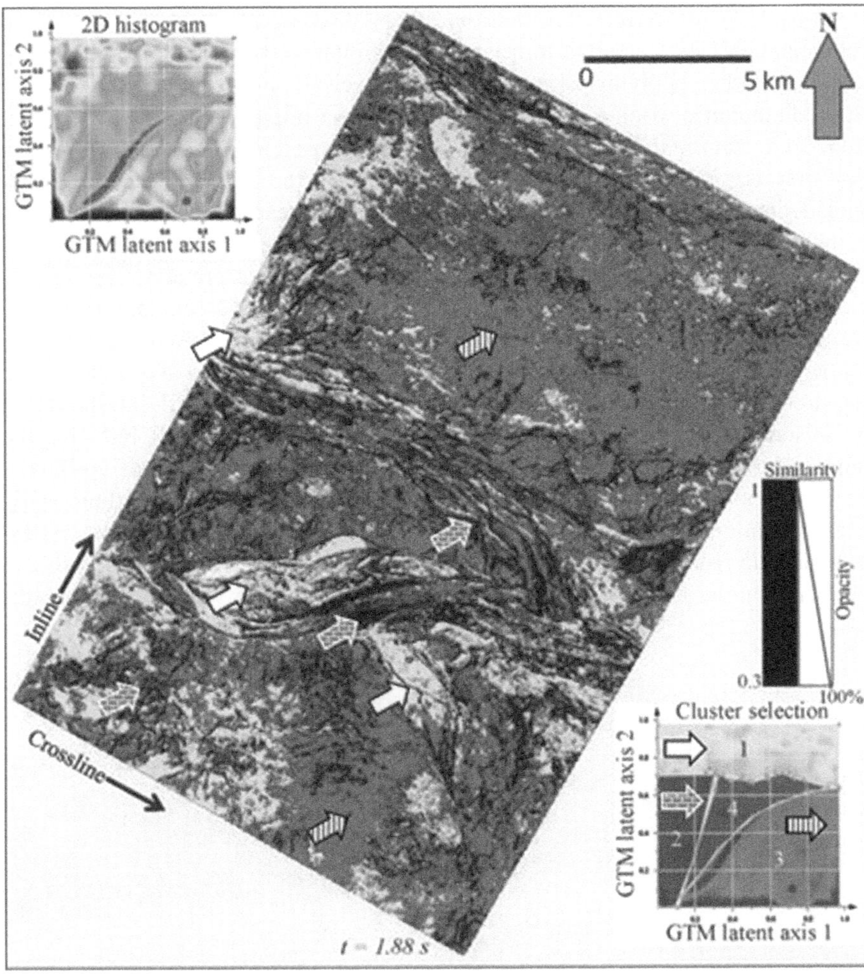

FIGURE 3.2 An example of seismic facies interpretation along a seismic time slice using a projection algorithm GTM followed by interactive class assignment. The four classes are defined on the cross-plot domain, representing (1) high-energy point bars and crevasse splays, (2) multistoried channels, (3) slope fans, and (4) "everything else." Classes 1, 2, and 3 are marked by the respective arrows in the facies map. Data are from the Canterbury Basin, offshore New Zealand. Modified from Zhao et al. (2015).

In recent years, deep learning has gained tremendous popularity for seismic facies interpretation. In addition to the ML models being much more complex, the fundamental difference between a deep learning method and a conventional ML method is the ability to take "raw" data as input and perform automatic feature extraction (LeCun et al., 2015). Due to its great success in computer vision problems, the convolutional neural network (CNN) methods (Krizhevsky et al., 2012) have almost become the synonym to deep learning for seismic facies interpretation. The first published work on seismic facies interpretation using deep learning is probably Zhao (2018), in which the author discussed both classification and segmentation CNNs applied to seismic facies interpretation. One of the biggest challenges when applying supervised CNNs to seismic facies interpretation is the limited amount of training data, which has a great impact on model generalization. To overcome this issue, Di et al. (2020) proposed a semi-supervised model that first uses a convolutional autoencoder to learn the most representative features from the unlabeled seismic data, then uses the learned feature extractor to generate features to be used in the supervise task. This approach greatly reduces the amount of required training data for seismic facies interpretation. Being special facies, faults are extremely important for building the regional structural framework as well as for reservoir characterization yet are challenging for automatic interpretation prior to the deep learning era. With the help of deep learning, researchers have made significant advancements in automatic fault interpretation on seismic data (Maniar et al., 2018; Wu et al., 2019; Zhao, 2019), reducing the man hours for performing fault interpretation from weeks to days, if not hours. Figure 3.3 shows an example from Wu et al. (2019), in which the authors apply a segmentation CNN model, which is pre-trained on synthetic data only, to field seismic data from the Campos Basin, offshore Brazil. The CNN model is able to delineate numerous closely spaced faults, which may take weeks if not months

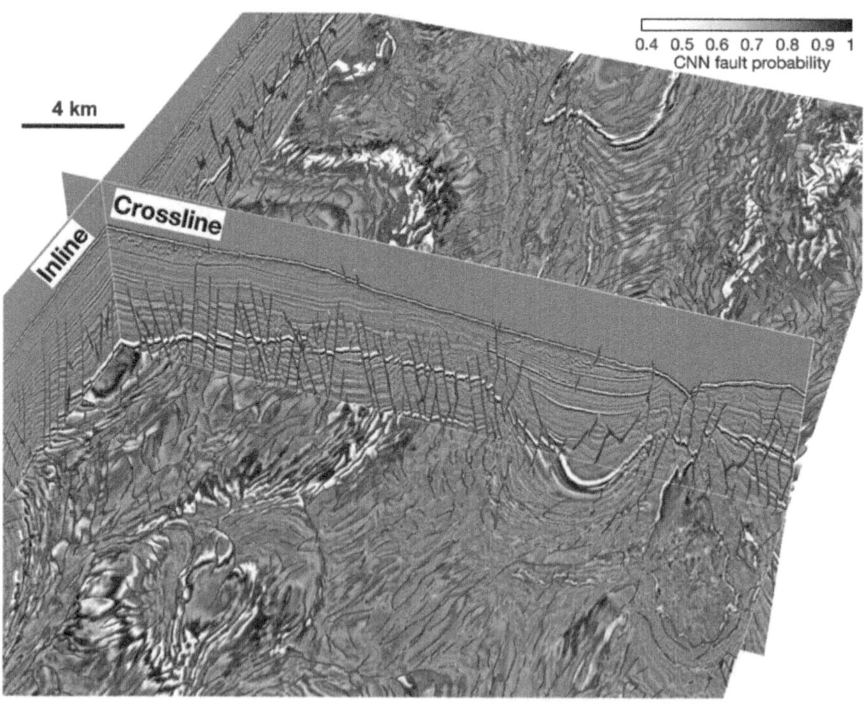

FIGURE 3.3 Fault probability computed using a segmentation CNN model, which is pre-trained on synthetic data. Modified from Wu et al. (2019).

to interpret in a traditional workflow. In the next section of this chapter, we present another ML-assisted workflow for rapid petrophysical QA/QC in challenging borehole conditions.

As a concluding remark for this section, we would also like to point our readers to more recent work on quantifying the sensitivity of seismic facies classification to seismic attribute selections. This work (Lubo-Robles et al. 2022) introduces the use of interpretable ML methods in studying sensitivity on seismic classification maps.

3.4 MACHINE LEARNING AIDING PETROPHYSICAL AUTOMATION

There has been an increasing need for rapid log QA/QC and reconstruction in challenging borehole conditions for petrophysicists to update models in an existing zone or assess new plays in time-sensitive workflows. Petroleum reservoirs are often associated with multiple target zones or a single zone adjacent to non-productive intervals. For example, in real-time geosteering, it is important to remain in a zone or to dynamically steer drilling and exploration toward a target. This requires knowledge of the petrophysical/rock mechanical properties of the rock surrounding the drill bit. Although logging while drilling can provide this information, a cost-effective and almost-real-time solution is lacking. In general, there is a depth lag and, therefore, a time delay between what the logging-while-drilling sub relays to the surface and the bit performance. Gupta et al. (2020) focused on relating drill-bit and drill string performance data in an ML workflow to predict the lithology at the bit while drilling.

The traditional workflows of log reconstruction involve the manual work of the petrophysicists, which results in subjective, repetitive, and labor-intensive QC edits. In Figure 3.4 a workflow to automate the log reconstruction in challenging borehole conditions with an ensemble of ML algorithms is shown. The purpose of this automated workflow is to reduce bias and variance between petrophysicists' interpretations and boost efficiency through process automation.

This specific automated workflow focuses on constructing models based on stratigraphic, spatial, and input features to make robust predictions for log reconstruction in challenging borehole conditions with mixed vintage data. A data QC was first performed using the interactive statistics and visuals, and then the compromised borehole conditions were identified to be removed from our training dataset. Then individual and ensemble ML algorithms, including random forest (Ho, 1995), support vector regression (Cortes and Vapnik, 1995), neural network, and XGBoost regression (Chen and Guestrin, 2016), are trained through cross-validation, tested against blind data, and undergo hyperparameter tuning. Evaluation of model metrics include a comparison of predicted values with actual measurements in good borehole conditions. This includes causal inference, residual analysis, and quantitative and qualitative metrics with geologic context. This work covers the life cycle of a model through data selection, data conditioning, hyperparameter tuning, model evaluation, and model selection and deployment.

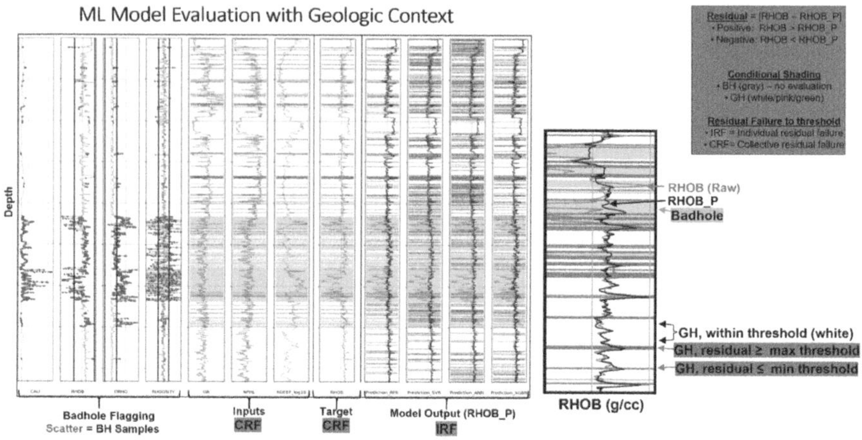

FIGURE 3.4 Automated well log reconstruction with identification of bad hole flagging.

In this work (presented by Jayaram et al.), the bad hole conditions were successfully identified, and the well logs were reconstructed. Candidate ML algorithms performed well within the internally defined tolerance set by the petrophysicists and the ensemble learning was able to further improve the accuracy. The result of this work was to use ML methods to automate key QA and QC steps for wireline logs. This, in turn, served to rapidly process well logs for providing critical decision support in field development and play assessment.

The workflow allows petrophysicists to overcome the "black box" impression of statistical models. Understanding the relationship between ML internal complexities and their output log predictions provides the opportunity for the petrophysicists to finesse model tuning and selection accounting for geologic context. The reference architecture implemented in this workflow has a modular design in which future algorithms and external data combined with geospatial information could be seamlessly integrated. In the current climate of consolidation, such workflow improves productivity and reduces turnaround times.

3.5 GEOSCIENCE AND ML-AIDED WELLBORE DIAGNOSTICS: A 360-DEGREE VIEW

In this section, we downscale from regional scale geophysical characterization to near wellbore reservoir characterization. The recent advances in seismic acquisition technology and the distributed fiber-optic measurements have provided game-changing insights into reservoir characterization. DAS is a new technology for seismic surveying in which the transmission of light through fiber-optic cables is used to monitor and record seismic energy, with unparallel spatial resolution compared with traditional techniques. Integrating DAS and DTS with microseismic and downhole pressure data provides critical insights into the physical processes taking place in the near- and far-field wellbore environments drilling and hydraulic fracturing (Jayaram et al., 2019). Processing and visualizing these integrated data both three dimensionally and temporally lead to an improved understanding of how the reservoir responds to stimulation, highlight processes relevant to optimizing completion designs, and enable better interpretation of each type of data.

In such studies, the oil and gas operators' objectives are to better understand the development of physical changes to the formation as wells were drilled and completed and to record far-field interactions between the fiber-optics-equipped horizontal monitor well and offsetting horizontal wells during stimulation.

In typical science well studies, a fiber-optic cable was cemented in place on the outside of the monitor well's casing. The well also was equipped with downhole pressure gauges and conventional downhole microseismic geophone arrays. The monitoring well and adjacent horizontal wells were then stimulated in a zipper frac sequence while recording data in the monitoring well to observe both near- and far-field effects of hydraulic fracturing.

Because it is coupled to the rock by cement, the fiber-optic cable translates stress changes of the rock as strain changes within the line. The temporal strain changes on the DAS are recorded and displayed over multiple frequencies. As the hydraulic fracture tip propagates, it introduces deformation and slippage along the fracture face and creates pressure and strain increases to the surrounding rock. The deformation process may result in mechanical slippage over a broad range of frequencies, which are recorded as stress drops. Microseismic recording of these stress drops tracks the development of the stress envelope through time.

Some of the recent works (Zhang et al., 2020) show how modern diagnostic techniques can be leveraged to provide insight into the near wellbore region. Specifically, these diagnostics combine interpretations from both fiber-optic DAS and external downhole pressure gauges (BHG) combined with ML to improve the characterization of the near-wellbore region (NWR). The ML algorithms were utilized to estimate the friction pressure induced by near-wellbore tortuosity. In Figure 3.5, we show a typical geometrical illustration of a 360-degree view of a near-wellbore response. This work was executed during the stimulation of a horizontal well located in the Wolfcamp formation within the Midland Basin.

3.6 CONCLUSIONS

In this chapter, we discussed various ML methods applied to different geophysical data modalities. First, using seismic data, we reviewed some state-of-the-art algorithms and workflows for seismic facies interpretation and fault detection. We then visited well log QC using an ensemble of ML models on wireline measurements. Finally, we introduced applying ML on DAS data for wellbore diagnostics.

While the chapter focused on showcasing ML being applied to various geophysical data modalities, it also provides a flavor to the multi-resolution scale at which it is applied and adapted. These scales vary from a regional scale to near wellbore studies. Oil field data now provides petabytes of rich, heterogeneous data from a diverse set of sources due to field technology and data capture modernization. Advances in ML and data science in the oil and gas industry have provided advanced-level automation for the scientific teams, quick insights from voluminous data, and operational efficiency.

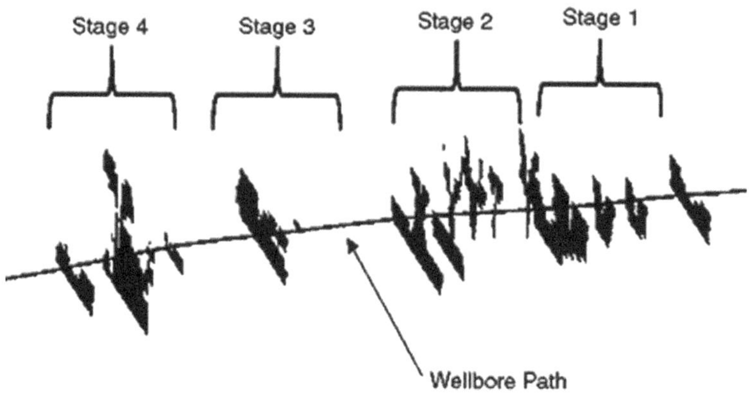

FIGURE 3.5 An illustration of a 360-degree view of a near wellbore response.

In spite of these maturing endeavors, the sources of data are still disparate and challenging to derive subsurface characterization and insights. Much of it is attributed to the operational and data silos created during the well life cycle.

3.6.1 Future Directions

In the next 10 years, oil and gas operators and their technical/scientific teams shall start adopting advanced-level data culture breaking away from geosciences and petroleum engineering silos. Challenges to maintain high-quality stewardship under environmental, social, and governance (ESG) aspects will become a prime focus. Such a culture shall constantly (relentlessly) seek to understand the true data drivers of surface and subsurface phenomena. Secondly, there should be accountability to measurable goals that are good proxies of the oil and gas operator's mission and values (focus on ESG). Responsible operations (respecting the air, land, and water), thoughtful planning and innovation in field development, producing low-carbon fuel, and creating a NETZERO pathway utilizing oil field data shall continue to be the mainstay for the energy industry. Lastly, expect ML workflows to do fast iteration and experimentation leading to valuable and actionable data insights and the confluence of collaborations between the Geosciences and Engineering communities.

REFERENCES

Araya-Polo, M., J. Jennings, A. Adler, and T. Dahlke, 2018, Deep-learning tomography, The Leading Edge, **37**, 58–66.

Barnes, A. E., and K. J. Laughlin, 2002, Investigation of methods for unsupervised classification of seismic data, 72nd Annual International Meeting, SEG, Expanded Abstracts, 2221–2224.

Bergen, K. J., P. A. Johnson, V. Maarten, and G. C. Beroza, 2019, Machine learning for data-driven discovery in solid Earth geoscience, Science, **363**, 6433.

Bishop, C. M., M. Svensén, and C. K. Williams, 1998, GTM: The generative topographic mapping, Neural Computation, **10**, 215–234.

Carrassi, A., M. Bocquet, L. Bertino, and G. Evensen, 2018, Data assimilation in the geosciences: An overview of methods, issues, and perspectives, Wiley Interdisciplinary Reviews: Climate Change, **9**, e535.

Chen, T., and C. Guestrin, 2016, XGBoost: A scalable tree boosting system, Proceedings of the 22nd ACM SIGKDD International Conference on Knowledge Discovery and Data Mining, 785–794.

Chopra, S., and K. J. Marfurt, 2005, Seismic attributes—A historical perspective, Geophysics, **70**, 3SO–28SO.

Coleou, T., M. Poupon, and K. Azbel, 2003, Unsupervised seismic facies classification, A review and comparison of techniques and implementation, The Leading Edge, **22**, 942–953.

Cortes, C., and V. Vapnik, 1995, Support-vector networks, Machine Learning, **20**, 273–297.

Di, H., Z. Li, H. Maniar, and A. Abubakar, 2020, Seismic stratigraphy interpretation by deep convolutional neural networks: A semisupervised workflow, Geophysics, **85**, WA77–WA86.

Duda, R. O., P. E. Hart, and D. G. Stork, 2001, Pattern classification, 2nd ed.: John Wiley & Sons.

Gupta, I., N. Tran, D. Devegowda, V. Jayaram, C. Rai, C. Sondergeld, and H. Karami, 2020, Looking ahead of the bit using surface drilling and petrophysical data: Machine-learning-based real-time geosteering in Volve Field, SPE Journal, **25**, 990–1006.

Hampson, D. P., B. H. Russell, and B. Bankhead, 2005, Simultaneous inversion of pre-stack seismic data, 75th Annual International Meeting, SEG, Expanded Abstracts, 1633–1637.

Hegde, C., H. Daigle, H. Millwater, and K. Gray, 2017, Analysis of rate of penetration (ROP) prediction in drilling using physics-based and data-driven models, Journal of Petroleum Science and Engineering, **159**, 295–306.

Ho, T. K., 1995, Random decision forests, Proceedings of the 3rd International Conference on Document Analysis and Recognition, 278–282.

Hu, W., Y. Jin, X. Wu, and J. Chen, 2019, A progressive deep transfer learning approach to cycle-skipping mitigation in FWI, 89th Annual International Meeting, SEG, Expanded Abstracts, 2348–2352.

Jayaram, V., Hull, R., Wagner, J., and Zhang, S., 2019, Hydraulic Fracturing Stimulation Monitoring with Distributed Fiber Optic Sensing and Microseismic in the Permian Wolfcamp Shale Play: URTEC-2019-291-MS, presented at the SPE/AAPG/SEG Unconventional Resources Technology Conference, Denver, Colorado, USA, July 2019.

Justice, J. H., D. J. Hawkins, and D. J. Wong, 1985, Multidimensional attribute analysis and pattern recognition for seismic interpretation, Pattern Recognition, **18**, 391–399.

Kohonen, T., 1982, Self-organized formation of topologically correct feature maps, Biological Cybernetics, **43**, 59–69.

Krizhevsky, A., I. Sutskever, and G. E. Hinton, 2012, ImageNet classification with deep convolutional neural networks, Advances in Neural Information Processing Systems, **25**, 1097–1115.

LeCun, Y., Y. Bengio, and G. Hinton, 2015, Deep learning, Nature, **521**, 436–444.

Lubo-Robles, D., D. Devegowda, V. Jayaram, H. Bedle, K. J. Marfurt, M. J. Pranter, 2022, Quantifying the sensitivity of seismic facies classification to seismic attribute selection: An explainable machine learning study, Interpretation, **10**, 1–90.

MacQueen, J. B., 1967, Some methods for classification and analysis of multivariate observations, Proceedings of 5th Berkeley Symposium on Mathematical Statistics and Probability, **1**, 281–297.

Maniar, H., S. Ryali, M. S. Kulkarni, and A. Abubakar, 2018, Machine-learning methods in geoscience, 88th Annual International Meeting, SEG, Expanded Abstracts, 4638–4642.

Mitchum, R. M., P. R. Vail, and J. B. Sangree, 1977, Seismic stratigraphy and global changes in sea level, part 6: stratigraphic interpretations of seismic reflection patterns in depositional sequences, in Payton, C. E., ed., Seismic Stratigraphy and Applications to Hydrocarbon Exploration: AAPG Memoir, **26**, 117–133.

Ovcharenko, O., V. Kazei, M. Kalita, D. Peter, and T. Alkhalifah, 2019, Deep learning for low-frequency extrapolation from multioffset seismic data, Geophysics, **84**, R989–R1001.

Sonneland, L., 1983, Computer aided interpretation of seismic data, 53rd Annual International Meeting, SEG, Expanded Abstracts, 546–549.

Strecker, U., and R. Uden, 2002, Data mining of 3D poststack attribute volumes using Kohonen self-organizing maps, The Leading Edge, **21**, 1032–1037.

Sun, H. and L. Demanet, 2020, Extrapolated full-waveform inversion with deep learning, Geophysics, **85**, R275–R288.

van der Baan, M., and C. Jutten, 2000, Neural networks in geophysical applications, Geophysics, **65**, 1032–1047.

Virieux, J., A. Asnaashari, R. Brossier, L. Métivier, A. Ribodetti, and W. Zhou, 2017, An introduction to full waveform inversion, Encyclopedia of Exploration Geophysics, Society of Exploration Geophysicists, R1-1–R1-40.

Virieux, J. and S. Operto, 2009, An overview of full-waveform inversion in exploration geophysics, Geophysics, **74**, WCC1–WCC26.

West, P. B., R. S. May, E. J. Eastwood, and C. Rossen, 2002, Interactive seismic facies classification using textural attributes and neural networks, The Leading Edge, **21**, 1042–1049.

Wu, X., L. Liang, Y. Shi, and S. Fomel, 2019, FaultSeg3D: Using synthetic data sets to train an end-to-end convolutional neural network for 3D seismic fault segmentation, Geophysics, **84**, IM35–IM45.

Yang, F., and J. Ma, 2019, Deep-learning inversion: A next-generation seismic velocity model building method, Geophysics, **84**, R583–R599.

Yao J., L. Guasch, M. Warner, T. Lin, and E. Percak-Dennett, 2020, Geophysical data and gradient translation using deep neural networks, 90th Annual International Meeting, SEG, Expanded Abstracts, 1725–1729.

Zhang, S., H. Tang, R. Hurt, V. Jayaram, and J. Wagner, 2020, Joint interpretation of fiber optics and downhole gauge data for near wellbore region characterization, SPE Hydraulic Fracturing Technology Conference, SPE-199715-MS.

Zhao, T., 2018, Seismic facies classification using different deep convolutional neural networks, 88th Annual International Meeting, SEG, Expanded Abstracts, 2046–2050.

Zhao, T., 2019, 3D convolutional neural networks for efficient fault detection and orientation estimation, 89th Annual International Meeting, SEG, Expanded Abstracts, 2418–2422.

Zhao, T., A. Abubakar, X. Cheng, and L. Fu, 2020, Augment time-domain FWI with iterative deep learning, 90th Annual International Meeting, SEG, Expanded Abstracts, 850–854.

Zhao, T., V. Jayaram, A. Roy, and K. J. Marfurt, 2015, A comparison of classification techniques for seismic facies recognition, Interpretation, **3**, SAE29–SAE58.

4 Machine Learning to Discover, Characterize, and Produce Geothermal Energy

V. Vesselinov[1], *M. Mudunuru*[2], *B. Ahmmed*[1], *S. Karra*[1], *and D. O'Malley*[1]

[1]Computational Earth Science Group,
Earth and Environmental Sciences Division,
Los Alamos National Laboratory,
Los Alamos, New Mexico, USA

[2]Watershed and Ecosystem Science,
Pacific Northwest National Laboratory,
Richland, Washington, USA

CONTENTS

4.1 INTRODUCTION AND MOTIVATION

Geothermal energy [Brown et al. 2012, Mudunuru et al. 2021] extracted from Earth's heat is a secure, reliable, flexible, and nearly inexhaustible renewable resource beneath our feet. It offers answers to many of the world's and the United States' essential energy needs across residential, commercial, and industrial sectors

DOI: 10.1201/9781003207009-6

[GeoVision Report 2019]. Some U.S. Department of Energy (DOE) mission-critical needs include energy grid stability, affordable and reduced carbon footprint energy options (e.g., efficient heating and cooling solutions), and extraction of critical materials such as lithium and hydrogen [Hamm 2021]. As of 2021, the United States has more than 13 GW of installed geothermal electrical-generating capacity [Robins et al. 2021], which is enough to power approximately 2.5 million homes [SEIA 2021]. According to resource estimates and assessments from the U.S. Geological Survey (USGS) in 2008 [Williams et al. 2008], Blackwell's geothermal heat flow map in 2011 [Blackwell et al. 2011], and the National Renewable Energy Laboratory (NREL) study in 2016 [Augustine 2016], more than 100 GW (~19.2 million homes) of new geothermal energy can be extracted through enhanced geothermal systems (EGS) and hydrothermal resources. Figure 4.1 represents geologic regions and basins in the United States [Stanton et al. 2017] and temperature profile [Blackwell et al. 2011] at a depth of approximately 6 km. From the figure, it is evident that the geothermal potential is primarily concentrated in the western basins of the United States.

Extracting this renewable energy at depth has various technical and non-technical barriers, which contributed to slow progress toward exploring and deploying geothermal technology [Hamm 2021]. However, if we overcome these barriers, geothermal technology would allow for nearly inexhaustible energy supply from this renewable resource [Hamm 2021]. Figure 4.2 shows the obstacles that need to be overcome to efficiently explore, develop, and manage a geothermal resource efficiently. High-risk challenges include pre-survey, exploration, and drilling [Gehringer and Loksha 2012, Hamm 2021]. Based on the literature survey [GeoVision 2019, Mudunuru et al. 2021], we envision that the recent advancements in machine learning (ML) [Beardsmore 2014, Holtzman et al. 2018, Buster et al. 2021, Cavur et al. 2021, Siler and Pepin 2021, Siler et al. 2021] can help us overcome the challenges and barriers that are in our way (see Figure 4.3) to discover, characterize, and produce geothermal energy widely at an affordable option.

4.1.1 Need for Advanced Geothermal Technologies

In 2019, the DOE developed a comprehensive report called GeoVision analysis [Hamm et al. 2019, Hamm 2021, GeoVision 2019] to assess and discuss geothermal energy opportunities in the United States. A significant finding from this analysis is that advanced technologies are needed to reduce geothermal exploration and development costs and risks. An improved understanding of the subsurface is crucial to achieving better exploration technology for assessing geothermal prospectivity, energy affordability, and resource development through efficient heat extraction. Improved subsurface characterization (e.g., fractures, faults, heat sources) is critical to geothermal deployment [Brown et al. 2012, Mudunuru et al. 2021]. For geothermal resource development to be transformative, the GeoVision analysis confirmed that subsurface characterization is vital to the following:

- Exploration of hidden geothermal systems [Dobson 2016].
- Unlocking the potential of EGS [Olasolo et al. 2016].
- Capturing undiscovered hydrothermal resources [Williams et al. 2008].

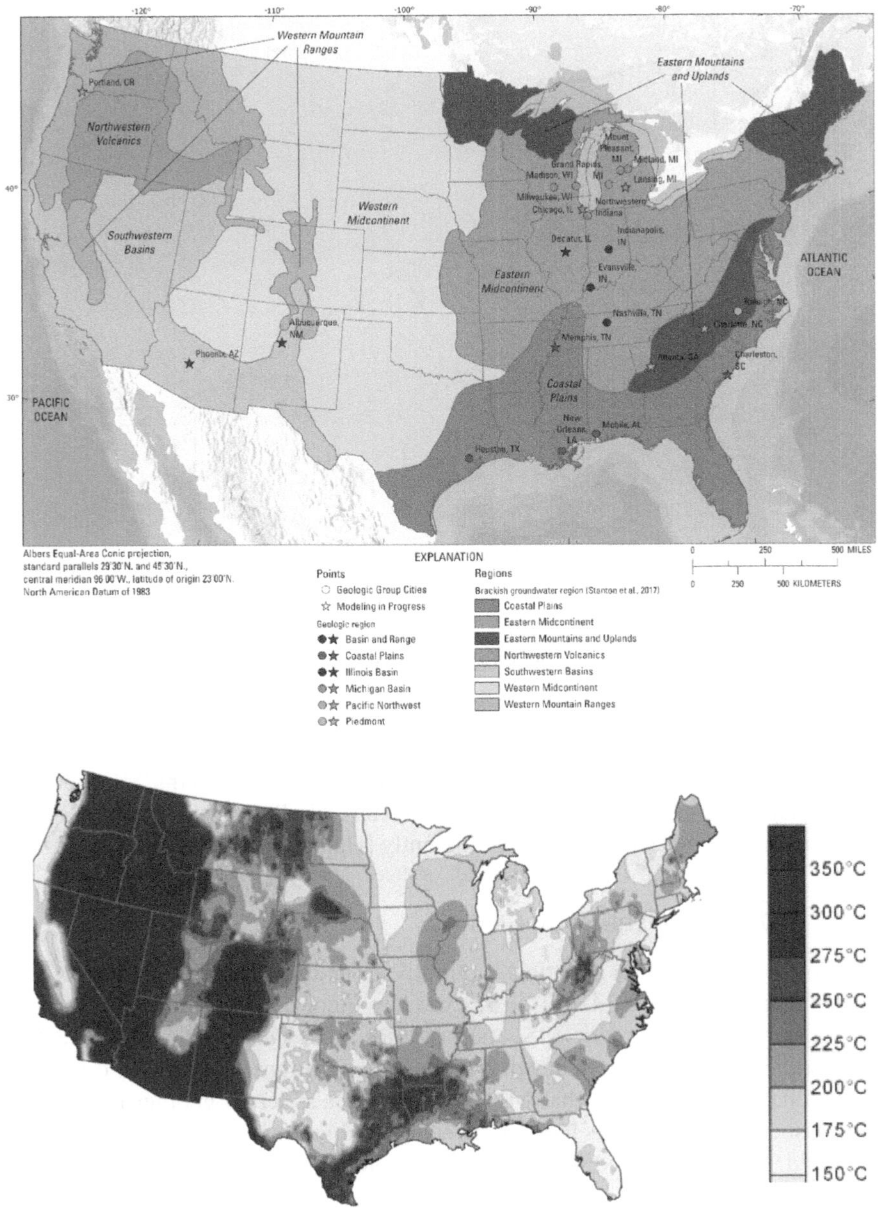

FIGURE 4.1 A representation of the geologic regions, basins, and temperature profile at depth of geothermal resources within the United States. The top figure shows the geologic regions (e.g., basin) corresponding to previously identified geologic units with structural controls [Credit: Stanton et al. 2017]. The bottom figure shows the temperature profile at a 6-km depth [Credit: Blackwell et al. 2011], which illustrates the potential of geothermal resources within these basins.

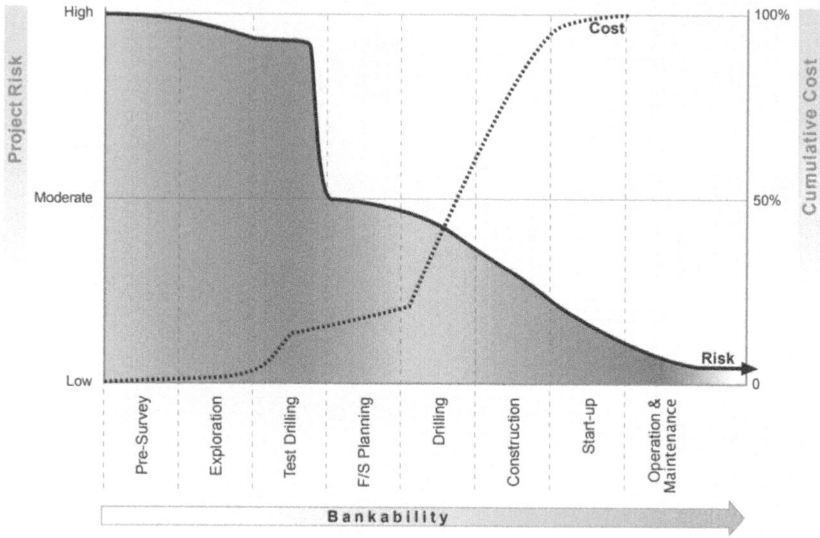

FIGURE 4.2 A pictorial description of project cost and risk profile at different phases in geothermal resource exploration, development, and deployment [Credit: Gehringer and Loksha 2012].

The GeoVision study identified that new exploration tools utilizing ML methods could enable the geothermal community to identify undiscovered resources better and develop prospective EGS more efficiently [Hamm et al. 2019, Hamm 2021].

4.1.2 Outline of the Book Chapter

Section 4.1 provides the introduction, motivation, and the need for ML for geothermal energy. Section 4.2 discusses the key ingredients for developing accurate and reliable ML models. Section 4.3 describes case studies that use ML for geothermal exploration, development, and operations. Section 4.4 discusses some challenge problems that ML can solve to accelerate geothermal energy and associates the barriers toward the successful forward deployment of ML in the field. Finally, concluding remarks and potential future work is provided in Section 4.5.

4.2 MACHINE LEARNING FOR GEOTHERMAL ENERGY

In a broader perspective, ML is a subset of artificial intelligence (AI) [Jordan and Mitchell 2015], which provides the ability to automatically learn and improve from experience without knowledge being explicitly programmed into a system. Recent advances in ML [Goodfellow et al. 2016, Innes et al. 2019, Schoenholz and

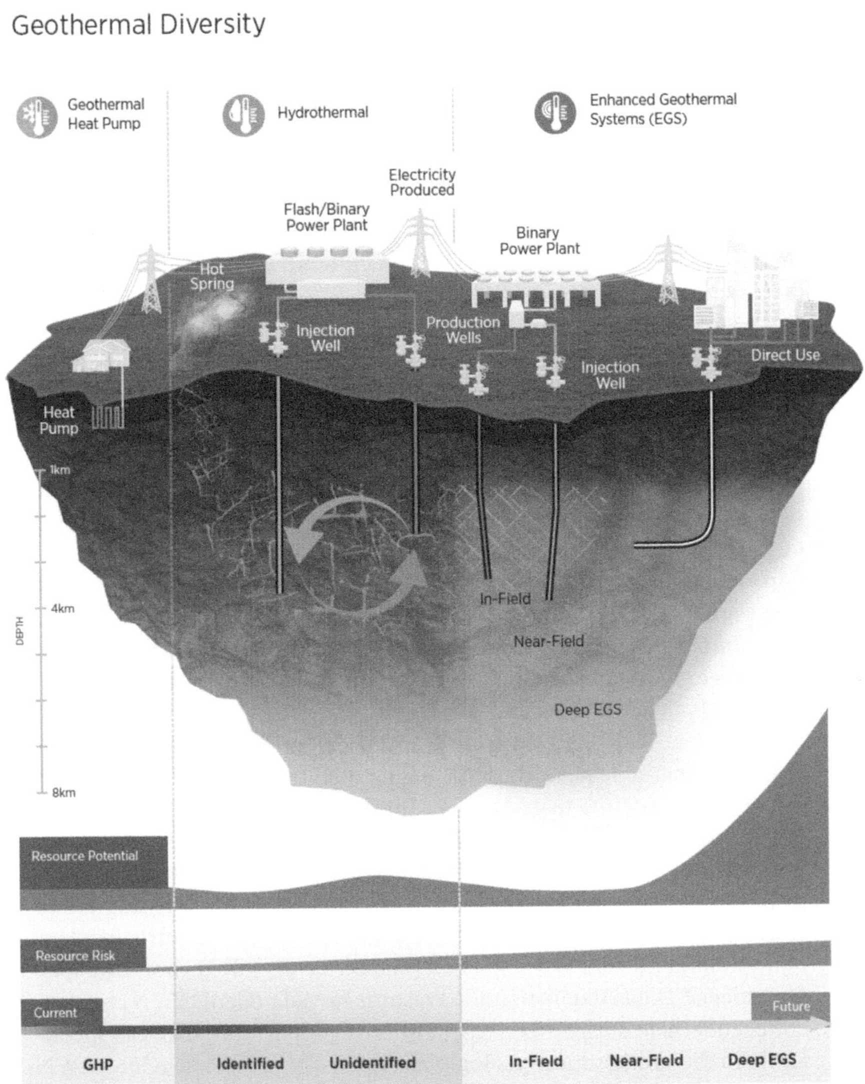

FIGURE 4.3 A pictorial description on the development of geothermal resource. From a geothermal heat pump (GHP) to deep EGS [Credit: GeoVision 2019].

Cubuk 2020] provide examples of use cases unlocking the potential to address many scientific, technological, and economic problems our society faces. They can also be applied to discover, characterize, develop, and produce thermal power from hidden geothermal, EGS, or hydrothermal systems [Mudunuru et al. (2017a, b), Siler and Pepin 2021, Siler et al. 2021]. This section primarily focuses on two main ingredients needed to create ML models. These include data and algorithms or methods to build models [Goodfellow et al. 2016].

4.2.1 Data

The cornerstone for developing accurate and robust ML models is informative data [Brunton and Kutz 2019]. We can generate this data in primarily three different ways:

- ***Lab-scale and local field-scale data:*** We can conduct different experiments in lab and field experiments to generate data for ML analyses and model development. Experiments [Brown et al. 2012, Mudunuru et al. (2017a, b), Kneafsey et al. 2018, Moore et al. 2019] performed in the field (e.g., EGS Colab, Utah FORGE, Fenton Hill) provide realistic data on geothermal conditions, geological factors influencing production, fracture networks, faults, and other constraints (e.g., metadata) to guide predictive ML models. Core samples collected in the field can be analyzed in a lab setting [Kneafsey et al. 2018] to provide complementary data to field experiments. However, the cost and effort (e.g., workforce, time) to perform lab-scale and local field-scale experiments are generally high (e.g., drilling wells) [Gehringer and Loksha 2012] because of project costs and risks associated with geothermal exploration and development (see Figure. 4.2 in Section 4.1).
- ***Distributed field-scale data:*** Various remote sensing and geophysical methods exist to collect data over large regions (e.g., different spatial resolutions) [van der Meer et al. 2014, Cavur et al. 2021]. Popular techniques include electrical (e.g., time-lapse electrical impedance tomography), electromagnetic (e.g., airborne, magnetotellurics), gravity, magnetic, seismic, and satellite (e.g., LiDAR, InSAR) [Wright et al. 1985, Brown et al. 2012, van der Meer et al. 2014]. The main advantage is that these methods are non-invasive or minimally invasive. They can be applied to improve the understanding of the near subsurface (e.g., geothermal features such as fractures, faults, thermal hot spots) toward geothermal exploration and development [Faulds et al. 2015]. For example, Geoscience Data Acquisition for Western Nevada (GeoDAWN) aims to collect new near surface data (e.g., airborne geophysical, LiDAR) specifically in the Walker Lane geologic zone in western Nevada [GeoDAWN 2020]. This distributed data collection objective is to identify highly prospective geothermal resources hidden in this region. However, a potential downside is that the collected data are generally noisy and may need extensive pre-processing. Statistical and unsupervised ML can be used for this pre-processing [Vesselinov et al. 2020, Cavur et al. 2021, Smith et al. 2021, Siler et al. 2021] to denoise the remote and geophysical data before they are ingested into physics-based and data-driven models. In addition, the models developed using the data from a single source can be non-unique and ambiguous for interpretation. As a result, multiple geophysical and remote sensing datasets are combined (e.g., multi-sensor data fusion approaches) to create better models from the collected data and

reduce the uncertainty associated with their understanding [Mudunuru et al. (2017a, b)].

- ***Synthetic data:*** To overcome the bottleneck of generating data for ML model development, process model simulations using simulators such as FEHM [https://fehm.lanl.gov/], PFLOTRAN [https://www.pflotran.org/], STOMP [https://www.pnnl.gov/projects/stomp], and GeoDT [Frash 2021] of different complexity and resolution (e.g., low-fidelity, high-fidelity) can be used to develop synthetic datasets. The model inputs are generally conditioned on the site specifics to make such simulation applicable for modeling a region of interest [Mudunuru et al. (2017a, b)]. Such a dataset allows us to overcome the cost, time, and risk challenges of collecting lab-scale and field data. However, the synthetically generated data might not accurately represent the site-specific conditions. As a result, the conceptual uncertainty in the generated synthetic data needs to be accounted for in the ML process [Witter et al. 2019].

There are various resources to obtain geothermal datasets (e.g., lab-scale, field-scale, synthetic, raw, and curated) for training ML models. Typically, the generated data from lab, field, and model simulations are in a raw digital format (e.g., raster, GIS, VTK, SEG-Y). These raw data are curated and processed for ML analysis [Pepin 2019, Cavur et al. 2021]. Commonly, the datasets may need to be restructured, rescaled, hardcoded, transformed (e.g., log-based), and normalized [Vesselinov et al. 2020]. This is necessary as the performance of most of the ML algorithms is heavily dependent on how the data are curated. Note that both the raw and curated geothermal datasets are generally available in some open-source data repositories. Examples of some data sharing repositories and tools are included in this Table 4.1

TABLE 4.1
Examples of Geothermal Data Repositories

Data and Informatics Tools	Brief Description
National Geothermal Data Systems (NGDS)	Documents and datasets about geothermal resources in the United States [Anderson et al. 2013]
Geothermal Data Repository (GDR)	Extensive database of raw and curated experimental and field data, simulation data, and analysis code [GDR, 2022; Weers and Anderson 2016]
Geothermal Prospector	Visualization and download of geospatial datasets curated by universities and government agencies [Getman and Anderson 2015]
Renewable Energy (RE) Data Explorer	Provides information to support data-driven analyses for policymakers to inform critical decisions needed for renewable energy (e.g., geothermal) deployment [Cox et al. 2020]

4.2.2 State-of-the-Art ML Methods

ML can be divided into purely data-driven and physics (or science)-informed [Brunton and Kutz 2019, Innes et al. 2019] categories. The data-driven methods learn everything from the provided data (e.g., collected from field or lab, physics-based, or conceptual model simulation outputs) [Beardsmore 2014, Holtzman et al. 2018, Buster et al. 2021, Cavur et al. 2021, Siler and Pepin 2021, Siler et al. 2021]. The physics (or science)-informed methods use not only the provided data but also existing knowledge about the solved problem (e.g., physics-informed neural networks) [Raissi et al. 2019, Tartakovsky et al. 2020]. The knowledge can be any form, such as constitutive relationships (e.g., permeability-fracture correlations), conservation laws (e.g., flow, transport, energy), non-negative constraints on chemical species and temperature, and mathematical models [He and Barajas-Solano 2020]. This a priori domain knowledge can be directly embedded in the ML process (e.g., simulators embedded in neural networks through differentiable programming or adjoints) or provided as optimization constraints (e.g., penalty functions using Lagrange multipliers) [Innes et al. 2019, Karra et al. 2021, Mitusch et al. 2021]. Furthermore, the ML algorithms can be categorized into supervised, unsupervised (or self-supervised), and semi-supervised learning. We briefly describe these methods and their applicability for geothermal energy in the following paragraphs.

- ***Supervised ML:*** Supervised methods are a subset of ML algorithms [Singh et al. 2016, Duplyakin et al. 2022] that map inputs (e.g., geothermal attributes such as temperature and permeability) to outputs (e.g., quantities such as thermal power produced, or labels of geothermal potential such as high, intermediate, and low) [Gangwani et al. 2020, Liu et al. 2021]. Typically, the output quantities or labels are generated through process model simulations, field data (e.g., thermal power output), or inputs by subject-matter experts (SMEs) [Gudmundsdottir and Horne 2020]. For example, SMEs can identify locations with high-, intermediate-, and low-temperature geothermal prospects or specific geologic features representing hidden systems (e.g., fault offsets) [Faulds et al. 2020]. In the case of simulation outputs, the supervised ML methods are then applied to the pairs of input-output datasets to learn geothermal potential. We note that the successful training of supervised ML algorithms requires massive datasets, which may not be available for geothermal exploration. There are also approaches to automate the labeling process in the pre-processing data. Table 4.2 provides popular supervised ML (some of these methods can also be applied for unsupervised ML).
- ***Unsupervised (or self-supervised) ML:*** On the other hand, unsupervised ML algorithms identify patterns and extract hidden signals from geothermal data without any prior labeling from SMEs [Vesselinov et al. 2020, Cavur et al. 2021, Siler et al. 2021, Siler and Pepin 2021, Smith et al. 2021].

TABLE 4.2
Popular ML Methods for Geothermal Applications

Popular Supervised ML Methods	Geothermal Application – Discovery, Characterization, and Production
Deep neural networks (DNNs)	**Characterization and production** – DNNs are used to characterize and model geothermal reservoirs to predict thermal power production [Gudmundsdottir and Horne 2020, Pandey and Singh 2021]. **Discovery** – ML-based PFA analysis to assess geothermal potential in the Great Basin region of Nevada [Faulds et al. 2020] also used DNNs
Convolutional neural networks (CNNs)	**Characterization** – CNNs have been used to delineate geothermal field faults and predict geothermal reservoirs [Gao et al. 2021]. They have been used to determine the state of stress changes at Coso geothermal field in California, using surface deformation data [Roberts et al. 2021]
Recurrent neural networks (RNNs)	**Production –** RNNs were used to predict temperature outputs in geothermal systems using the historical data from injection wells and production mass flow rates [Jiang et al. 2020, Shi et al. 2021]
Support vector machines (SVMs)	**Characterization** – SVMs are used to learn the mapping between temperature and pressure data to subsurface permeability [Başer and Sert 2021]
Random forests (RFs)	**Characterization** – RFs have been used to estimate permeability, monitor subsurface, and identify regions of geothermal potential [Kiran and Salehi 2020, Başer and Sert 2021]

The identified patterns and extracted low-dimensional features are post-processed by SMEs to associate a physical meaning (e.g., signatures that correspond to resource type) [Vesselinov et al. 2020]. Popular unsupervised ML methods include singular value decomposition (SVD), principal component analysis (PCA), independent component analysis (ICA), non-negative matrix and tensor factorization (NMF/NTF), various clustering methods including *k*-means, and Gaussian mixture models (GMMs; aka kriging and co-kriging) [Gentleman and Carey 2008, Cichocki et al. 2009, Vesselinov et al. 2019]. SVD decomposes the geothermal data into several component matrices to gain insights into the system state [Beltrami 2001]. PCA is a linear dimensionality reduction technique to project data into lower-dimensional space. It is used to decompose a multivariate geothermal dataset into a set of successive orthogonal components that explain a maximum amount of the variance. PCA has been used to reveal geologic controls on energy production at Brady's geothermal field [Siler and Pepin

2021]. The ICA method allows us to separate multivariate geothermal data into additive subcomponents that are maximally independent. The *k*-means clustering enables us to partition the data into *k*-different clusters. ICA and clustering analysis have been used to analyze remote sensing satellite imagery to prospect hydrothermal systems [Abubakar et al. 2019]. GMMs are probabilistic models, which assume that all the data are generated from a mixture of a finite number of Gaussian distributions with unknown parameters. They have been used to characterize hydrothermal alteration in two different wells in the Hachimantai geothermal field in Japan [ESI, 2021, Ishitsuka et al. 2021].

Recently, a new type of unsupervised ML method called NMFk/NTFk has successfully analyzed geothermal data [Vesselinov et al. 2020, Siler et al. 2021]. The methods combine the power of NMF/NTF to extract hidden signals from the data [Alexandrov and Vesselinov 2014, Vesselinov et al. 2019]. Customized *k*-means clustering (NMF*k*/NTF*k*) produces optimal latent space. The NMFk/NTFk was used to identify geologic factors that control flow in Brady's geothermal field in Nevada [Siler et al. 2021]. Additionally, this technique was used to discover signals corresponding to geothermal resources in play fairway analysis (PFA) data. As a result, NMFk found signatures that can be applied to PFAs to improve the detection of hidden geothermal resources, thereby reducing play development risk.

Furthermore, recent unsupervised ML advances in application of deep neural network (DNN) autoencoders have shown promise for extracting latent features from geothermal data. These methods leverage DNNs and impose bottlenecks in the network to extract compressed presentation of the inputs. As a result, autoencoders learn the structure in the original geothermal data through this representation learning. They have been used for modeling geothermal systems and predicting energy production [Gangwani et al. 2020].

- ***Semi-supervised (or semi-unsupervised) ML:*** Semi-supervised ML falls between supervised and unsupervised learning. Typically, if an ML algorithm is given any labels or metadata or auxiliary (including physics-based or science-based) information, it is generally referred to this category. For example, algorithms learn by combining a small number of labeled data samples or metadata with large unlabeled data samples in this method. Such a methodology can improve learning accuracy when labeling training datasets is infeasible or costly, and acquiring unlabeled data is relatively inexpensive. Geothermal applications in this category include Smale et al. [2018] and Cavur et al. [2021], where they utilize semi-supervised learning to analyze geothermal fields in the United States (e.g., Brady's field in Nevada) and New Zealand. Smale and co-workers [Smale et al. 2018] collected geothermal data attributes such as vegetation, soil chemical fertility, and subsurface soil temperature at 38 sites on 15 geothermal fields in the Taupō Volcanic Zone, central

North Island, New Zealand. Semi-supervised clustering was used to partition these geothermal data attributes. Then these clustered data were ingested in a fuzzy classification framework to identify the main environmental drivers of geothermal vegetation composition. This classification identified 16 vegetation associations in eight structural classes unique to geothermal fields.

Cavur et al. [2021] used self-organizing map (SOM) algorithms on remote sensing data to analyze the direction and patterns of displacement observed in Brady's geothermal field in Nevada. Specifically, they curated and analyzed 70 Sentinel-1A Synthetic-Aperture Radar (SAR) images using persistent scatterer interferometry (PSInSAR) to estimate vertical displacement at this geothermal site. The SOM algorithms are then used on this curated PSInSAR data to classify zones affected by displacement due to geothermal activity. The SOM-based ML analysis showed that uplift occurred on the northeast part of Brady's field, where the injection wells are located. On the other hand, the semi-supervised ML revealed that subsidence is primarily concentrated on the southwest part of the field where the production wells are located.

4.3 POPULAR CASE STUDIES

Over the past decade, the DOE Geothermal Technologies Office has developed various initiatives to accelerate the geothermal project life cycle [Hamm 2021]. These projects have provided and curated diverse datasets (see Section 4.2.1 Data), made available through open-source repositories such as Geothermal Data Repository (GDR), National Geothermal Data Systems (NGDS), and geothermal prospector. ML showed promise to fully utilize the data available to better understand how physical processes impact subsurface geothermal conditions and how they are represented in observations. For instance, Vesselinov [2021] discussed how novel ML methods and tools that can (1) analyze large field datasets, (2) assimilate model simulations (large inputs and outputs), (3) process sparse datasets, (4) perform transfer learning (between sites with different exploratory levels), (5) extract hidden geothermal signatures in the field and simulation data, (6) label geothermal resources and processes, (7) identify high-value data acquisition targets, and (8) guide geothermal exploration and production by selecting optimal exploration, production, and drilling strategies. In the following subsections, we provide some popular applications of ML for geothermal energy and their status. The focus is on ML for geothermal exploration, development (e.g., EGS, drilling), and operations (e.g., power plant).

4.3.1 Exploration and Development

Building on the techniques from petroleum industries, the concept of PFA has allowed us to develop maps of geothermal potential toward identifying prospective locations of hidden geothermal systems. Various PFAs were conducted primarily in the western United States (e.g., basins in the southwest and northwest), resulting in rich and diverse data for ML. Examples include great basins, atlas cascades, Appalachian basins, southwest New Mexico (SWNM), Hawaii, and the Snake River plain [PFA, GTO 2015]. In SWNM play type, Pepin [2019] used PCA to gain insights into geothermal resource development (e.g., where to explore next). This analysis identified promising regions (e.g., medium and high-temperature resources). Building on this work, Siler and co-workers [Siler et al. 2021, Siler and Pepin 2021] have used PCA and NMFk to reveal geologic controls on geothermal production at Brady geothermal field in Nevada (see Figure 4.4). Their analysis indicated that macro-scale faults and a local step-over in the fault system preferentially occur along the production wells compared with injection and non-productive wells. This better identification of features within the data instills confidence in combining unsupervised ML techniques with SMEs to improve understanding of the subsurface processes necessary for geothermal exploration.

Building on the PFA conducted in Person et al. [2015] in SWNM, Vesselinov [2021] analyzed the curated data resulting in better identification of low- and medium hydrothermal systems. They have used NMFk on the PFA data and found dominant attributes and spatial distribution of extracted hidden hydrothermal signatures (see Figure 4.5). These signatures were then related to different regions in the basin, resulting in enhanced predictions of resources within a regional physiographic province. Other PFA datasets analyzed using NMFk were Tularosa Basin in New Mexico, Hawaii, the Utah FORGE region, and the Great Basin [Vesselinov 2021]. NMFk identified geothermal signatures from the data sampled at 120 locations within the Tularosa Basin. Then, it associated these signatures with low-, medium-, and high-temperature hydrothermal systems. In Hawaii, NMFk analyzed PFA data in four different islands to associate the grade of the hydrothermal system. A detailed prospectivity analysis was performed in Utah FORGE to identify future drilling locations using geological, geochemical, and geophysical attributes curated from the GDR repository [Ahmmed and Vesselinov, 2021]. Temperature and heat flow maps were constructed at different depths based on NMFk analysis that utilized InSAR, gravity, and seismic data. These prospectivity maps provide more insights into the drilling locations for future geothermal exploration within this region. NMFk was also used to assess geothermal potential in the Texas region. This analysis showed that the western part of the regional area had a higher potential for geothermal exploration at lower depths than central and eastern areas. Such a detailed ML analysis resulted in enhanced PFA or ePFA as shown in Figure 4.6. ePFA workflow digests site data, extracts hidden features, and assesses geothermal site prospectivity.

ML was applied extensively to study the Great Basin region, which shows promise in identifying prospective locations of hidden systems [Smith 2021]. Vesselinov [2021] has analyzed the PFA data and extracted hidden geothermal signatures associated with low-, medium-, and high-temperature hydrothermal systems. Their analysis provided insights into the dominant characterization attributes and spatial distribution of the resource type within the study area

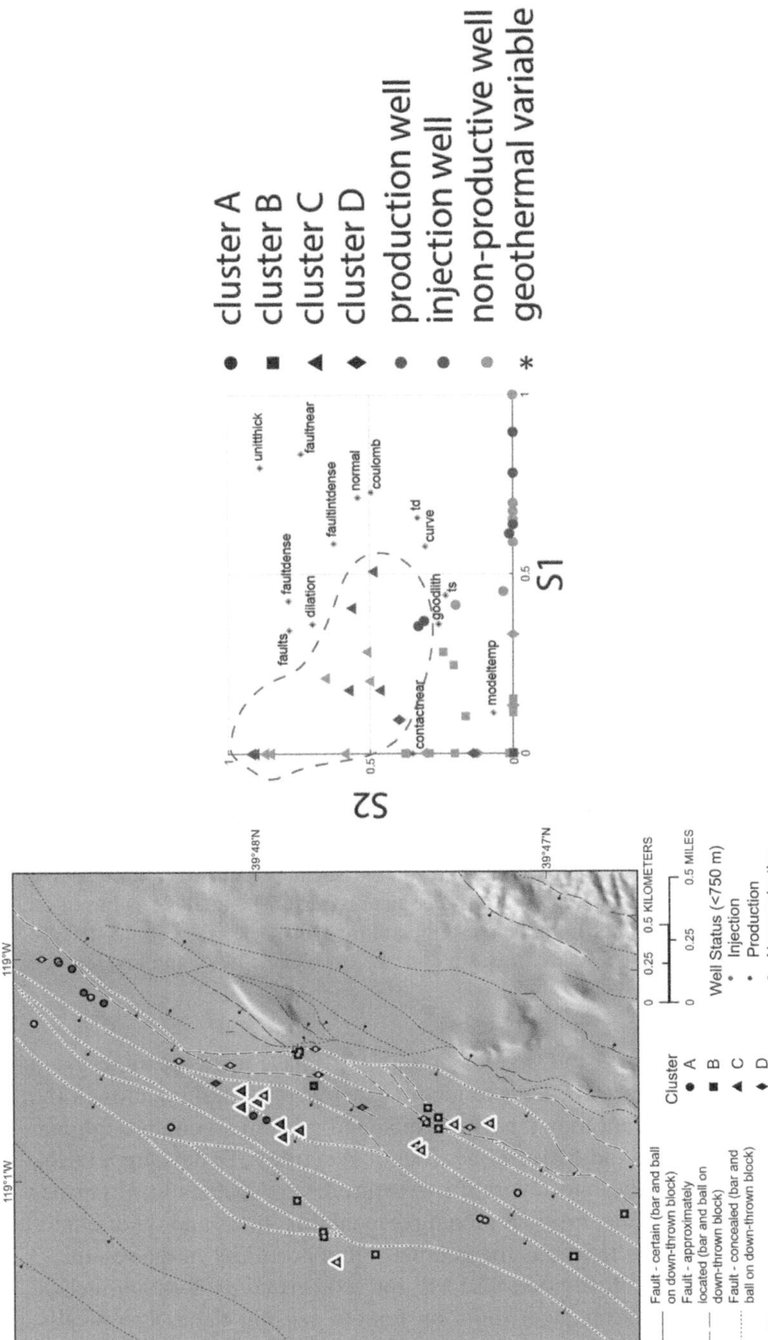

FIGURE 4.4 This figure shows the identification of producing, non-producing, and injecting wells using NMFk [Credit: Siler et al. 2021]. The left figure shows a map of Brady's geothermal field in Nevada, along with well clusters identified based on the NMFk method. The right figure shows a biplot of hidden signatures. It indicates that NMFk was able to identify the variables that control the separation of producing wells from non-producing and injection wells.

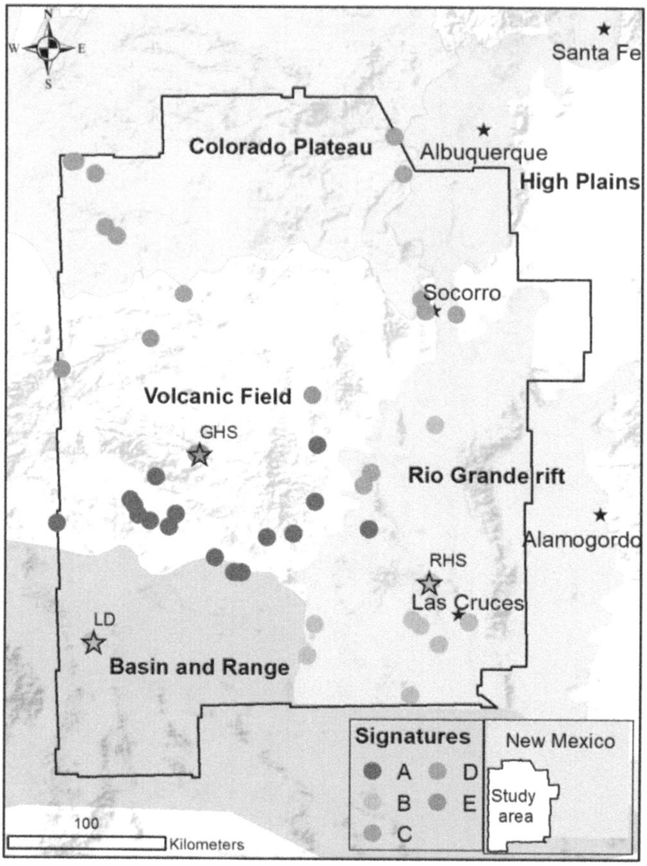

FIGURE 4.5 This figure shows ML signatures and associates them with the resource type and province [Credit: Vesselinov 2021, technical report]. The analysis is based on NMFk, which was able to identify moderate-temperature hydrothermal locations that are represented by open stars. These are in Basin and range, volcanic field, and Rio Grande rift provinces.

(Figure 4.7). On the other hand, Faulds et al. [2020] have used the supervised ML technique to develop workflows to identify new geothermal systems in the Great Basin. The PFA data they a nalyzed included geologic, geophysical, and geochemical parameters indicative of geochemical activity. To develop a DNN, they used different geothermal favorability feature layers such as local permeability, intermediate permeability, regional permeability, heat, and other exploration metadata as inputs to assess the geothermal opportunity in the region.

Other success stories of unsupervised ML for geothermal exploration include the application of clustering algorithms on remote sensing data. Specifically, hyperspectral image data were used [Cavur et al. 2021] to assess geothermal energy potential at Basin and range in the western United States. They have used

SOM clustering algorithms to analyze 70 Sentinel-1A SAR images for Brady's geothermal site. The resulting clusters from the ML method were used to determine subsidence, uplift, and stable regions automatically. The coupling of the displacement data from InSAR and the SOM algorithms was shown to effectively analyze the direction and pattern of the displacements observed in Brady's field. Matzel et al. [2021] has combined geological, 3D seismic, and magnetotellurics data with unsupervised ML to understand and define high permeability areas within the Raft River geothermal resource area. They used *k*-means clustering on the fused dataset to distinguish lithologies and verify against the known lithologic boundaries. They show a clear correspondence between the geophysical measurements and the known geology.

From a supervised ML perspective, deep learning models have been applied on lab-scale and field-scale seismic data to understand different aspects of geothermal fields. For example, Marone [2021] developed and curated microearthquake (MEQ) data (e.g., waveforms) to image geothermal reservoir properties and forecast seismic events. These data represent the evolution of 149 MEQs between 2012 and 2013 at the Newberry EGS site for analysis with the DNNs. In addition to field data analysis, Marone and co-workers [Alireza et al. 2021] have developed various ML models to understand lab-scale frictional data and

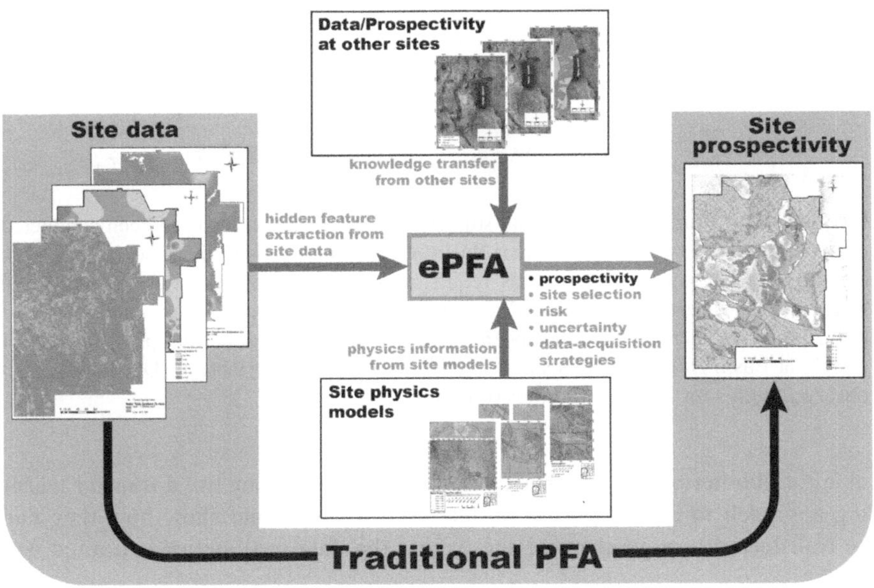

FIGURE 4.6 This figure compares the ML-enhanced PFA (ePFA) with traditional PFA analysis. In a conventional PFA, site data are only utilized to develop geothermal prospectivity maps, making it difficult to transfer knowledge across sites. On the other hand, ePFA combines site data, its extracted features, prospectivity from other sites, and physics model simulations to assess the site's geothermal potential. This ePFA workflow allows us to transfer knowledge across sites with minimal retraining.

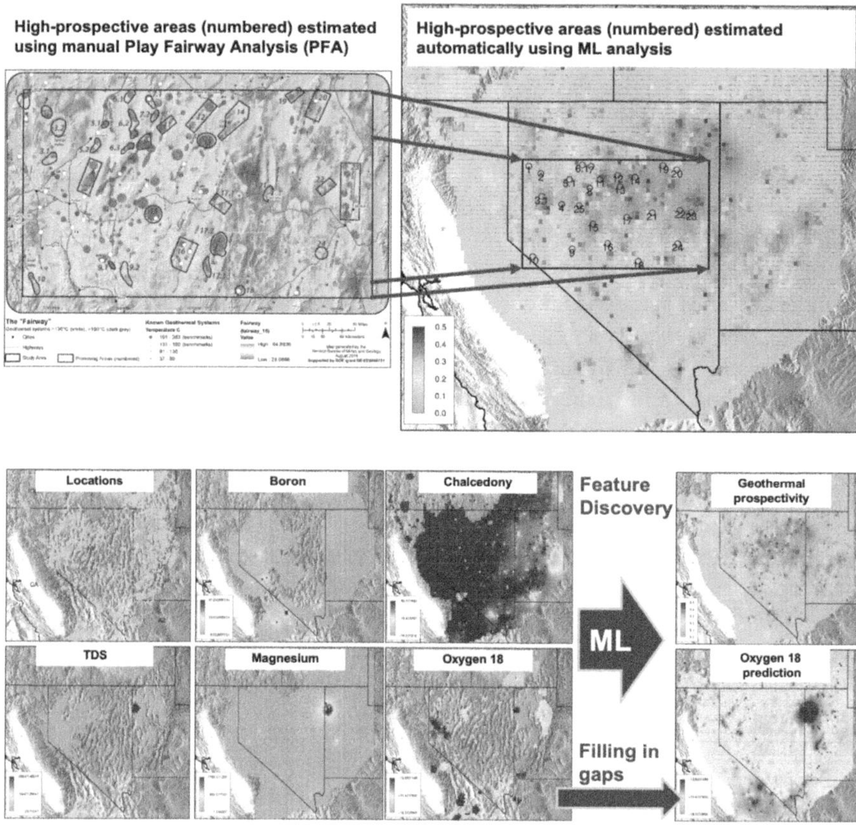

FIGURE 4.7 Numerous geothermal studies and data acquisitions have been conducted over the last 50 years with the Great Basin. Geothermal prospectivity so far has been evaluated through a manual process requiring advanced subject-matter expertise (top left). Now, GeoThermalCloud can achieve similar results with reduced computational cost (top right). GeoThermalCloud has been applied to discover geothermal prospectively of the Great Basin from limited, sparse geochemical data (18 attributes; TDS, B, Mg, etc.) observed with gaps and uncertainties (bottom).

monitor geothermal reservoir evolution. Specifically, they use a transfer learning approach to create models on lab-scale datasets and show how they can be transferred to a geothermal site, where the data availability is limited and sparse.

Similarly, Li [2021] have used MEQs to understand EGS. They used a hybrid ML model to predict the permeability evolution of intermediate-scale (~10 m) hydraulic stimulation experiments at the Sanford Underground Research Facility (SURF) [Kneafsey et al. 2018]. An ML model on MEQ data and injection histories (e.g., wellhead pressure and flow rate) was used to develop a 3D map of in situ permeability evolution for two stimulation episodes at the SURF site. The 3D map details average reservoir permeability evolution over time and local fracture permeability distribution within the evolved geothermal system. They showed that the average permeability predicted by ML is consistent and in good agreement with field observation, resulting in identifying the potential fluid flow path within the geothermal reservoir.

The previously mentioned case studies show considerable success in applying ML to geothermal studies using seismic data. The primary focus of such ML analysis is an interpretation of diverse, complex, and multi-terabytes of waveform datasets [Gao et al. 2021, Zheng et al. 2021]. Supervised ML is attractive in such applications because of the availability of seismic sensor networks (e.g., dense arrays) at various locations, collected and curated label data, and the technique's power in lowering risks associated with drilling and in building a reasonably good reservoir model in different stages of exploration, development, and well optimization.

4.3.2 Operations

There are recent success stories of applying ML to improve thermal energy production efficiencies. Many geothermal power plants could realize significant improvements in operational efficiency using ML-enhanced modeling software. The focus is through the ML analysis of operational power plant and reservoir data (e.g., downhole sensors, wellhead and other surface sensors, company databases, and power plant historians). Buster et al. [2021] have developed a new framework for geothermal operational optimization with machine learning (GOOML). Through a hybrid data-driven thermodynamics approach, GOOML can accurately predict the real-world performance characteristics of as-built geothermal systems. Their framework is adaptable and can readily be integrated into the more significant ML ecosystem for proper state-of-the-art optimization of geothermal operations. The GOOML framework has already been applied to several geothermal power plants because of its success in substantial improvements in operational efficiency [see Section 4 in Buster et al. 2021]. Another advantage of GOOML is that geothermal developers can use it to develop digital twins that will help maximize outputs, optimize processes, and sustainably manage their plant operations.

Another success story using ML and advanced analytics for efficiency and automation in geothermal operations is by Jafarpour and co-workers [Jiang et al. 2020, Liu et al. 2021]. They have developed a deep learning approach to predict and detect faults (abnormal events) in geothermal operations. Typically, measurements collected from geothermal power plants are often high dimensional, highly cross-correlated, and autocorrelated, and influenced by disturbances and subsequent adjustments of operational settings. They have used an autoencoder to extract latent representations to develop a deep learning model. Using historical monitoring and performance data from the binary cycle geothermal power plant, they demonstrated the applicability of ML for abnormal event detection during power plant operations.

4.4 CHALLENGE PROBLEMS – OUR VISION

For ML to have a genuine and transformative impact on the geothermal industry, the primary barrier that needs to be overcome is the availability of informative geothermal data. This data can come from multiple sources such as lab-scale data, physics simulations, and field data. As a step toward overcoming this barrier, the U.S. DOE has supported a part of a larger community-wide Geothermal Code Comparison effort, which resulted in the development of challenge problems [White et al. 2016, 2018]. The issues discussed by the community formed the benchmark problems for modeling EGS leading to various success stories applications [e.g., White et al. 2016, 2018, Mudunuru et al. (2017a, b)]. For example, Figure 4.8 shows an ML workflow for the Fenton Hill reservoir, which used the limited site and extensive physics data to predict thermal power output. We believe that such a similar effort can be undertaken to discover, characterize, and analyze hidden geothermal resources in the Great Basin region. For instance, Figure 4.9 shows the area of interest in which DOE and USGS collect next-generation data (called GeoDAWN) to discover new hydrothermal systems. Specifically, airborne geophysical and 3DEP LiDAR surveys over Nevada and California are acquired to find geothermal resources in the western Great Basin and the Walker Lane region. For example, LiDAR provides information on the fault patterns, which are needed for the discovery and development of geothermal reservoirs. The airborne geophysical surveys provide field-scale distributed data on the subsurface permeability (a couple of tens of meters below the surface), necessary for better characterization of geothermal reservoirs. A challenge we envision is how efficiently the GeoDAWN data can be utilized to discover and characterize blind geothermal systems. When combined with physics, we believe ML on this data can transform how we explore geothermal resources (e.g., discovering hidden hydrothermal signatures) in the Western United States.

Another challenge we envision is the development of ML/AI theory and methods, specifically for geothermal energy applications. Typically, subsurface data availability is limited, sparsely sampled, noisy (e.g., geophysical, remote sensing), and has data gaps. However, such a scenario of sparsely available data and uncertainty in governing physics are less pronounced in computer vision and natural language processing fields, and are primarily data-driven. Therefore, direct utilization of methods and models developed in these fields may not apply to geothermal applications. Hence, there is a need for ML methods that adhere to the underlying physics and collected data to make transformative advances in geothermal science [Baker et al. 2019, Stevens et al. 2020].

4.5 CONCLUSIONS

This chapter provided our perspective on using ML to discover, characterize, and accelerate the utilization of geothermal energy resources. Our view is based on the literature survey of recent advances in using ML for geothermal exploration, development, and operations. First, we discussed the barriers that stood in the

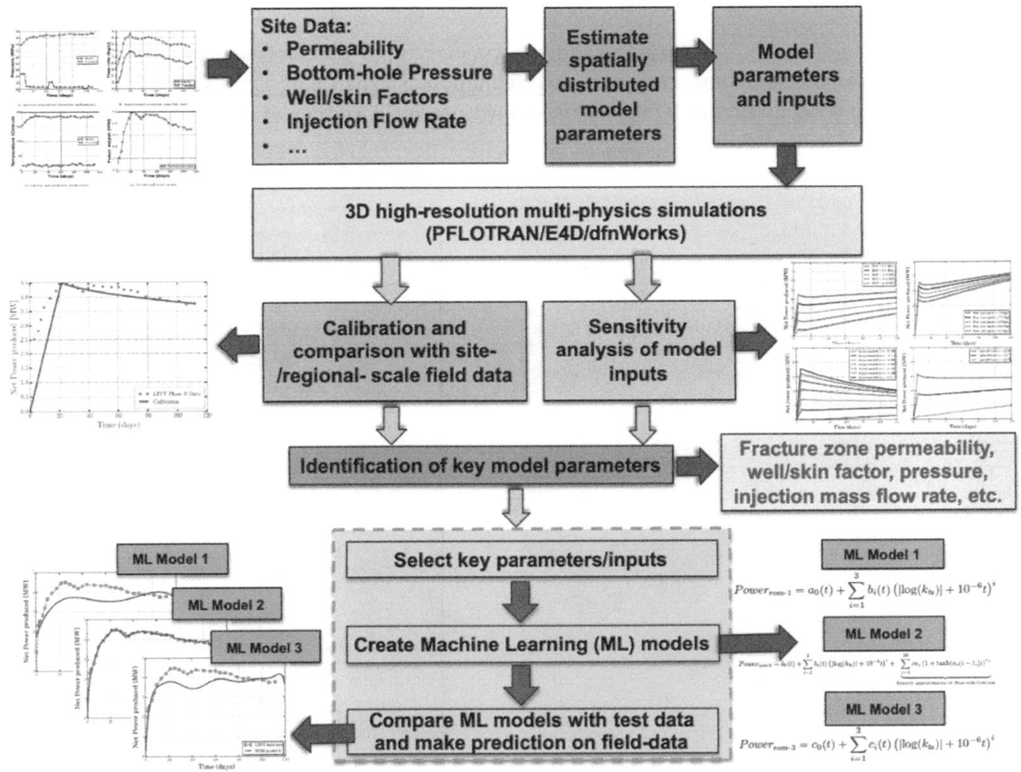

FIGURE 4.8 This figure shows a workflow to develop ML models to predict the thermal energy production at Fenton Hill geothermal reservoir [Credit: Mudunuru et al. 2021, Stanford Geothermal Workshop]. The workflow was an effort of larger wide geothermal community project supported by the DOE to overcome the challenges in modeling EGS reservoirs.

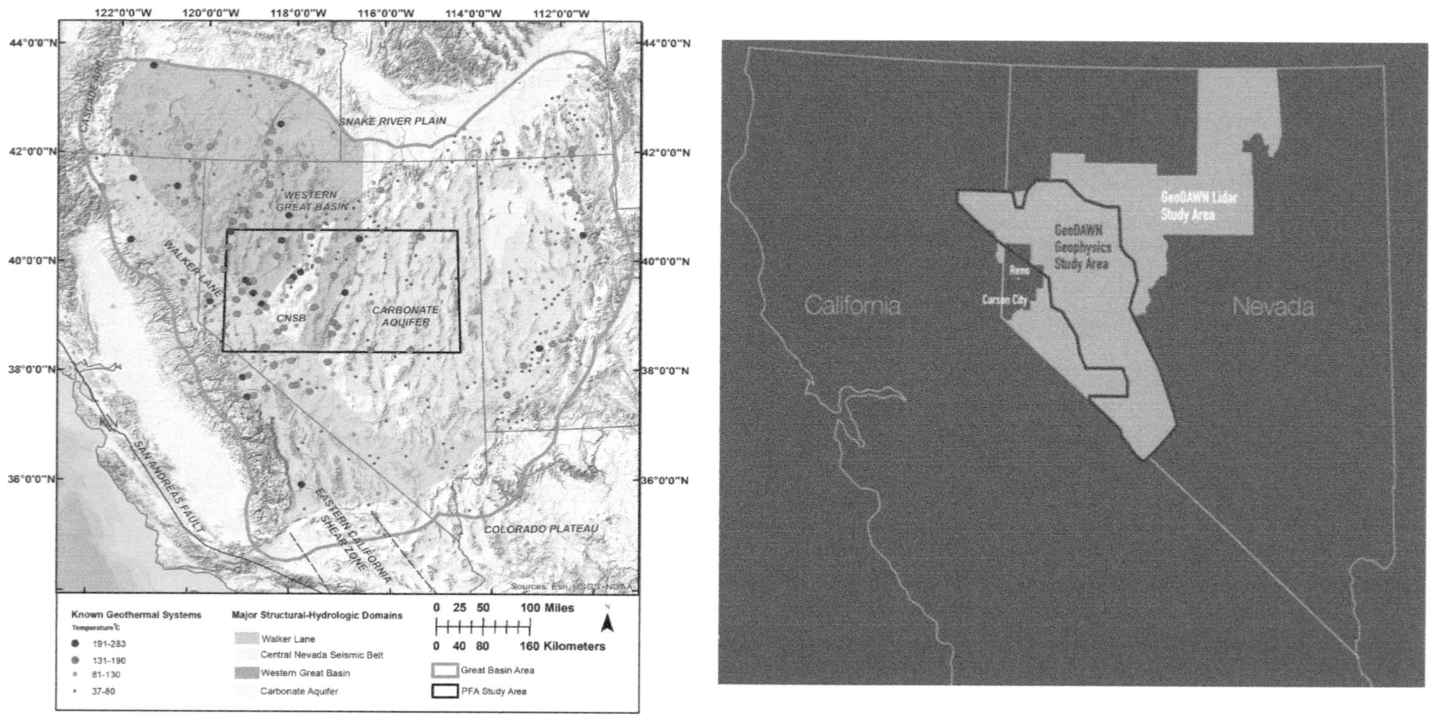

FIGURE 4.9 Next-generation data collection to identify hydrothermal resources. The left figure shows a map of existing high temperature geothermal systems in the Great Basin area [Credit: Smith 2021, dissertation]. The right figure shows the area where geophysical and LiDAR data are collected to accelerate the discovery of hidden systems [Credit: GeoDAWN 2021, website].

way of the widespread deployment of geothermal energy. Then, we presented state-of-the-art ML algorithms and data used to train those models. Next, we discussed popular case studies of using ML for exploration, development, and operations. These examples show the promise of ML to accelerate the widespread deployment of geothermal energy. Toward the end, we summarized a critical geothermal resource exploration challenge related to data availability to successfully develop accurate and robust ML models. We envision that ML on GeoDAWN for exploration of hydrothermal resources can be transformative in making geothermal energy a cost-competitive, widely available, and geographically diverse component of U.S. energy.

ACKNOWLEDGMENTS AND DISCLAIMER

This research is based on work supported by the U.S. Department of Energy's (DOE) Office of Energy Efficiency and Renewable Energy (EERE) under the Geothermal Technology Office (GTO) Machine Learning (ML) for Geothermal Energy funding opportunity, Award Number DE-EE-3.1.8.1. LANL is operated by Triad National Security, LLC, for the National Nuclear Security Administration of the U.S. DOE (Contract No. 89233218CNA000001). PNNL is operated for the DOE by Battelle Memorial Institute under contract DE-AC05-76RL01830. The views and opinions of authors expressed herein do not necessarily state or reflect those of the United States Government or any agency thereof.

REFERENCES

Abubakar, A.J.A., Hashim, M. and Pour, A.B., 2019. Remote sensing satellite imagery for prospecting geothermal systems in an aseismic geologic setting: Yankari Park, Nigeria. International Journal of Applied Earth Observation and Geoinformation, 80, pp.157–172.

Ahmmed, B. and Vesselinov, V.V., 2021. Prospectivity analyses of the Utah FORGE site using Unsupervised Machine Learning. GRC Transactions, 45, p. 213.

Alexandrov, B.S. and Vesselinov, V.V., 2014. Blind source separation for groundwater pressure analysis based on non-negative matrix factorization. Water Resources Research, 50(9), pp.7332–7347.

Alireza Sepehrinezhad, A.S., Srisharan Shreedharan, S.S., Marone, C. and Parisa Shokouhi, P.S., 2021, June. Learning from Laboratory Earthquakes: A Transfer Learning Approach to Active Source Seismic Monitoring. Paper presented at the 55th U.S. Rock Mechanics/Geomechanics Symposium, Virtual, Paper Number: ARMA-2021-2027.

Anderson, A., Blackwell, D., Chickering, C., Boyd, T., Horne, R., MacKenzie, M., Moore, J., Nickull, D., Richard, S. and Shevenell, L.A., 2013. NGDS geothermal data domain: assessment of geothermal community data needs. Boise State Univ., ID (United States).

Augustine, C., 2016. Update to Enhanced Geothermal System Resource Potential Estimate (No. NREL/CP-6A20-66428).

Baker, N., Alexander, F., Bremer, T., Hagberg, A., Kevrekidis, Y., Najm, H., Parashar, M., Patra, A., Sethian, J., Wild, S. and Willcox, K., 2019. Workshop report on basic research needs for scientific machine learning: Core technologies for artificial intelligence. USDOE.

Başer, A. and Sert, M., 2021, August. Permeability Estimation for Natural State Modeling of Geothermal Fields with Machine Learning. In 2021 INISTA (pp. 1–5). IEEE.

Beardsmore, G., 2014. Data fusion and machine learning for geothermal target exploration and characterisation. Technical report, National ICT Australia Limited (NICTA), Australia.

Beltrami, H., 2001. Surface heat flux histories from geothermal data: inferences from inversion. Geophysical Research Letters, 28(4), pp.655–658.

Blackwell, D., Richards, M., Frone, Z., Batir, J., Ruzo, A., 2011. Temperature-at-depth maps for the conterminous US and geothermal resource estimates (No. GRC1029452).

Brown, D.W., Duchane, D.V., Heiken, G. and Hriscu, V.T., 2012. Mining the earth's heat: hot dry rock geothermal energy. Springer Science & Business Media.

Brunton, S.L. and Kutz, J.N., 2019. Data-driven science and engineering: Machine learning, dynamical systems, and control. Cambridge University Press.

Buster, G., Siratovich, P., Taverna, N., Rossol, M., Weers, J., Blair, A., Huggins, J., Siega, C., Mannington, W., Urgel, A., Cen, J., Quinao, J., Watt R. and Akerley, J. 2021. A new modeling framework for Geothermal Operational Optimization with Machine Learning (GOOML). Energies, 14(20), p.6852.

Cavur, M., Moraga, J., Duzgun, H.S., Soydan, H. and Jin, G., 2021. Displacement Analysis of Geothermal Field Based on PSInSAR and SOM Clustering Algorithms A Case Study of Brady Field, Nevada—USA. Remote Sensing, 13(3), p.349.

Cichocki, A., Zdunek, R., Phan, A.H. and Amari, S.I., 2009. Non-negative matrix and tensor factorizations: applications to exploratory multi-way data analysis and blind source separation. John Wiley & Sons.

Cox, S., Lee, N. and Jacobson, M., 2020. RE Data Explorer: Use of the Tool to Support Renewable Energy Project Development (No. NREL/TP-6A20-76573).

Dobson, P.F., 2016. A review of exploration methods for discovering hidden geothermal systems. GRC Transactions, 40, pp.695–706.

Duplyakin, D., Beckers, K.F., Siler, D.L., Martin, M.J, Johnston, H.E., 2022. Modeling subsurface performance of a geothermal reservoir using machine learning. Energies, 15, p.967.

Faulds, J., Hinz, N., Coolbaugh, M., 2015, September. Discovering geothermal systems in the Great Basin region: An integrated geologic, geochemical, and geophysical approach for establishing geothermal play fairways. No. DOE-UNR-06731-06.

Faulds, J.E., Brown, S., Coolbaugh, M., DeAngelo, J., Queen, J.H., Treitel, S., Fehler, M., Mlawsky, E., 2020. Preliminary report on applications of machine learning techniques to the nevada geothermal play fairway analysis. SGW, SGP-TR-216 (Vol. 1).

FEHM. https://github.com/lanl/FEHM; PFLOTRAN: https://www.pflotran.org/

Frash, L., 2021. Geothermal Design Tool (GeoDT). SGW Conference, Stanford, CA.

Gangwani, P., J. Soni, H. Upadhyay and S. Joshi, 2020. A deep learning approach for modeling of geothermal energy prediction. International Journal of Computer Science and Information Security, 18(1).

Gao, K., Huang, L., Lin, R., Hu, H., Zheng, Y. and Cladohous, T., 2021a. Delineating faults in the Soda Lake geothermal field using machine learning. Proceedings of 46th Workshop on Geothermal Reservoir Engineering Stanford University, Stanford, California, February 16–18, 2021, SGP-TR-216.

Gao, K., Huang, L. and Zheng, Y., 2021b. Fault detection on seismic structural images using a nested residual U-Net. IEEE Transactions on Geoscience and Remote Sensing, 60, pp. 1–15, doi: 10.1109/TGRS.2021.3073840.

GDR: Geothermal Data Repository. 2022, https://gdr.openei.org/, Accessed 01/30/2022

Gehringer, M. and Loksha, V., 2012. Geothermal Handbook: Planning and Financing Power Generation a Pre-launch. ESMAP Technical Report 002/12. World Bank, Washington, DC. ©World Bank. License: CC BY 3.0 IGO. https://openknowledge.worldbank.org/handle/10986/23712

Gentleman, R. and Carey, V.J., 2008. Unsupervised machine learning. In Bioconductor case studies (pp.137–157). Springer.

GeoDAWN 2020. https://www.usgs.gov/media/images/geodawn-geoscience-data-acquisition-western-nevada

GeoVision: Harnessing the Heat Beneath Our Feet, 2019.

Getman, D., Anderson, A. 2015. Geothermal Prospector: Supporting geothermal analysis through spatial data visualization and querying tools. Geothermal Resource Council Bulletin, 39, pp.977–986.

Goodfellow, I., Bengio, Y. and Courville, A., 2016. Deep learning. MIT press.

Gudmundsdottir, H. and Horne, R.N., 2020, February. Prediction modeling for geothermal reservoirs using deep learning. SGW Workshop (pp.229–34).

Hamm, S.G., 2021. Geothermal Energy R&D: An Overview of the U.S. Department of Energy's Geothermal Technologies Office. Journal of Energy Resources Technology, 143(10), p.100903.

Hamm, S.G., Augustine, C.R., Tasca, C. and Winick, J., 2019. An Overview of the US Department of Energy's GeoVision Report.

He, Q., Barajas-Solano, D., 2020. Physics-informed neural networks for multiphysics data assimilation with application to subsurface transport. Advances in Water Resources, 141, p.103610.

Holtzman, B.K., Paté, A., Paisley, J., 2018. Machine learning reveals cyclic changes in seismic source spectra in Geysers geothermal field. Science Advances, 4(5), p.eaao2929.

Innes, M., Edelman, A., Fischer, K., Rackauckas, C., Saba, E., Shah, V.B. and Tebbutt, W., 2019. A differentiable programming system to bridge machine learning and scientific computing. arXiv preprint arXiv:1907.07587.

Jiang, A., Zhen, Q., Jafarpour, B., 2020, Recurrent neural networks for predicting the dynamic response of geothermal reservoirs from monitoring data. American Geophysical Union Abstract H166–0031.

Jordan, M.I. and Mitchell, T.M., 2015. Machine learning: Trends, perspectives, and prospects. Science, 349(6245), pp.255–260.

Karra, S., Ahmmed, B. and Mudunuru, M.K., 2021. AdjointNet: Constraining machine learning models with physics-based codes. arXiv preprint arXiv:2109.03956.

Kiran, R. and Salehi, S., 2020, October. Assessing the Relation between Petrophysical and Operational Parameters in Geothermal Wells: A Machine Learning Approach. Proceedings of the 45th Workshop on Geothermal Reservoir Engineering Stanford University, Stanford, California, February 10–12, 2020, SGP-TR-216.

Kneafsey, T.J. et al., EGS Colab Project, 2018.

Li Z. Hybrid machine learning model to predict 3D in-situ permeability evolution, 2021. Proceedings of the Stanford Geothermal Workshop, February 16–18, Stanford, CA.

Liu, Y., Ling, W., Young, R., 2021, February. Deep learning for prediction and fault detection in geothermal operations. GRE, Stanford, CA, USA (pp.15–17).

Marone, C. 2021. Data arrays for microearthquake (MEQ) monitoring using deep learning for the Newberry EGS sites. United States Geothermal Data Repository. doi:10.15121/1787546.

Matzel, E., Magana-Zook, S., Mellors, R.J., 2021. Looking for permeability on combined 3d seismic and magnetotelluric datasets with machine learning. Proceedings of the 45th Workshop on Geothermal Reservoir Engineering Stanford University, Stanford, California, February 10–12, 2020, SGP-TR-216.

Mitusch, S.K., Funke, S.W. and Kuchta, M., 2021. Hybrid FEM-NN models: Combining artificial neural networks with the finite element method. arXiv preprint arXiv: 2101.00962.

Moore, J., McLennan, J., Allis, R., 2019, February. The UTAH FORGE: An International Laboratory for Enhanced Geothermal System Technology Development. In Proceedings of the 44th Workshop on Geothermal Reservoir Engineering Stanford University, Stanford, California, February 11–13, 2019, SGP-TR-214.

Mudunuru, M.K., Carey, J.W., Chen, L., 2021. Subsurface energy: Flow and reactive-transport in porous and fractured media. In Handbook of Porous Materials: Volume 4-Porous Materials for Energy Conversion and Storage (pp.323–395). worldscientific.com

Mudunuru, M.K., Karra, S., Harp, D.R., Guthrie, G.D. and Viswanathan, H.S., 2017a. Regression-based reduced-order models to predict transient thermal output for enhanced geothermal systems. Geothermics, 70, pp.192–205.

Mudunuru, M.K., Karra, S., Makedonska, N. and Chen, T., 2017b. Sequential geophysical and flow inversion to characterize fracture networks in subsurface systems. Statistical Analysis and Data Mining: The ASA Data Science Journal, 10(5), pp.326–342.

Olasolo, P., Juárez, M.C., Morales, M.P. and Liarte, I.A., 2016. Enhanced geothermal systems (EGS): A review. Renewable and Sustainable Energy Reviews, 56, pp.133–144.

Pandey, S. N. and Singh, M., 2021. Artificial neural network to predict the thermal drawdown of enhanced geothermal system. Journal of Energy Resources Technology, 143(1), 010901.

Pepin, J. 2019. New approaches and insights to geothermal resource exploration and characterization. PhD thesis, New Mexico Institute of Mining and Technology.

Person, M., Kelley, S., Kelley, R., Karra, S., Harp, D., Witcher, J., 2015. Hydrogeologic windows. GRC Transactions, 39, pp.751–759.

PFA GTO, 2015. Play Fairway Analysis, Department of Energy.

Raissi, M., Perdikaris, P. and Karniadakis, G.E., 2019. Physics-informed neural networks: A deep learning framework for solving forward and inverse problems involving nonlinear partial differential equations. Journal of Computational Physics, 378, pp.686–707.

Roberts, S., Delorey, A., Johnson, C.W., Guyer, R., Alfaro-Diaz, R. and Johnson, P., 2021. Using Surface Deformation and Machine Learning to Determine State of Stress Changes at the Coso Geothermal Field, California USA. Proceedings of the 46th Workshop on Geothermal Reservoir Engineering Stanford University, Stanford, California, February 15–17, 2021, SGP-TR-218.

Robins, J.C., Kolker, A., Flores-Espino, F., Pettitt, W., Schmidt, B., Beckers, K., Pauling, H. and Anderson, B., 2021. 2021 US Geothermal Power Production and District Heating Market Report (No. NREL/TP-5700-78291).

Schoenholz, S. and Cubuk, E.D., 2020. Jax MD: a framework for differentiable physics. Advances in Neural Information Processing Systems, 33.

SEIA, 2021. https://www.seia.org/initiatives/whats-megawatt

Shi, Y., Song, X. and Song, G., 2021. Productivity prediction of a multilateral-well geothermal system based on a long short-term memory and multi-layer perceptron combinational neural network. Applied Energy, 282, p.116046.

Siler, D.L. and Pepin, J.D., 2021. 3-D Geologic controls of hydrothermal fluid flow at Brady Geothermal Field, Nevada, USA. Geothermics, 94, p.102112.

Siler, D.L., Pepin, J.D., Vesselinov, V.V., Mudunuru, M.K. and Ahmmed, B., 2021. ML to identify geologic factors associated with production in geothermal fields: a case-study using 3D geologic data, Brady geothermal field, Nevada. Geothermal Energy, pp.1–17.

Singh, A., Thakur, N. and Sharma, A., 2016, March. A review of supervised machine learning algorithms. 3rd International Conference on Computing for Sustainable Global Development (INDIACom), pp. 1310–1315.

Smale, M.C., Wiser, S.K., Bergin, M.J. and Fitzgerald, N.B., 2018. A classification of the geothermal vegetation of the Taupō Volcanic Zone, New Zealand. Journal of the Royal Society of New Zealand, 48(1), pp.21–38.

Smith, C.M., 2021. Machine Learning Techniques Applied to the Nevada Geothermal Play Fairway Analysis (Doctoral dissertation, University of Nevada, Reno).

Smith, C.M., Faulds, J.E., Brown, S., Coolbaugh, M., Lindsey, C.R., Treitel, S., 2021. Characterizing Signatures of Geothermal Exploration Data with Machine Learning Techniques: GRE conference paper.

Stanton, J.S., Anning, D.W., Brown, C.J., Moore, R.B., McGuire, V.L., Qi, S.L., Harris, A.C., Dennehy, K.F., McMahon, P.B., Degnan, J.R. and Böhlke, J.K., 2017. Brackish groundwater in the United States (No. 1833). U.S. Geological Survey.

Stevens, R., Taylor, V., Nichols, J., Maccabe, A.B., Yelick, K. and Brown, D., 2020. AI for Science (No. ANL-20/17). Argonne National Lab. (ANL), Argonne, IL (United States).

STOMP. https://www.pnnl.gov/projects/stomp

Tartakovsky, A.M., Marrero, C.O., Perdikaris, P., Tartakovsky, G.D. and Barajas-Solano, D., 2020. Physics-informed deep neural networks for learning parameters and constitutive relationships in subsurface flow problems. Water Resources Research, 56(5), p.e2019WR026731.

Tester, J., Reber, T., Beckers, K., Lukawski, M., Camp, E., Aguirre, G.A., Jordan, T. and Horowitz, F., 2015, April. Integrating geothermal energy use into re-building American infrastructure. In Proceedings of the World Geothermal Congress, April 19–22.

van der Meer, F., Hecker, C., van Ruitenbeek, F., van der Werff, H., de Wijkerslooth, C. and Wechsler, C., 2014. Geologic remote sensing for geothermal exploration: A review. International Journal of Applied Earth Observation and Geoinformation, 33, pp.255–269.

Vesselinov, V.V., 2021. Cloud fusion of big data and multi-physics models using machine learning for discovery, exploration, and development of hidden geothermal resources. Los Alamos National Laboratory Internal Report.

Vesselinov, V.V., Mudunuru, M.K., Ahmmed, B., Karra, S. and Middleton, R.S., 2020. Discovering signatures of hidden geothermal resources based on unsupervised learning (No. LA-UR-20-21030).

Vesselinov, V.V., Mudunuru, M.K., Karra, S., O'Malley, D. and Alexandrov, B.S., 2019. Unsupervised machine learning based on non-negative tensor factorization for analyzing reactive-mixing. Journal of Computational Physics, 395, pp.85–104.

Weers, J. and Anderson, A., 2016. The DOE geothermal data repository and the future of geothermal data (No. NREL/CP-6A20-65967).

White, M., Fu, P., McClure, M., Danko, G., Elsworth, D., 2018. A suite of benchmark and challenge problems for enhanced geothermal systems. Geomechanics, pp.79–117.

White, M.D., Podgorney, R., Kelkar, S.M., McClure, M.W., Danko, G., Ghassemi, A., Fu, P., Bahrami, D., Barbier, C., Cheng, Q. and Chiu, K.K., 2016. Benchmark problems of the geothermal technologies office code comparison study (No. PNNL-26016).

Williams, C.F., Reed, M.J., Mariner, R.H., 2008. Assessment of moderate-and high-temperature geothermal resources of the United States (USGS No. 2008-3082).

Witter, J.B., Trainor, W.J. and Siler, D.L., 2019. Uncertainty and risk evaluation during the exploration stage of geothermal development: A review. Geothermics, pp.233–242.

Wright, P.M., Ward, S.H., Ross, H.P. and West, R.C., 1985. State-of-the-art geophysical exploration for geothermal resources. Geophysics, 50(12), pp.2666–2696.

Zheng, Y., Li, J., Lin, R., Hu, H., Gao, K. and Huang, L., 2021. Physics-guided machine learning approach to characterizing small-scale fractures in geothermal fields. In Proceedings, GRE Conference, Stanford University.

Section III

Drilling Operations Applications

5 Real-Time Drilling and Completion Analytics

From Cloud Computing to Edge Computing and Their Machine Learning Applications

Dingzhou Cao[1], Jingshuang Xue[2], and Yu Sun[3]
[1]Devon Energy,
Oklahoma City, Oklahoma
[2]Pioneer Natural Resources,
Dallas, Texas
[3]Our Lady of the Lake University,
San Antonio, Texas

CONTENTS

DOI: 10.1201/9781003207009-8

5.1 REAL-TIME DRILLING AND COMPLETION SYSTEM

Real-time analytics nests in and acts as the heart of the real-time system. Without discussing the real-time system, it is pointless to discuss the real-time analytics part. In this chapter, the real-time system will be discussed first, followed by various real-time machine learning (ML) applications. Real-time technology was first applied in the drilling domain, which gave birth to the real-time drilling (RTD) system and later was expanded to cover completion operations (RTC). In this chapter, the focus of RTD/RTC and their ML applications will be limited to the shale operations, not the offshore and deep water operations in conventional reservoirs.

5.1.1 Brief History of RTD Systems

The RTD system has been available for decades. It evolved from the stand-along rig-site drilling data monitoring system (Kastrop 1971; Kennedy 1973). With satellite communication (Veenkant and Vitali 1986), microwave communication (Isaacs and Bobo 1984), and fiber-optic communication (Booth 2009, 2010), the stand-along rig-site drilling data monitoring systems were connected to develop the RTD monitoring/analytics systems. They became part of the real-time drilling operations centers (RTDOCs) starting from the 80s up to today, either built/maintained by service companies, operators, or a collaboration of them. Booth (2009, 2010) provided a good literature overview of RTDOCs. This kind of RTDOC was usually staffed with drilling subject-matter experts (SMEs) to provide 24/7 support for the drilling operations, and the cost of building and maintaining such an RTDOC was very high. This is especially true in the 80s, when computer systems were not as popular as they are today and only some big players could afford it, not to mention the 24/7 staffing labor cost. As stated by Booth (2009, 2010), except Superior's RTD data center, all other RTDOCs mentioned in his papers could not survive during oil and gas downtime cycles, or merges/acquisitions. Nowadays, thanks to the popularity of computer systems and the internet, building such RTDOCs is not as costly, but running them is still not cheap due to the 24/7 staff cost. Because of the high daily rate, such RTDOCs usually are only used to support high-cost offshore/deep water well drilling operations. In contrast, for low-budget unconventional land wells, drilling teams usually are reluctant to pay the high daily rate of an RTDOC to use such services. After 2010, thanks to the popularity of cloud services (i.e., the pay as you go business model and very low upfront cost), low cost of PC hardware, and the availability of open-sources software, the barrier to build a real-time system is not as high as the RTDOC, especially without the 24/7 support staff. Thus, a lightweight real-time system became popular within unconventional drilling domains. This system pulled real-time data from Wellsite Information Transfer Standard Markup Language (WITSML) servers, analyzed the data in the cloud, and finally pushed the real-time analysis results to the web-based User Interface (UI) or smartphone. Such a lightweight RTD system can run automatically with minimal information technology (IT) support. The lightweight RTD system has the following features: (1) unlike the RTDOC, which is usually built/maintained by big operators or service companies, the lightweight RTD is usually built by unconventional drillers (Behounek 2017; Liu et al. 2019; Cao 2017, 2021; Cao et al. 2018, 2019) or newborn startups; (2) without 24/7 support SMEs, the primary goal of a lightweight RTD is to deliver real-time results via smartphones or computers anytime and anywhere to people who need it, including drillers in the field, and drilling engineers/managers in the office; and (3) commercial automation systems usually are deployed in three modes including shadow mode, advisory mode, and closed-loop mode. This lightweight RTD usually runs in shadow mode or advisory mode. However, the ultimate goal of such a system is to evolve into the closed-loop mode and hook up with a control system to materialize partial drilling/completion automation (i.e., auto drilling/fracturing). The discussion of real-time systems in this chapter refers to the lightweight real-time systems in the unconventional reservoir community.

5.1.2 Stream Processing Technology Selection and Discussion

With the evolution of sensors and big data technology, more data is available for analyzing and the data needs to be processed as soon as possible. To avoid adding latency to the real-time system, the timing of processing the data is becoming more demanding, especially with the goal of the real-time system being upgraded to closed-loop mode from advisory mode. There are two ways to process real-time data: data at rest and data in motion. For data at rest, as data flows in, it is saved to a database first and is subsequently pulled out of the database for processing. It is inevitable that the latency will be added to the real-time system as saving and pulling data in and out of the database takes time. The other way of processing the data in motion would allow data to be computed immediately as data was received, namely stream processing. Stream processing enables processing data with minimal latency, while still being able to process a large amount of data. The stream processing engine is a key element of any modern real-time system. Figure 5.1 illustrates the most common stream processing engines available in the market. The lower left open-source

FIGURE 5.1 Stream processing engines (not complete list).

(After Cao 2017)

ones are more flexible and almost free; the lower right lists the cloud services, and the upper right shows the off-the-shelf products. The license fee of off-the-shelf products should be a consideration. The closed-source ones also have the vendor dependency issue: once a system is built within certain services or products, the system is dependent on those services/products. Boom and bust cycles are very common in the oil and gas industry. To increase the survival rate of a system during bust cycles, the selection of stream processing technology is very important. A solution is to use the open-source technology with the available commercial version to avoid vendor dependency.

Examples of vendor dependency are the RTD and RTC systems in these studies (Cao et al. 2018, 2019; Ben et al. 2020a). The RTD was built on top of the Tibco StreamBase (now Tibco Streaming) stream processing engine and the RTC was built with Google Cloud Services (GCP). However, like the majority of the RTDOC discussed in the previous section, the RTC and RTD systems cannot avoid a similar fate: they cannot survive in the merger/acquisition. For RTD, the cost is the main consideration as the Tibco Streaming annual license fee is not cost-effective. For RTC, it was built on top of the GCP, but the new company's main cloud provider is not GCP. Due to the acquisition, the RTC needs to be either rebuilt within another cloud service or totally shut down.

5.1.3 From Cloud Computing to Edge Computing

The bar to build the real-time system is not high due to the popularity of cloud services. Thus, the RTD/RTC systems in the industry are almost all cloud based. However, as a real-time system, from a data transmission perspective, the dataflow in a cloud-based system is like a detour (wellsite to cloud and back to wellsite) and it is not a true real-time system. Cloud-based systems are also highly dependent on internet connectivity, which is usually an issue as the wells are almost always drilled in remote areas. For the real-time system, both the advisory mode and the closed-loop mode require edge computing. A popular analogy of edge computing is autonomous vehicles. Imagine if an autonomous vehicle collects the real-time driving data, sends it to the cloud for computing, and the analytics results are sent back to the vehicle to guide the vehicle to drive. If latency or network disconnect occurs, the auto-driving vehicle may have a car incident. The same scenario also applies to cloud-based real-time systems. Thus, there is a trend that moves the real-time analytics out of clouds and into wellsites (Cao 2021; Jacobs 2021; Korolov 2021; Violine 2021), namely edge computing. An example of such a system is the legacy WPX Energy's edge computing system (Cao 2021). With all the lessons learned, in 2020, legacy WPX Energy (now Devon Energy) started to build the InnFrac®, a real-time edge computing system, for its hydraulic fracturing operations with a brand new edge stream processing architecture. When the solution was designed, it tried to address the limitations of the previous system (Cao 2017; Cao et al. 2018, 2019) and was the architecture for future analytics needs. Two primary highlights of the solution are as follows. (1) The data source is agnostic. The InnFrac® architecture is the fundamental architecture for legacy WPX's real-time analytics strategy; it would be expanded to cover its drilling operation (InnDrill®) (Korolov 2021; Violine 2021) and production operation (InnProd®) in the future. (2) The cloud provider is agnostic. The open-source technology Apache Kafka was used as the backbone of InnFrac®, which frees InnFrac® from vendor dependency. It takes a few days, rather than weeks or months, for the migration of InnFrac® from one cloud provider to the other. This is an important factor to take into consideration when designing the system, mainly because the oil and gas industry is a very cyclical industry, and M/A happens along with the cycle of commodity price. To increase the survival rate of the system, the portability of the technology is very important.

5.2 REAL-TIME ANALYTICS

Real-time analytics could provide rapid actionable information to operations to make decisions and change operations from reactive to proactive. In drilling and completion, a lot of human know-how and interpretation is involved in decision-making. Real-time analytics can transform those human know-how and operational expertise into algorithms to achieve repeatable and consistent operations performance. Compared with non-real-time analytics, real-time analytics has several characteristics: (1) the real-time analytics model can only use the real-time data up to the current timestamp, unlike the historical data analytics where an amount of previously existing data can be used and (2) the model should be computation cost-effective. The model inference time should be less than the inference period. For example, if the model will be inferenced every 1 second and it is 1-Hz real-time data, the model inference time must be less than 1 second; otherwise, there will be data congestion in the real-time system. Let's use a toll highway analogy as an example. A real-time system is like a toll highway, and the car depicts the real-time data point. The model inference is like the toll gate; every second, there is a car entering the highway. If the model inference time is more than 1 second, then most likely it takes more than 1 second for the car to pass the toll gate. In this way, more and more cars will be stuck in the toll gate and there will be traffic jams as time goes on. The same scenario applies to real-time systems. If the model inference time is longer than the model inference period, then there will be data traffic in the system and the latency will be added to the real-time system. For most ML applications discussed in the next several sections, these studies claimed to be real-time applications but did not address the model computation cost, which should be discussed. Real-time analytics is a broad concept including descriptive analytics (what happens), diagnostic analytics (why it happened), predictive analytics (what will happen), and prescriptive analytics (how we make something happen), which contains ML (a general term for when computers learn from data and conduct predictive and prescriptive analytics) as part of it (Mishra et al. 2021). Deep learning (DL) is a subset of ML methods based on artificial neural networks with representation learning (Wikipedia 2003, 2011). In this chapter, the discussion focuses on specific use cases when the ML/DL model has been deployed in production or in a place where it can be very easily deployed in a real-time setting.

5.2.1 Real-Time Drilling Machine Learning Applications

The common drilling issues, such as axial/lateral/torsional vibration, excessive torque and drag, bit wear, bad hole cleaning, stuck pipe incidents, loss of circulation, wellbore instability, and washout, are the main causes of tool failures and drilling dysfunction. Many drilling operations are reactive in nature, in that the dysfunction first occurs and then action is taken to remedy the situation. Real-time analytics and monitoring could provide rapid actionable information to operations to make decisions and change operations from reactive to proactive. Three main focuses in RTD analytics are discussed in the following sections: drilling key performance indicators (KPI) tracking and efficiency improvement; dysfunction prediction to eliminate or mitigate the potential risk; and downhole conditions prediction using surface data, such as bit wear prediction and downhole vibration prediction.

5.2.1.1 KPI Tracking and Drilling Efficiency Improvement

5.2.1.1.1 Rig State Classification for KPI Tracking

Accurate rig state classification is the foundation to calculate KPI, analyze invisible lost time, and is the key input to develop real-time advanced analytics to predict drilling dysfunction and improve efficiency. For example, drilling and tripping connection times can be calculated based on rig states, and any slight improvement to reduce connection time during repetitive operations can potentially have a substantial performance gain on multiple wells and multiple rigs. Ben et al. (2019) used a rule-based model and three ML models of Random Forest (Wikipedia 2005b), convolutional neural networks (CNN), and a hybrid CNN/recurrent neural network (RNN) model (Wikipedia 2005a, 2013) to classify rig states. The architectures of three models can be found in Figure 5.2. ML models were far superior compared with the rule-based model, and two DL

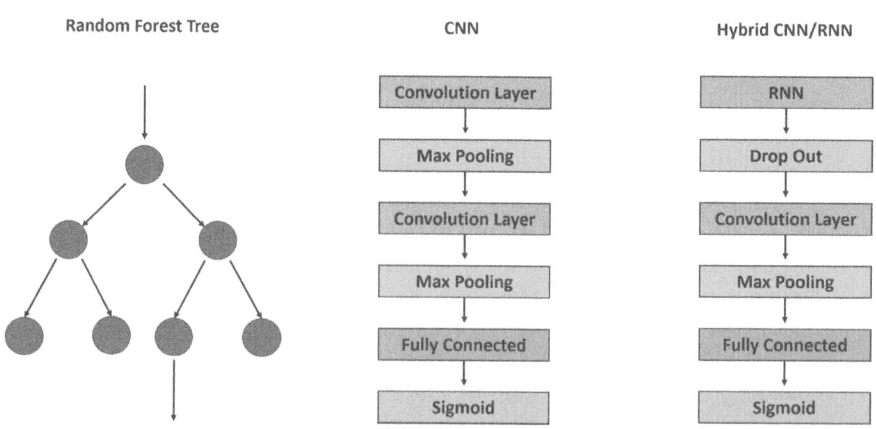

FIGURE 5.2 Random forest, CNN, and hybrid CNN/RNN architectures.

(After Ben et al. 2019)

models are more capable of handling complex drilling patterns compared with Random Forest. The CNN model is finally chosen to deploy into a real-time system because its inference time is less than the hybrid CNN/RNN model. Ben et al. (2020c) proposed a reliable, accurate, scalable solution using the rig state prediction model as a case study. It proposed to run an offline model utilizing semantic segmentation and batch data aggregation, a more complicated model with a better accuracy, to continuously monitor the performance of the online model. The offline model accuracy is periodically compared with the online model, and the retraining process can be triggered if the online model accuracy is deviated from offline model by a certain threshold (the second opinion mechanism). This workflow utilizing cloud pipeline can be easily adopted in other real-time ML models in the drilling. How to retrain the model effectively and automatically at the right time with minimal manual effort is another key in a real-time setting because the field operation may change over time and the previously trained model may not be accurate for the current operation.

5.2.1.1.2 Directional Drilling Efficiency

Directional drilling is essential for horizontal wells to ensure better reservoir exposure and less wellbore collision risk. Faster and reliable data-driven decisions to improve drilling efficiency and optimize well path is desired for horizontal drilling. Mud motor and rotary steerable systems (RSS) are mainly used in directional drilling, and the decision to utilize RSS or mud motor is based on economical and operational considerations.

In sliding using a mud motor, it is critical to determine the optimal operation control sequence of the motor. Yu et al. (2021) trained an automatic agent for motor directional drilling using reinforcement learning to interact with an environment through choices of controls in a sequence. The environment is affected by the agent's actions and returns corresponding rewards (positive or negative) to the agent. A drilling simulator with layered formation and bottom hole assembly (BHA) directional responses was used to train this directional-drilling agent, and the agent can successfully drill to target in the simulated environment through decisions of sliding and rotating actions to follow a planned trajectory. This paper presents the prototype of an automated directional drilling decision-making system and would be easily integrated with the real-world dataset and applied in the field.

Mud pulses "downlinking" in RSS are used as commands to steer the mud motor while drilling. Identifying the duration of downlinking is very important to calculate high-quality operational KPIs and evaluate its performance and economics; it is also helpful to reduce false positives, which may present in trend analysis and drilling dysfunction predictions. Cao et al. (2020a) developed a real-time DL model using the dynamic U-Net concept (Ronneberger et al. 2015), the pre-trained ResNet-34 (He et al. 2016), and transfer learning, to detect and estimate the duration of downlinking based on standpipe pressure (SPP). Again, it was treated as an image segmentation problem and SPP time-series data was segmented based on a stand of pipe that was drilled and then was fed into a

DL model. The model was trained on 10 onshore wells drilled with mud motors and validated on additional tests using RSS wells; the model performance had a 99.93% accuracy.

5.2.1.2 Dysfunction Prediction

5.2.1.2.1 Washout Detection

Washout is caused by small holes in drill pipes that quickly erode, and it can result in severe and expensive failure such as fishing events, long non-productive time (NPT), and a sidetrack in the worst scenario if not detected at an early stage. An early signal of washout would enable the field engineers to make the right decision to pull out of the hole for a washout check instead of continuing drilling to avoid potential cost. Jeong et al. (2020) built a hybrid deep neural network that embeds physics models with ML workflow that can be trained by available data. Physics models have great interpretability, and the power of ML can capture the hidden patterns of data that the traditional physics model may not catch. The hybrid model contained a physics model utilizing the Darcy-Weisbach equation to catch the relationship between SPP and mud flow velocity for this specific washout problem; the parameter network for estimating model parameters to learn the system behavior based on the embedded physics model; and finally, the residue network for predicting regression or classification results. The paper applied Dep Kalman filter as a deep neural network model for integrating the parameter network outputs and the physics model in the residue network. Once trained, the hybrid model can control the parameters of the physics model and predict the SPP trend in real time. The result showed the hybrid model had better predictability and sensitivity to washout detection compared with the data-driven approach without a physics model, and it provided a general-purpose framework for combining physics models with ML models for many drilling problems.

5.2.1.2.2 Standpipe Pressure Prediction

High-pressure spike monitoring and prediction is very important for safer and more optimized drilling operations, and it is a good indicator for many drilling issues such as circulation. Erge and van Oort (2021) proposed a hybrid physics-based and data-driven modeling for SPP prediction. Its physics-based model used a new shear-rate and time-dependent thixotropy model, and its data-driven model utilizing DL neural networks was trained with flow rate, rotation rate, block velocity, bit depth, and hole depth to predict SPP. After obtaining the results from the physics-based and DL model, a rule-based stochastic decision-making algorithm by developing a hidden Markov model was developed to combine these model results. It gave higher weight and trust to the data-driven model if the physics-based model was inaccurate compared with the data-driven model and vice versa. The hybrid model combines the best features of data-driven and physics-based approaches and yields better prediction of pressure compared with using standalone physics-based or data-driven modeling.

5.2.1.3 Downhole Condition Prediction

5.2.1.3.1 Bit Wear Prediction

Monitoring bit performance and making predictions on cutter wear conditions to optimize the decision of when to pull out of the hole for a new bit is very important to improve drilling performance and reduce NPT. Liu et al. (2018) used 40 bit runs including bit dull grade, bit records, BHA information, and surface drilling parameters to quantify the downhole bit wear condition. The paper first identified wear factor, defined as const*WOB/(ROP/bit RPM), was most sensitive to cutter wear and its value variation can be quantified to distinguish influence from formation change and cutter wear damage. Then a decision tree algorithm was trained on the historical data to determine the relationship between wear factor and bottom drilling hours to define the range to trigger each state of bit wear. This model can establish effective bit pull criteria for the field guidance and only requires real-time surface drilling data to be implemented in real time. The final application based on the model successfully rolled out onto multiple rigs in west Texas and had a very high prediction accuracy.

Witt-Doerring et al. (2021) proposed a similar way to calculate the wear metric using surface drilling data to predict bit wear in real time. This paper identified dysfunction from surface data such as stick-slip, bit/BHA whirl, and downhole motor fail and used bit dull photos to do forensics to understand the mode and cause of cutter damage. Surface drilling data and bit dull photos were combined to build the relationship between the wear metric and observed bit wear. The final recorded wear metric value can be equated to the severity of the bit damage when not related to tool or downhole failure. The presented methodology and wear metric were validated using nine wells and 51 lateral production runs. Final results can separate the wear severity into four separate categories; all relevant sets of dull photos were split into these categories and pull criteria were established.

5.2.1.3.2 Downhole Vibration Prediction

Downhole vibration is a major part of downhole tool failure and NPT in drilling. Direct downhole vibration measurement relies on downhole sensors but it's neither feasible to install a downhole sensor on every well due to high cost nor transfer downhole data in real time to the surface due to bandwidth constraints of mud telemetry. It would be great to predict downhole vibration from surface drilling data by utilizing advanced analytics.

Zha and Pham (2018) proposed a DL model consisting of a parallel combination of CNN and RNN to classify drilling data at the surface and identify downhole vibration from surface data (Figure 5.3). It was trained on multi-channel surface data with simultaneously recorded downhole data to recognize patterns in the surface drilling data relating to severe downhole lateral and torsional vibrations. Applying CNN to drilling data learned essential patterns closely related to the stick-slip vibration process, and RNN learned essential sequential relationships within the data. Due to the heavily imbalanced nature of the training dataset, the data augmentation was applied by randomly stretching and adding noise

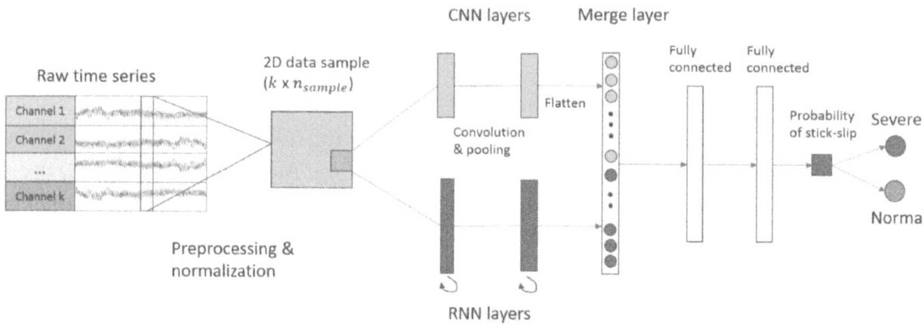

FIGURE 5.3 DL workflow for downhole vibration predication.

(After Zha and Pham, 2018)

to the training data. The validation dataset shows an accuracy of 99% and precision of 97%. This paper demonstrated the model's capability in recognizing surface data patterns that correspond to severe downhole stick-slip vibration, and this model can be easily deployed on a drilling rig to provide automated real-time diagnostics and advisory.

5.2.2 Real-Time Completion Machine Learning Applications

In the real-time completion analytics, most ML applications can be divided into two categories: (1) fracturing events recognition such as the recognition of stage start/end and (2) pressure predictions such as the wellhead pressure predictions and the screen-out predictions. Due to space limitations, only a few typical ML applications are discussed. These applications are related to each other rather than independent. For example, the prediction of wellhead pressure can be used to predict screen-out, as well as to adjust the hydraulic fracturing pumping schedule to conduct cost optimization in real time.

5.2.2.1. Real-Time Fracturing Events Recognition

5.2.2.1.1 Stage Start/End Recognition

Stage start and end are the key flags for stagewise analysis and the fundamental element for further advanced hydraulic fracturing analysis. However, manual selection of stage start/end is often very time-consuming and prone to inaccuracies due to the lack of uniform selection and interpretation methods across the industry, and human bias. Shen et al. (2020a) first formulated the stage start/end flags prediction problem as a near real-time time-series classification problem. The rule-based models and the CNN were applied to predict the stage flags. The flag-wise accuracy metric was first proposed and defined to evaluate the model performance. A total of six different scenarios were analyzed, and the combination of the median filter and the CNN model gave the best overall performance with a 0.95 F1 score. Compared with the rule-based method, the CNN model achieves higher accuracy, and is much more robust against noise and outliers in the data. However, the CNN model needed to buffer a certain amount of data, which resulted in a delay in the prediction. Thus, the CNN model could be deployed as a near real-time application in the real-time system.

5.2.2.2 Real-Time Pressure Prediction

5.2.2.2.1 Screen-Out Prediction

During hydraulic fracturing, screen-out is a problem that may cause the delay of placement of subsequent stages, the potential failure of wellbore or wellhead equipment, loss production, and safety concerns. Sun et al. (2020) tried to predict the screen-out in real time using a combination of physics-based methods and ML models. The physics-based inverse slope method was easy to implement for real-time applications. The issue is that it might lead to a large amount of false positive screen-out predictions. Thus, Sun et al. (2020) used the weight average to build an ensemble model based on the inverse slope model, and the CNN-long short-term memory (LSTM) model, to provide advanced warning of screen-out in real time. Compared with the individual model, the ensemble model improves the accuracy of prediction very well. However, the authors did not discuss the computing cost of the CNN-LSTM model in the real-time setting, as this may be a concern as the inference time of the LSTM type model is usually longer than other models like CNN.

5.2.2.2.2 Hydraulic Fracturing Pressure Prediction

The hydraulic fracturing pressure (wellhead pressure) is a very important measurement during the hydraulic fracturing jobs because engineers must monitor the wellhead pressure and adjust the pumping schedule in real time to avoid screen-out, optimize the proppant and fluid amounts, and minimize cost at the same time. Thus, the prediction of wellhead pressure in real time would bring huge business values. Ben et al. (2020a) used continuous learning mechanisms and several neural network architectures including the CNN, RNN, and multi-layer perceptron (MLP), to predict the wellhead pressure. Continuous learning means being able to update the prediction model with newly available data and keep the model evergreen. In a real-time fashion, using the 2-minute forecasting window as an example, for every 2 minutes, the past 2 minutes of data was added to the training dataset to retrain the model and the new trained model was used to predict the next 2 minutes of wellhead pressure. The timing of prediction results available to engineers was very important, thus the computation cost (also named inference time) of the model was a key factor for the model to be deployed in the real-time system. In Ben's (2020a) study, the computation cost of the proposed ML models was discussed and MLP was concluded as the lightest model with the least computation cost and acceptable accuracy. It was recommended for deploying to the cloud-based RTC system, although for such real-time applications it was not a perfect solution. As discussed in the previous sections, the cloud-based real-time systems have the risk of internet instability and latency issues and would not guarantee the prediction results would be available to engineers at the wellsite in real time. The perfect place for deploying such models was the edge computing system. With the acknowledgement of such limitations of cloud-based systems, the co-author of this chapter, Cao, re-architected the real-time system from cloud to edge and proposed a new edge computing system (the legacy WPX Energy's Edge Stream Processing Solution), which was already discussed in the previous section.

5.2.2.2.3 Real-Time Completion Cost Optimization

One of the final goals of predicting the wellhead pressure is to conduct real-time completion cost optimization. Based on the predicted wellhead pressure, Ben et al. (2020b) used the model predictive control (MPC) mechanism to optimize the completion cost subject to real-time operational constraints. The principle of MPC was to use a model to predict the system behavior and find the optimal actions to manipulate the control inputs to achieve a certain target. The optimization workflow was proposed like this: It started with the wellhead pressure model to predict the next several minutes of wellhead pressure. If there was a potential screen-out, then the action was to "cut sand." If not, the MPC optimizer would provide optimal actions to manipulate the three control variables (slurry rate, proppant concentration, and friction reducer concentration) to control the actual wellhead pressure. This proposed solution was ideally to be deployed in a latency-free edge computing system, rather than the cloud-based real-time system.

5.3 CHALLENGES AND FUTURE DIRECTIONS

There are several challenges in RTD and completion analytics, such as lacking good quality of data, the life-cycle model management of real-time ML models (aka MLOps), and the model adoption issue.

The data quality/integrity issue is an old topic. Garbage in garbage out. Without the good quality of data, analytics would become useless. In drilling, real-time surface drilling data with the standard WITSML format is usually available through electronic drilling recorder (EDR). However, not all the data needed for real-time application are available. For example, the real-time equivalent circulating density model required real-time mud properties for high accuracies; however, usually only a few mud property measurements are available each day if the real-time mud skid isn't utilized. Non-sensor-based channels in EDR may require manual input from the field, which is highly prone to human input error. Some ML applications required additional well metadata such as drill string and casing components to get a good prediction. Cao et al. (2020b) demonstrated an automated and sustainable digital transformation to address well data digitization issues. To enhance the quality and access of valuable data, it usually requires redesign and adjustment of current workflow, which may take a relatively long time to complete. Another challenge related to drilling data is enough volume of good data to be trained. Take drilling dysfunction prediction, for example, historical dysfunction events usually take only a very small fraction of the whole dataset. For some rare dysfunction events, very limited data are available for training. In completion domain, the data quality issue is more significant as hydraulic fracturing doesn't have similar industry standards like WITSML and aggregators in the drilling domain. Imaging multiple data sources collected from various service companies, without a designated aggregator to conduct the tasks like timestamp syncing and channel mapping, are hardly used for any analytics. The solution to this is analogous to what happens in the drilling domain; a designated on-site aggregator would conduct the aggregation tasks and make sure that the data is clean before being sent to the analytics models. Some start-ups already can provide such data aggregation services in frac sites. However, establishing an industrywide data collection protocol is not foreseen in the near term, although there were already some joint industry project (JIP) discussions among shale operators and service companies.

For real-time ML applications, life-cycle model management is another challenge. The models deployed in the real-time system may need to be retrained periodically to adapt to the new data. On the other hand, for real-time analysis, especially like the real-time series classification, there is no ground truth available to check the accuracy of the model. How to retrain the model effectively and automatically at the right time with minimal manual effort is not easy to address in a real-time setting. A possible solution to this challenge is as proposed by Cao et al. (2019b) and Ben et al. (2020c), which is to build an ML pipeline that consists of auto detection of accuracy deviation and model retrain mechanism (the second opinion mechanism). For ML algorithm selection, real-time applications' model accuracy and fast inference speed are equally important, whereas look-back/historical study cares more for accuracy. It gives extra constraints on selecting the algorithms. A very complicated model even with higher accuracy is usually not favored in the real-time setting.

The last significant challenge is the adaptation of the analytics/ML models. The final goal of the real-time analysis is to evolve into the closed-loop mode to materialize a certain degree of automation in drilling and completion operations. However, none of the ML applications discussed in the previous sections are closed-loop applications. They are either in shadow mode or in advisory mode. When talking about automation in drilling or completion, autonomous driving is always a good analogy. However, unlike autonomous driving, which heavily relies on closed-loop ML algorithms, there are almost no closed-loop ML applications in literature in the drilling and completion domain. Even for the non-ML closed-loop use cases, there are only several of them available in literature (Israel et al. 2017, 2018; Lai et al. 2021). There is a long way to go for the ML applications to evolve from advisory mode to closed-loop mode. One of the reasons is the model's adaption, either due to the difficulty of explaining ML models because of its black box nature or job security threat the field personnel think the ML technology would bring to them. How to get buy-in from the field so that engineers are willing to use the ML results to aid their decision making, and finally get their trust so that they feel comfortable to be hands off in the control-loop is the key. A possible way to do this may be similar to the human factor engineering and change management strategy Behounek et al. (2017, 2019) discussed in their studies when they wanted to introduce automation tools to rigs. The trust and adoption of ML models sometimes is a culture problem. Early involvement with the field and effective communication of how algorithms work throughout the whole process will help to build the trust and relationship.

REFERENCES

Behounek, Michael, Blake Millican, Brian Nelson, Matthew Wicks, Eugene Rintala, Matthew White, Taylor Thetford, Pradeepkumar Ashok, and Dawson Ramos. 2019. "Change Management Challenges Deploying a Rig-Based Drilling Advisory System." Paper presented at the *SPE/IADC International Drilling Conference and Exhibition, The Hague, The Netherlands*, March 2019. doi: https://doi.org/10.2118/194184-MS

Behounek, Michael, Taylor Thetford, Lisa Yang, Evan Hofer, Matthew White, Pradeepkumar Ashok, Adrian Ambrus, and Dawson Ramos. 2017. "Human Factors Engineering in the Design and Deployment of a Novel Data Aggregation and Distribution System for Drilling Operations." Paper presented at the *SPE/IADC Drilling Conference and Exhibition, The Hague, The Netherlands*, March 2017. doi: https://doi.org/10.2118/184743-MS

Ben, Yuxing, Chris James, and Dingzhou Cao. 2019. "Development and Application of a Real-Time Drilling State Classification Algorithm with Machine Learning." Paper presented at the *SPE/AAPG/SEG Unconventional Resources Technology Conference, Denver, Colorado, USA*, July 2019. doi: https://doi.org/10.15530/urtec-2019-253

Ben, Yuxing, Michael Perrotte, Mohammadmehdi Ezzatabadipour, Irfan Ali, Sathish Sankaran, Clayton Harlin, and Dingzhou Cao. 2020a. "Real-Time Hydraulic Fracturing Pressure Prediction with Machine Learning." Paper presented at the *SPE Hydraulic Fracturing Technology Conference and Exhibition, The Woodlands, Texas, USA*, February 2020. doi: https://doi.org/10.2118/199699-MS

Ben, Yuxing, Sathish Sankaran, Clayton Harlin, and Michael Perrotte. 2020b. "Real-Time Completion Cost Optimization Using Model Predictive Control." Paper presented at the *SPE Hydraulic Fracturing Technology Conference and Exhibition, The Woodlands, Texas, USA*, February 2020. doi: https://doi.org/10.2118/199688-MS

Ben, Yuxing, Weilu Han, Chris James, and Dingzhou Cao. 2020c. "Building a General and Sustainable Machine Learning Solution in a Real-Time Drilling System." Paper presented at the *IADC/SPE International Drilling Conference and Exhibition, Galveston, Texas, USA*, March 2020. doi: https://doi.org/10.2118/199603-MS

Booth, J. Ernest. 2009. "Drilling Operations Centers: A History of Functionality and Organizational Purpose-The First Generation." Paper presented at the *SPE Digital Energy Conference and Exhibition, Houston, Texas, USA*, April 2009. doi: https://doi.org/10.2118/123225-MS

Booth, J. Ernest. 2010. "Real-Time Drilling Operations Centers: A History of Functionality and Organizational Purpose—The Second Generation." Paper presented at the *SPE Intelligent Energy Conference and Exhibition, Utrecht, The Netherlands*, March 2010. doi: https://doi.org/10.2118/126017-MS

Cao, Dingzhou. 2017. "Rapid Development of Open Real-Time Drilling Analytics System." YouTube Video, 19:17, October 25, 2017. https://www.youtube.com/watch?v=tMwMGWY58Co.

Cao, Dingzhou. 2021. "Design for TRUE Real-Time Analytics-Driven Decision-Making, the edge Computing." YouTube Video, 53:29, April 28, 2021. https://www.youtube.com/watch?v=P9U4xybPzLc.

Cao, Dingzhou, Yuxing Ben, Chris James, and Kate Ruddy. 2019. "Developing an Integrated Real-Time Drilling Ecosystem to Provide a One-Stop Solution for Drilling Monitoring and Optimization." Paper presented at the *SPE Annual Technical Conference and Exhibition, Calgary, Alberta, Canada*, September 2019. doi: https://doi.org/10.2118/196228-MS

Cao, Dingzhou, Don Hender, Sam Ariabod, Chris James, Yuxing Ben, and Micheal Lee. 2020b. "The Development and Application of Real-Time Deep Learning Models to Drive Directional Drilling Efficiency." Paper presented at *the IADC/SPE International Drilling Conference and Exhibition, Galveston, Texas, USA*, March 2020. doi: https://doi.org/10.2118/199584-MS

Cao, Dingzhou, Don Hender, Sam Ariabod, Kate Ruddy, and Chris James. 2020a. "Digital Transformation Strategy Enables Automated Real-Time Torque-and-Drag Modeling." Paper presented at the *IADC/SPE International Drilling Conference and Exhibition, Galveston, Texas, USA*, March 2020. doi: https://doi.org/10.2118/199670-MS

Cao, Dingzhou, Chad Loesel, and Sanjay Paranji. 2018. "Rapid Development of Real-Time Drilling Analytics System." Paper presented at the *IADC/SPE Drilling Conference and Exhibition, Fort Worth, Texas, USA*, March 2018. doi: https://doi.org/10.2118/189595-MS

Erge, Oney, and Eric van Oort. 2021. "Hybrid Physics-Based and Data-Driven Modeling for Improved Standpipe Pressure Prediction." Paper presented at the *SPE/IADC International Drilling Conference and Exhibition, Virtual*, March 2021. doi: https://doi.org/10.2118/204094-MS

He, Kaiming, Xianyu Zhang, Shaoqing Ren, and Jian Sun. 2016. "Deep Residual Learning for Image Recognition," *2016 IEEE Conference on Computer Vision and Pattern Recognition (CVPR), 2016*, pp. 770–778, doi: 10.1109/CVPR.2016.90.

Isaacs, W.R., and J.E. Bobo. 1984. "Design and Impact of a Real-Time Drilling Data Center." Paper presented at the *SPE Annual Technical Conference and Exhibition, Houston, Texas*, September 1984. doi: https://doi.org/10.2118/13109-MS

Israel, Riaz, Julian Farthing, Hamish Walker, Rodrigo Gallo Covarrubias, Jason Bryant, and Christian Vahle. 2017. "Development to Delivery - A Collaborative Approach to Implementing Drilling Automation." Paper presented at the *SPE/IADC Drilling Conference and Exhibition, The Hague, The Netherlands*, March 2017. doi: https://doi.org/10.2118/184695-MS

Israel, Riaz, Doug McCrae, Nathan Sperry, Brad Gorham, Jacob Thompson, Kyle Raese, Steven Pink, and Andrew Coit. 2018. "Delivering Drilling Automation II – Novel Automation Platform and Wired Drill Pipe Deployed on Arctic Drilling Operations." Paper presented at the *SPE Annual Technical Conference and Exhibition, Dallas, Texas, USA*, September 2018. doi: https://doi.org/10.2118/191574-MS

Jacobs, Trent. 2021. "As road to Real-time Fracturing Optimization is Paved, Shale Producers Approach On-Ramps Cautiously." Accessed December 21, 2021. https://jpt.spe.org/as-road-to-real-time-fracturing-optimization-is-paved-shale-producers-approach-on-ramps-cautiously.

Jeong, Cheolkyun, Yingwei Yu, Darine Mansour, Velizar Vesslinov, and Richard Meehan. 2020. "A Physics Model Embedded Hybrid Deep Neural Network for Drillstring Washout Detection." Paper presented at the *IADC/SPE International Drilling Conference and Exhibition, Galveston, Texas, USA*, March 2020. doi: https://doi.org/10.2118/199629-MS

Kastrop, J. E. 1971. "Automatic Drilling Data Acquisition, Processing and Telemetry." Petroleum Engineer (Aug. 1971): 36–62.

Kennedy, J. L. 1973. "Data Monitoring on Today's Rig." The Oil and Gas Journal (Sept. 24, 1973): 119–126.

Korolov, Maria. 2021. "Edge devices improve drilling efficiency for energy company." Accessed December 21, 2021. https://www.networkworld.com/article/3620072/energy-company-deploys-edge-devices-to-improve-drilling-efficiency.html

Lai, Stephen W., James Ng, Aaron Eddy, Sergey Khromov, Dan Paslawski, Ryan van Beurden, Lars Olesen, Gregory S. Payette, and Benjamin J. Spivey. 2021. "Large-Scale Deployment of a Closed-Loop Drilling Optimization System: Implementation and Field Results." SPE Drill & Completion 36 (2021): 47–62. doi: https://doi.org/10.2118/199601-PA

Liu, Yu, Justin Kibbey, Yanbin Bai, and Xianping Wu. 2018. "Real-Time Bit Wear Monitoring and Prediction Using Surface Mechanics Data Analytics: A Step Toward Digitization Through Agile Development." Paper presented at the *IADC/SPE Drilling Conference and Exhibition, Fort Worth, Texas, USA*, March 2018. doi: https://doi.org/10.2118/189602-MS

Liu, Yu, Xianping Wu, Theresa Baumgartner, Douglas Zimpfer, Denver Wedgwood, Cody Teff, Jiaqi Zhu, Yuying Jia, Yucheng Zhao, and Lei Tao. 2019. "Real-Time Drilling Advisor Appstore – An Agile Development and Deployment Program." Paper presented at the *SPE/IADC International Drilling Conference and Exhibition, The Hague, The Netherlands*, March 2019. doi: https://doi.org/10.2118/194100-MS

Mishra, Srikanta, Jared Schuetter, Akhil Datta-Gupta, and Grant Bromhal. 2021. "Robust Data-Driven Machine-Learning Models for Subsurface Applications: Are We There Yet?" Journal of Petroleum Technology 73 (2021): 25–30. doi: https://doi.org/10.2118/0321-0025-JPT

Ronneberger, O., P. Fischer, and T. Brox. 2015. U-Net: Convolutional Networks for Biomedical Image Segmentation. In: Navab, N., Hornegger, J., Wells, W., Frangi, A. (eds) Medical Image Computing and Computer-Assisted Intervention – MICCAI 2015. MICCAI 2015. *Lecture Notes in Computer Science*, vol 9351. Springer, Cham. https://doi.org/10.1007/978-3-319-24574-4_28

Shen, Yuchang, Dingzhou Cao, Kate Ruddy, and Luis Felipe Teixeira de Moraes. 2020a. "Near Real-Time Hydraulic Fracturing Event Recognition Using Deep Learning Methods." *SPE Drill & Completion* 35 (2020): 478–489. doi: https://doi.org/10.2118/199738-PA

Sun, Jianlei John, Arvind Battula, Brandon Hruby, and Paymon Hossaini. 2020. "Application of Both Physics-Based and Data-Driven Techniques for Real-Time Screen-Out Prediction with High Frequency Data." Paper presented at the *SPE/AAPG/SEG Unconventional Resources Technology Conference, Virtual*, July 2020. doi: https://doi.org/10.15530/urtec-2020-3349

Veenkant R., and J. D. Vitali. 1986. "Satellite Communications Change Drilling Operations and Procedures." Paper presented at the SPE 1986 International Meeting on Petroleum Engineering, Beijing, China, March 17–20, 1986.

Violine, Bob. 2021. "3 enterprise AI success." Accessed December 21, 2021. https://www.infoworld.com/article/3615449/3-enterprise-ai-success-stories.html

Wikipedia. 2003. "Machine learning." Last modified May 11, 2022. https://en.wikipedia.org/wiki/Machine_learning.

Wikipedia. 2005a. "Random forest." Last modified May 4, 2022. https://en.wikipedia.org/wiki/Random_forest.

Wikipedia. 2005b. "Recurrent neural network." Last modified April 30, 2022. https://en.wikipedia.org/wiki/Recurrent_neural_network.

Wikipedia. 2011. "Deep learning." Last modified May 09, 2022. https://en.wikipedia.org/wiki/Machine_learning.

Wikipedia. 2013. "Convolutional neural network." Last modified April 28, 2022. https://en.wikipedia.org/wiki/Convolutional_neural_network.

Witt-Doerring, Ysabel, Paul Pastusek, Pradeepkumar Ashok, and Eric van Oort. 2021. "Quantifying PDC Bit Wear in Real-Time and Establishing an Effective Bit Pull Criterion Using Surface Sensors." Paper presented at the *SPE Annual Technical Conference and Exhibition, Dubai, UAE*, September 2021. doi: https://doi.org/10.2118/205844-MS

Yu, Yingwei, Wei Chen, Qiuhua Liu, Minh Chau, Velizar Vesselinov, and Richard Meehan. 2021. "Training an Automated Directional Drilling Agent with Deep Reinforcement Learning in a Simulated Environment." Paper presented at the *SPE/IADC International Drilling Conference and Exhibition, Virtual*, March 2021. doi: https://doi.org/10.2118/204105-MS

Zha, Yang, and Son Pham. 2018. "Monitoring downhole drilling vibrations using surface data through deep learning." Paper presented at the SEG International Exposition and Annual Meeting, *Anaheim, California, USA*, October 2018. doi: https://doi.org/10.1190/segam2018-2964198.1

6 Using Machine Learning to Improve Drilling of Unconventional Resources

Ruizhi Zhong
University of Queensland,
Brisbane, Australia

CONTENTS

6.1 DRILLING OF UNCONVENTIONAL RESOURCES

Unconventional resources (tight oil and gas resources or coal seam gas [CSG]) have increased contribution in the energy supply. In the United States, over 60% of natural gas is produced from unconventional gas resources (Aminzadeh and Dasgupta 2013). Unlike conventional resources that are stored in porous and permeable rocks, unconventional resources are trapped in low-permeability rocks. To economically produce these unconventional resources, drilling techniques such as horizontal drilling, hydraulic fracturing, and underbalanced drilling have been implemented.

Shale is an organically rich sedimentary rock formed in low permeability layers (in the order of nano-Darcy). It contains tight oil and gas that cannot be extracted by conventional technology. With revolutionary techniques of horizontal drilling

DOI: 10.1201/9781003207009-9

and hydraulic fracturing, oil and gas production has been greatly improved in these ultra-tight formations since the 1950s (Boyer et al. 2006). CSG (also known as coalbed methane [CBM]) is another common type of unconventional resource, which is stored in naturally fractured coal seams. It experienced rapid development in Australia and China over the last two decades (Al-Jubori et al. 2009). For instance, Australia has been the largest liquefied natural gas (LNG) export country due to the increased exploration and production of CSG in the Surat and Bowen basins. The depth of coal seams is usually less than 1000 m, which is shallower than that of shales (1500–4000 m). The CSG wells are usually vertical, but recently there are more horizontal wells drilled to increase gas production. For hydraulic fracturing, it is also commonly used in CSG wells. In addition, underbalanced drilling is frequently used to reduce formation damage because there are multiple coal seams in a production well.

6.2 REVIEW OF MACHINE LEARNING STUDIES IN UNCONVENTIONAL RESOURCES DRILLING

This section briefly reviews some machine learning studies in the following areas: drilling hydraulics, drilling dynamics, drilling problems, drilling target identifications, and miscellaneous drilling applications. These machine learning studies are either related to horizontal drilling and underbalanced drilling in unconventional reservoirs drilling, or direct drilling applications in shale or CSG reservoirs. In general, a huge volume of high-frequency drilling data is routinely recorded by service companies. These drilling data combined with high-resolution petrophysical logs have great potentials, which can reveal rock properties, solve some drilling problems, and improve drilling activities.

Drilling hydraulics. Knowledge of drilling hydraulics is important to main safe drilling pressure (avoiding borehole collapse or tensile fracturing), cementing displacement, and cuttings transport. Machine learning studies are mainly focused on the estimation of pressure loss and cuttings concentration. For the cuttings transport study, Ozbayoglu et al. (2002) used an artificial neural network (ANN) to predict cuttings bed height in horizontal and highly inclined wells. Experimental data from a large-scale flow loop are used, and ANN outperforms the least squares method (LSM) with an error rate percentage less than 10%. Rooki et al. (2014) used ANN and multiple linear regression (MLR) to predict cuttings concentration for foam drilling. The foam drilling can be used in underbalanced drilling, which is advantageous for CSG drilling as it can minimize the formation damage. The foam properties (e.g., foam velocity and quality), subsurface conditions (pressure and temperature), and pipe rotation are input parameters. However, this study is limited because only 60 samples are used and cuttings conditions (size and density) are not varied. Nevertheless, ANN has the best performance (absolute average percentage error [AAPE] of 3.2%) compared with the MLR model and mechanistic model. In terms of pressure loss for horizontal wells, Rooki (2015) developed an ANN for Herschel-Bulkley drilling fluids for horizontal annulus, using wellbore information and rheological properties as inputs. The model achieves an extremely high R^2 value of 0.998. Barati-Harooni et al. (2016) used a genetic algorithm (GA) to predict pressure loss in inclined annuli during underbalanced drilling. The model deals with the multiphase flow (gas and liquid) and the results are excellent as R^2 value of 0.99 has been achieved. Considering that analytical solutions exist for pressure loss during drilling, these machine learning models have little value and usually have very good performance. However, machine learning can model the more complex multiphase flow problem (possibly the combination of fluid, solids, and gas) during drilling as conventional methods are difficult to handle this type of problem.

Drilling dynamics. Drilling parameters (e.g., rate of penetration [ROP], torque, rotations per minute [RPM], weight-on-bit [WOB], etc.) are routinely recorded during drilling. These parameters help drillers monitor the drilling and make real-time decisions to optimize drilling or detect anomalies. Recently, there has been increased interest in using machine learning to predict ROP and further optimize drilling, especially for deep offshore wells. Drilling dynamics is complex due to the involvement of fluid mechanics, solid mechanics, and rock mechanics. Hegde et al. (2017) and Soares and Gray (2019) conducted studies to compare the performance of ROP prediction using traditional physics-based models and data-driven models (ANN, random forest [RF], support vector machines [SVMs]). Machine learning models have significantly better accuracies on ROP prediction. Later Hegde et al. (2018) studied some optimization algorithms for a well drilled in the Bakken Shale and found that the eyeball method and random-search method are plausible for real-time drilling optimization. Over 20% ROP improvement was achieved with a computational run time of less than 100 ms. The study also tested some metaheuristic algorithms; however, the longer computational run time makes them undesirable for real-time optimization.

Drilling problems. Drilling problems (e.g., stuck pipe, lost circulation, well control, etc.) are commonly encountered during drilling. Machine learning can potentially save billions of dollars each year by detecting these drilling problems. For stuck pipe, Miri et al. (2007) used data from 64 sidetracked and horizontal wells in Persian Gulf Offshore Iran with two types of ANNs. It was found that differential pressure, fluid loss, plastic viscosity, and gel strength are important to pipe sticking. Jahanbakhshi et al. (2012) compared the performance of SVM and ANN and concluded that SVM outperformed the ANN on the stuck pipe prediction. In addition, some controllable parameters (i.e., mud properties) can be modified to optimize the drilling, which can potentially save drilling costs. Lost circulation has been one of the major drilling problems and can cause non-productive time (NPT). Behnoud far and Hosseini (2017) used an ANN to estimate lost circulation volume during underbalanced drilling. The study found that factors such as pump pressure, mud weight, and flow rate are important to loss circulation, and controlling these factors can minimize the lost circulation volume. Jahanbakhshi et al. (2014) used ANN to predict lost circulation volume and investigated the influence of geomechanical properties in naturally fractured reservoirs. It was found that minimum in-situ stress, natural fracture orientation, uniaxial compressive strength, tensile strength, and Young's modulus are important parameters.

Drilling target identification. Due to some unique characteristics of unconventional resources (e.g., low permeability, high or low gamma ray, low density), the drilling response combined with wireline logs can be used to identify these resources. For shales, Kormaksson et al. (2015) developed an automatic data-driven workflow to identify sweet spots in shale plays. The functional principal component analysis (PCA) is used on well logs with over 30 features. The results showed that the accuracy is 0.71 for oil predictions and 0.9 for gas predictions. For coal seams, Zhong et al. (2020b) used drilling and logging-while-drilling (LWD) data to identify low-density coals in six wells from Surat Basin, Australia. Five machine learning algorithms and two data manipulation techniques are used. The coal identification rate exceeds 90% in all wells. It was found that high ROP, low gamma ray, and large hole diameter are strong indicators of coal zones. In another study by Zhong et al. (2020a), pseudo density logs were generated and coals can be directly identified from generated density logs.

Miscellaneous drilling applications. Besides the common drilling applications mentioned earlier, some other novel applications are worth mentioning. For instance, some natural language processing (NLP) studies have been conducted to extract valuable information from reports. Castineira et al. (2018) and Ma et al. (2018) used NPL on drilling reports to identify the NPT and associated drilling problems. It was found that horizontal wells underperform in both drilling and production. The NPT can be reduced and the production rate of horizontal wells can be enhanced after NLP identifies the issues. AI-Yami et al. (2016) used a Bayesian network to develop drilling expert systems, which cover a range of applications including well control, underbalanced drilling, drilling fluids, completion, and cementing practices. For instance, optimal placement of horizontal wells has been conducted using expert opinions and real-time data in the Bayesian network. Similarly, Pollock et al. (2018) used reinforcement learning (RL) to develop an advisory system for directional drilling. By controlling tool face angle and guiding drill bit sliding during directional drilling, the efficiency of directional drilling has been improved as ROP is optimized and less personnel on board (POB) is required.

6.3 CASE STUDIES

In this section, two machine learning case studies are presented. The first study is a classification task to identify coal pay zones. The second study is a regression task to estimate the height of cuttings bed.

6.3.1 Case Study 1: Coal Identification

Over the past two decades, there has been rapid development in Australia and China to satisfy a growing gas demand in the Eastern Hemisphere (AI-Jubori et al. 2009). To maximize gas production, it is important to accurately identify coal pay zones. Conventional field methods of coal identification include coring and well logging. In CSG companies, it is a common practice to primarily use the density log to identify coals, which have a density cutoff of 1.8–2.0 g/cm^3.

However, these methods are relatively time and resource intensive. This study uses machine learning to identify the lithology (i.e., coal or non-coal), which can potentially be a validation method in addition to conventional coal identification methods or replace the density log to identify coals. Five supervised machine learning algorithms are used in this study: logistic regression (LR), SVM, ANN, RF, and XGBoost. Some algorithms (LR, SVM, ANN) require data preparation/normalization for the input data.

6.3.1.1 Data Manipulation Techniques

Many real-world classification problems are imbalanced, such as detection of fraudulent telephone calls (Fawcett and Provost 1996), medical diagnostics (Woods et al. 1993), and text classification (Zheng et al. 2004). The consequence of an imbalanced data problem is that machine learning algorithms usually perform well on the majority class but badly on the minority class. To solve this problem, some data manipulation techniques (e.g., oversampling or under-sampling techniques) can be applied to enhance the performance on the minority class. In this study, the non-coal is the majority class and the coal is the minority class because coal zones are much less than non-coal zones (generally <10% net-to-gross). As we are interested in the coal pay zones, using only machine learning algorithms would usually result in poor performance on coal predictions. Thus, two oversampling techniques are used to improve coal predictions. The first technique is naïve random oversampling (NROS). NROS (Chawla et al. 2004) simply generates new samples by randomly sampling the minority class. For coal identification, NROS randomly selects samples from coal zone samples and duplicates these samples multiple times (depending on the percentage of the minority class). The second technique is the synthetic minority oversampling technique (SMOTE). SMOTE (Chawla et al. 2002) is an oversampling method that generates new samples around the neighbors of a minority class. One advantage of SMOTE over NROS is that it can reduce the overfitting problem because SMOTE does not replicate the minority class samples.

6.3.1.2 Evaluation Metrics for Imbalanced Data

The coal identification is a classification task. Figure 6.1 shows the confusion matrix for the coal identification problem. Only true positive (TP) and true negative are correct predictions.

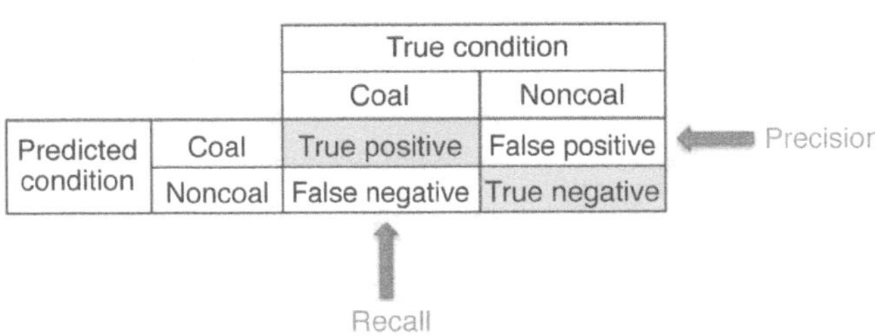

FIGURE 6.1 Confusion matrix for the coal identification problem. (After Zhong et al. 2020b, 1246.)

To better quantify the performance on the minority class (i.e., coal), precision, recall, and F1 score are used to evaluate the model performance:

$$\text{Precision} = \frac{\text{True positve}}{(\text{True positive} + \text{False positive})} \tag{6.1}$$

$$\text{Recall} = \frac{\text{True positve}}{\text{True positive} + \text{False negative}} \tag{6.2}$$

$$\text{F1 Score} = \frac{2}{\dfrac{1}{\text{Precision}} + \dfrac{1}{\text{Recall}}} = 2\frac{\text{Precision} \cdot \text{Recall}}{\text{Precision} + \text{Recall}} \tag{6.3}$$

Note that the recall is TP over the sum of TP and false negative, which represents the percentage of predicted coals in all true coals. Recall describes the ability to predict true coals from machine learning. F1 score is the harmonic mean of the precision and the recall, which considers the overall model performance.

6.3.1.3 Well Information

The coal identification techniques are illustrated on an example dataset from six wells in the Surat Basin, Australia. Table 6.1 shows a summary of the six wells. The measured depth (MD) of these wells is up to 1000 m and the true vertical depth (TVD) is up to around 500 m. These wells are inclined with final inclinations of about 70 degrees. The last column is the number of samples from each well. Well 4 has more samples than other wells because the resolution of the LWD log is higher.

Raw real-time drilling data and LWD data have been used. There are some preprocessing steps for data preparation, which include selecting effective drilling periods, extracting drilling data, and matching drilling data with LWD data. Usually there are several drilling periods from the received data; however, only long drilling periods that contain newly drilled formations are selected. Once the

TABLE 6.1
Summary of Study Wells

Well No.	TVD (m)	MD (m)	Number of Samples
1	95.9-478.0	96.1-950.5	8,510
2	184.5-483.6	190.1-950.7	7,607
3	97.7-495.0	98.0-949.9	8,520
4	99.6-346.5	99.8-480.4	14,159
5	363.3-510.2	544.4-1009.1	4,523
6	274.1-444.5	290.0-744.7	4,532

Source: Zhong et al. (2020b, 1246).
MD, measured depth; TVD, true vertical depth.

drilling periods are determined, the corresponding drilling data are extracted. The next step involves matching drilling data with LWD data. The drilling data are time-based and LWD data are depth-based. Depending on the resolution of LWD logs, the drilling data are processed to have the same resolution as the LWD data. For instance, most LWD logs in this study have a 0.1-m resolution; then the time-based drilling logs are converted to depth-based logs with 0.1-m resolution.

6.3.1.4 Procedure of Machine Learning

The experiment for multiple wells is suitable when drilling and log data from several nearby wells are available and coal zones in a new well can be predicted. In this case, the target well is not required to have the density log to place the slotted casings for the coal sections. The TVD is added as an input because the formation may be the same at the same TVD for closely spaced wells in the same field. The wells in this study are drilled in the same pad and have different well trajectories.

Figure 6.2 shows the flowchart of machine learning for multiple well experiments. The training data are from Wells 2–6. The training data can be categorized

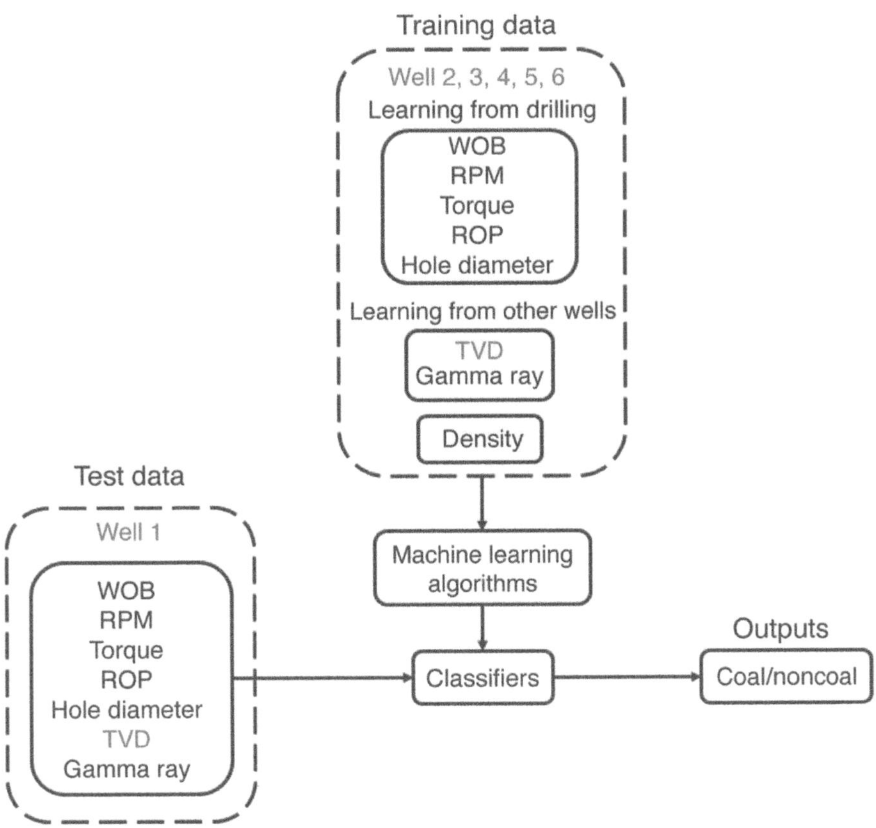

FIGURE 6.2 Flowchart of machine learning for coal identification in Well 1 using multiple wells. (After Zhong et al. 2020b, 1252.)

into learning from drilling response and learning from nearby lithology. Drilling-related parameters (WOB, RPM, torque, ROP, and hole diameter) reflect learning the drilling response of coal/non-coal from nearby wells. On the other hand, the TVD and gamma ray represent the learning of lithology from nearby wells. In the training process, the output is based on density logs from Wells 2–6 and the density log from Well 1 is no longer needed. The data from Wells 2–6 are fed into different machine learning algorithms to generate classifiers. The test data are processed from Well 1. Figure 6.2 shows the flowchart for Well 1 as the test well; however, the flowchart is also applicable to other wells. For instance, if Well 2 is the test well, Well 1 and Wells 3–6 are training wells.

Results and discussions. The results of coal prediction performance for multiple well experiments are shown in Table 6.2. Original algorithms (LR, SVM, ANN, and RF) have high precision and low recall. When coupled with NROS or SMOTE, these algorithms perform with a lower precision, higher recall, and lower F1 score. The XGBoost has the best F1 score with balanced precision and recall.

The results in Table 6.2 show that the scores of metrics are not very good using machine learning. However, this is because of the high resolution of logs (0.1 m). In reality, the resolution of the slotted casing is not 0.1 m. Usually, there are at least several meters for a slotted interval. To obtain the real coal identification performance in a field application, the criterion of slotted casing placement is shown in Figure 6.3. The basic idea is to retrieve coals with a thickness larger than 0.3 m within a certain interval (i.e., 5 m). If the coal zones are shorter than 0.3 m and there are no nearby coals within 5 m, the slotted casing will not be placed. Importantly, once an interval is targeted, the length of the slotted casing

TABLE 6.2
Results of Coal Prediction Performance for Multiple Well Experiments (Well 1) (After Zhong et al. 2020b, 1253)

Model	Precision	Recall	F1 Score
LR	0.84	0.34	0.48
LR+NROS	0.16	0.94	0.28
LR+SMOTE	0.17	0.94	0.29
SVM	0.81	0.39	0.53
SVM+NROS	0.31	0.89	0.46
SVM+SMOTE	0.30	0.88	0.44
ANN	0.78	0.48	0.59
ANN+NORS	0.38	0.84	0.53
ANN+SMOTE	0.37	0.81	0.5
RF	0.89	0.4	0.54
RF+NROS	0.34	0.82	0.48
RF+SMOTE	0.31	0.8	0.45
XGBoost	0.67	0.69	0.68

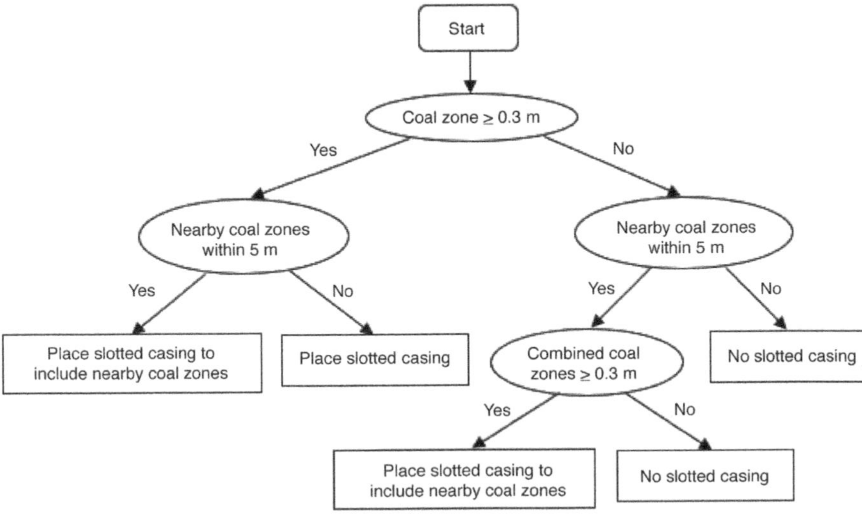

FIGURE 6.3 Flowchart of slotted casing placement. (After Zhong et al. 2020b, 1253.)

will extend 1.5 m above and below the targeted coal interval. The purpose of the length extension is to cover any possible missed coals nearby and take account of casing movement. To better describe the coal identification performance in a field glossary, we can define the coal identification rate as:

$$\text{Coal identification rate} = \frac{\text{Predicted coals covered by slotted casing}}{\text{Coals from the density log}} \tag{6.4}$$

Figure 6.4 shows an example of slotted casing placement for balanced precision and recall using XGBoost. This case represents the case with the best F1 score and balanced precision and recall. Figure 6.4a shows the original coals determined by the density log. In this study, a bulk density of less than 1.8 g/cm^3 from the density log is used as the criterion to determine the coal zones. The thickness of all coals is 31.8 m. Figure 6.4b shows the coals determined by the XGBoost. Following the criterion in Figure 6.3, the slotted casing placement is shown in Figure 6.4c. The dotted parts are the selected slotted casing sections. There are 12 sections, and only a 170.8- to 179.1-m section (marked in red) does not contain any coal. Some thin coals (0.1 or 0.2 m) in Figure 6.4b are not covered by the slotted casing based on the criterion (e.g., coals around 320 and 349 m). The slotted casing covers 26.4 of 31.8 m of coals in Figure 6.4a. Thus, the coal identification rate is 83%, which exceeds 14% as the recall is 0.69 (69%) using the XGBoost. We can expect the coal identification rate to be higher than the recall from the machine learning model.

Note that it is possible to have a higher coal identification rate if the recall is higher. For example, the coal identification rate is 94.7% for Well 1 if we use

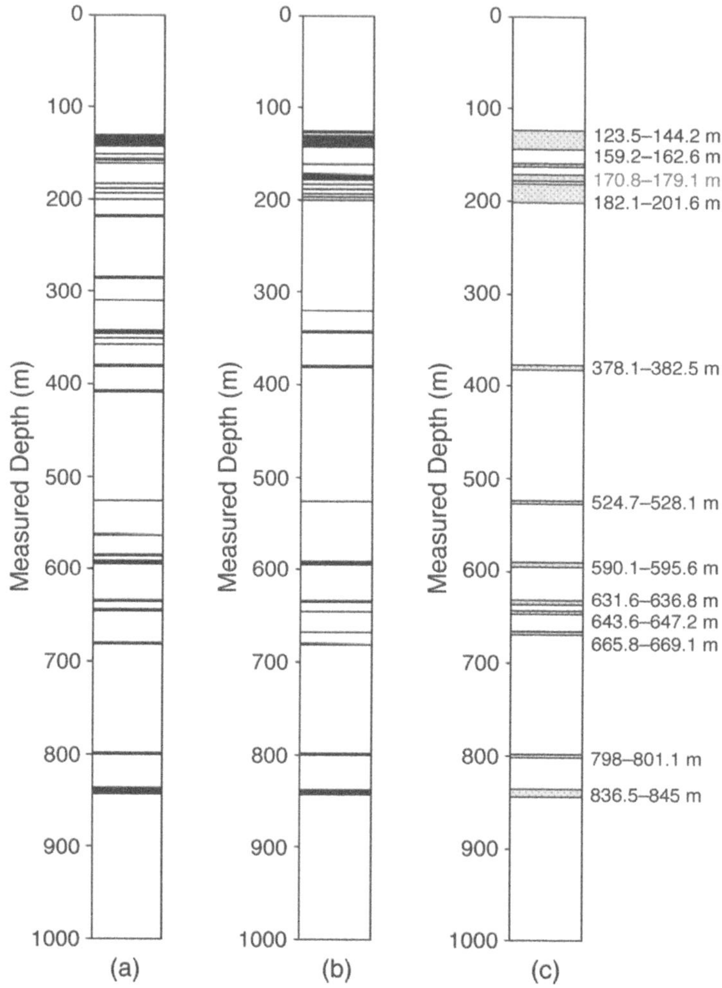

FIGURE 6.4 Coal zones and slotting casing placement for Well 1 using XGBoost (precision: 0.67, recall: 0.69, and F1 score: 0.68). (a) Coal zones based on the density log. (b) Coal zones determined by XGBoost. (c) Slotted casing placement based on results of XGBoost. The coal identification rate after slotted casing placement is 83%. (XGBoost: extreme gradient boosting). (After Zhong et al. 2020b, 1254.)

ANN with SMOTE. However, the tradeoff is the precision will be lower. In other words, there will be more misplaced slotted sections. Using a similar procedure in Figure 6.2, the experiments on the other five wells can be conducted. When one well is selected as the test well, the other five wells are the training wells. All wells have over a 90% coal identification rate and three of them (Wells 2, 5, and 6) have over 99% coal identification rate. Detailed results can be found in Zhong et al. (2020b, 1254–1255).

6.3.2 Case Study 2: Cuttings Bed Prediction

As mentioned in Section 6.1, unconventional resources are stored in low permeability rock layers. For instance, shale gas is trapped in the ultralow permeability (typically in the order of nano-Darcy) shales. Shale gas has been economically viable through the implementation of hydraulic fracturing and horizontal drilling (Zhong et al. 2014). During horizontal drilling, estimation of cuttings accumulated in the wellbore is important for controlling the bottom hole pressure and preventing stuck pipe. The second case study is from Ozbayoglu et al. (2002), which investigated the use of ANN to predict the height of cuttings bed. The results of machine learning were compared with results from a traditional regression method. The data were obtained from experimental studies of a large-scale flow loop.

Dimensionless groups. Hole cleaning is important during drilling because inadequate hole cleaning can cause problems such as bit wear, slow drilling rate, stuck pipe, etc. The investigation on cuttings transport can date back to the 1940s, which studied the terminal velocity determination for single-phase drilling fluid. Then other studies developed some semi-empirical equations and the effects of drilling parameters have been investigated (such as ROP, flow rate, fluid properties, mud density, inclination, cuttings size, eccentricity, etc.). Over the past several decades, some sophisticated mechanistic models (e.g., multiphase three-layered mechanistic models) with some assumptions were developed (Nguyen and Rahman 1996; Zhang et al. 2018). However, there is still a lack of understanding due to more complex fluids involved in the inclined and highly inclined wellbores (which represents complex boundary conditions).

To cover a wide range of conditions, dimensionless parameters are used in this study. Some independent parameters in the cuttings bed prediction include inclination angle (α), area (A), feed cuttings concentration (C_c), fluid density (ρ), apparent fluid viscosity (μ), average fluid velocity (v), and dimensions of the pipe and wellbore (D). Five dimensionless groups are developed as (Ozbayoglu et al. 2002):

$$\pi_1 = C_c = \frac{volume\ of\ cuttings}{volume\ of\ annulus} \tag{6.5}$$

$$\pi_2 = \alpha \tag{6.6}$$

$$\pi_3 = \frac{A_{bed}}{A_w} \tag{6.7}$$

$$\pi_4 = \frac{\rho v D}{\mu} = N_{Re} \tag{6.8}$$

$$\pi_5 = \frac{gD}{v^2} = \frac{1}{N_{Fr}} \tag{6.9}$$

where π_i, $i = 1, \ldots, 5$, represents the dimensionless parameter, N_{Re} is the Reynolds number, and N_{Fr} is the Froude number. In this study, the Reynolds number is defined as:

$$N_{Re} = \frac{8^{1-N} \rho v^{2-N} d_{hyd}^{N}}{K'} \quad (6.10)$$

where K' is the consistency index and N is the fluid behavior index.

Experimental work. The data used in the modeling were from experimental studies using a large-scale flow loop at The University of Tulsa Drilling Research Projects (TUDRP). A photo of the flow loop is shown in Figure 6.5. The test section is about 100 feet long. The diameter of the transparent casing is 8 inch and there is a 4.5-inch aluminum alloy drill pipe inside the casing. In this study, three different fluids are used: water, polymer mud, and foam. The mud system is composed of a centrifugal mud pump to supply liquid and a compressor to supply air. The cuttings are injected into the annular section from a 650-gallon tank. The flow loop can be inclined from 0 to 90° and there are 10 different locations to measure the height of developed cuttings bed. The data acquisition system can record data such as flow rates, rates of penetration, inclinations, fluid rheological properties, bed heights and eccentricities.

FIGURE 6.5 Photograph of the flow loop for the experimental study. (After Ozbayoglu et al. 2002.)

Evaluation metrics. The prediction of cuttings bed height is a regression problem. Thus, evaluation metrics for the regression problem are used. In this study, two evaluation metrics are used. The first one is the coefficient of determination (R^2), which has the following expression:

$$r = \frac{\sum_{i=1}^{n} (x_i - \bar{x})(y_i - \bar{y})}{\sqrt{\sum_{i=1}^{n} (x_i - \bar{x})^2} \sqrt{\sum_{i=1}^{n} (y_i - \bar{y})^2}} \tag{6.11}$$

$$R^2 = r^2 \tag{6.12}$$

where x_i and y_i are two parameters for correlation; $\bar{x}$ is the average of x_i; $\bar{y}$ is the average of y_i; n is the sample number; and r is the Pearson correlation coefficient. R^2 is simply the square of r and ranges from 0 to 1. Specifically, $R^2 = 1$ represents a perfect correlation between the predicted data and actual data.

To quantify the performance of prediction on each sample, another evaluation metric is the error percentage, which has the following expression:

$$\text{Error percentage} = \frac{\left|\widehat{y_i} - y_i\right|}{y_i} \tag{6.13}$$

where y_i is the original output and $\widehat{y_i}$ is predicted output.

Results and discussions. In this study, two methods are used to predict the cuttings bed height. The first method is LSM, which is a commonly used regression method for data fitting. It works by minimizing the sum of the squares of the residuals.

The dimensionless bed area (which can be used to determine the bed height) has a general expression (Ozbayoglu et al. 2002):

$$\frac{A_{bed}}{A_w} = k_1 (C_c)^{k_2} (N_{Re})^{k_3} (N_{Fr})^{k_4} \tag{6.14}$$

where k_i, $i = 1, \ldots, 4$, is the equation constant. Depending on the range of N, different fitting equations using LSM are obtained (Ozbayoglu et al. 2002). For $N \geq 0.9$:

$$\frac{A_{bed}}{A_w} = 4.1232(C_c)^{0.0035} (N_{Re})^{-0.2198} (N_{Fr})^{-0.2164} \tag{6.15}$$

For $0.6 < N < 0.9$,

$$\frac{A_{bed}}{A_w} = 0.7115(C_c)^{0.0697} (N_{Re})^{-0.0374} (N_{Fr})^{-0.0681} \tag{6.16}$$

For $N \leq 0.6$,

$$\frac{A_{bed}}{A_w} = 1.0484(C_c)^{0.0024}(N_{Re})^{-0.1502}(N_{Fr})^{-0.0646} \tag{6.17}$$

Figure 6.6 shows the performance of the empirical equations using LSM. Figure 6.6a shows the comparison between the predicted results using LSM (y-axis) and experimental results (x-axis). The data are plotted in different shapes based on the range of N (Eqs. 6.15–6.17). The dashed lines are ±20% error margin lines. When the dimensionless bed area is greater than 0.2, most of the points are within the 20% error margin. However, the results are poor when the ratio is smaller than 0.2. The

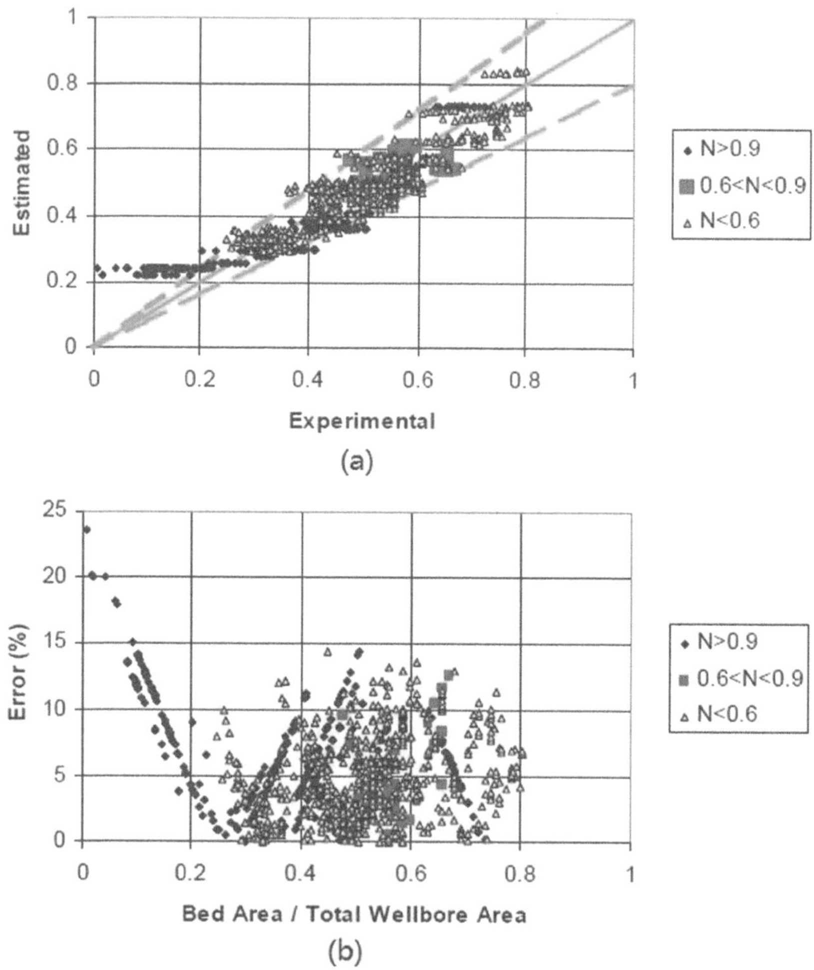

FIGURE 6.6 Results using the least square method (LSM). (a) Experimental results vs. the predicted results. (b) Error percentage. (After Ozbayoglu et al. 2002.)

predicted bed heights are overestimated. Figure 6.6b shows the results of error percentage for each sample. As expected, when the bed area ratio is smaller than 0.2, error percentages range from 0% to 25%. When the bed area ratio is greater than 0.2, all points have less than 15% error. Note that the error percentage in Figure 6.6b is not the same as the error margin in Figure 6.6a. The R^2 are 0.8612, 0.9318, and 0.7966 for $N \geq 0.9$, $0.6 < N < 0.9$, $N \leq 0.6$, respectively.

For the machine learning modeling, ANN is used. ANN is inspired by the biological neural networks that constitute animal brains, which consist of layers and nodes. In a typical ANN, there is one input layer, multiple hidden layers, and one output layer, and each layer has multiple nodes. The ANN works by performing mathematical computation between neurons from the input layer to the output layer. In this study, there is one input layer (3 nodes), one hidden layer (10 nodes), and one output layer. Three inputs are included: feed cuttings concentration, Reynolds number, and Froude number.

The performance of the ANN is presented in Figure 6.7. The dashed lines are ±15% error margin lines in Figure 6.7a. Most data points are within the dashed lines. Unlike the LSM, when the bed area ratio is smaller than 0.2, most points are still within the error margin. For the percentage errors shown in Figure 6.7b, all points have errors less than 10%. Thus, the ANN outperforms the LSM, especially when the bed area ratio is smaller than 0.2. The authors did not present results of R^2, but it is expected R^2 using ANN also outperforms that of LSM.

In summary, two different methods are used to estimate cuttings bed height for highly inclined and horizontal wells: a standard engineering correlation (LSM) and ANN. Both LSM and ANN can estimate cuttings bed height with an error of less than 15% when the bed area ratio is greater than 0.2. However, LSM has very poor performance when the bed area ratio is less than 0.2, and some points have more than 20% error percentage. On the other hand, the ANN method has consistent performance and has less than a 10% error percentage for all bed area ratios. It is expected that more quality data with different pipe and wellbore sizes, other cuttings sizes, and different fluids would improve the prediction accuracy and extend the model applicability.

6.4 CONCLUDING REMARKS

This chapter briefly reviews some machine learning studies for drilling unconventional resources. Drilling can create a huge volume of high-frequency drilling data, which is not available in the reservoir or production studies. Combined with high-resolution petrophysical logs, powerful tools can be created to reveal rock properties and improve drilling activities. However, one of the challenges is data quality control. The data from the top drive are usually noisy and require a good quality check. LWD or measurement while drilling (MWD) data have better quality, which are more desirable. Another challenge is real-time implementation, which is important for drilling optimization and detection of drilling problems. Real-time implementation requires proper selection of efficient algorithms, integration with hardware, and training of quality data.

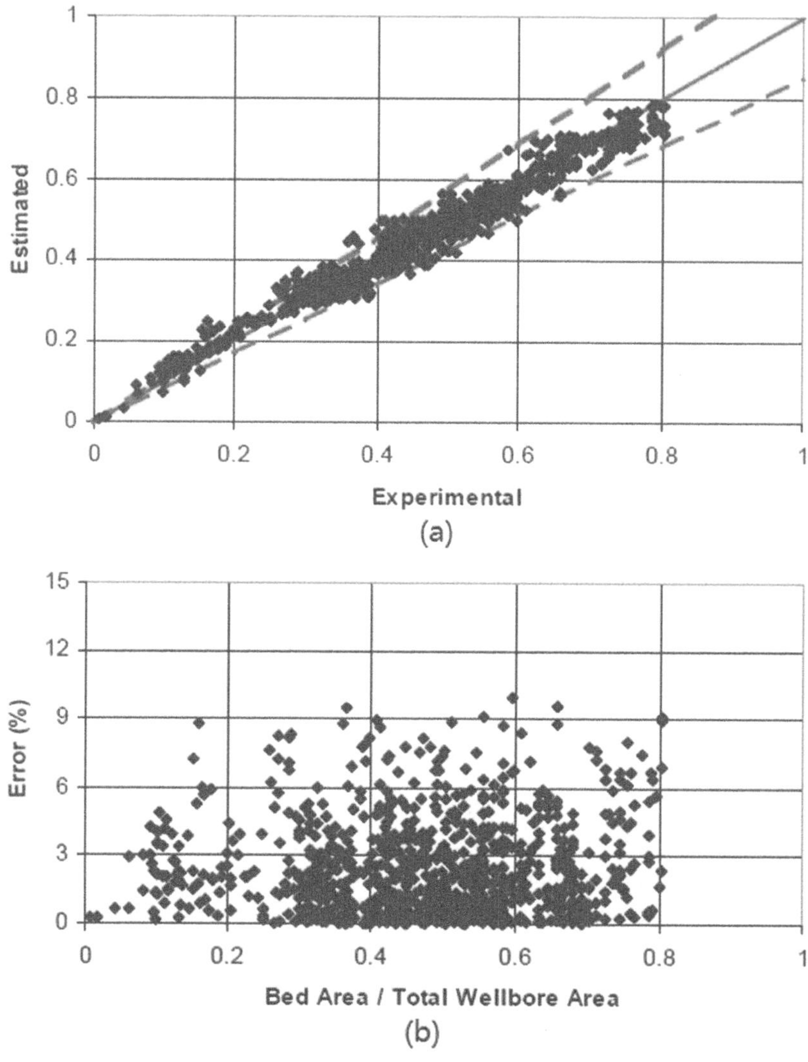

FIGURE 6.7 Results using the artificial neural network (ANN). (a) Experimental results vs. the predicted results. (b) Error percentage. (After Ozbayoglu et al. 2002.)

It is expected that machine learning studies in drilling will continue to grow, especially in drilling optimization and detection of drilling problems (e.g., stuck pipe, kick, and lost circulation). These studies not only have more economic benefits than other drilling applications but also reduce potential safety and environmental problems. For the detection of drilling problems, machine learning from anomaly detection will advance the development. For drilling optimization, more optimization studies in different fields and formations will be beneficial.

REFERENCES

Al-Jubori, Ahmed, Johnston, Sean, Boyer, Chuck, Lambert, Stephen W., Bustos, Oscar A., Pashin, Jack C., and Wray, Andy. 2009. "Coalbed methane: clean energy for the world." *Oilfield Review*, 2, 4–13.

Al-Yami, Abdullah S., Al-Shaarawi, Amjad, Al-Bahrani, Hussain, Wagle, Vikrant B., Al-Gharbi, Salem, and Al-Khudiri Musab B. 2016. "Using Bayesian network to develop drilling expert systems." Paper presented at SPE Heavy Oil Conference and Exhibition, Kuwait City, Kuwait, 6–8 December, SPE-184168-MS. https://doi.org/10.2118/184168-MS

Aminzadeh, Fred, and Dasgupta, Shivaji N. 2013. "Chapter 9: Geophysics for unconventional resources." *Developments in Petroleum Science*, 60, 247–271. https://doi.org/10.1016/B978-0-444-50662-7.00009-3

Barati-Harooni, Ali, Najafi-Marghmaleki, Adel, Tatar, Afshin, Arabloo, Milad, Phung, Le Thi Kim, Lee, Moonyong, and Bahadori, Alireza. 2016. "Prediction of frictional pressure loss for multiphase flow in inclined annuli during underbalanced drilling operations." *Natural Gas Industry B*, 3, 275–282. https://doi.org/10.1016/j.ngib.2016.12.002

Behnoud far, Pouria, and Hosseini, Pantea. 2017. "Estimation of lost circulation amount occurs during under balanced drilling using drilling data and neural network." *Egyptian Journal of Petroleum*, 26(3), 627–634. https://doi.org/10.1016/j.ejpe.2016.09.004

Boyer, Charles, Kieschnick, John, Suarez-Rivera, Roberto, Lewis, Richard E., and Waters, George. 2006. "Producing gas from its source." *Oilfield Review*, 18(3), 36–49.

Castineira, David, Toronyi, Robert, and Saleri, Nansen. 2018. "Machine learning and natural language processing for automated analysis of drilling and completion data." Paper presented at SPE Kingdom of Saudi Arabia Annual Technical Symposium and Exhibition, Dammam, Saudi Arabia, 23–26 April, SPE-192280-MS. https://doi.org/10.2118/192280-MS

Chawla, Nitesh V., Bowyer, Kevin W., Hall, Lawrence O., and Kegelmeyer, W. Philip. 2002. "SMOTE: synthetic minority over-sampling technique." *Journal of Artificial Intelligence Research*, 16, 321–357. https://doi.org/10.1613/jair.953

Chawla, Nitesh V., Japkowicz, Nathalie, and Kotcz, Aleksander. 2004. "Editorial: special issue on learning from imbalanced data sets." *ACM SIGKDD Explorations Newsletter*, 6(1), 1–6. https://doi.org/10.1145/1007730.1007733

Fawcett, Tom, and Provost, Foster. 1996. "Combining data mining and machine learning for effective user profiling." Paper presented at the Second International Conference on Knowledge Discovery and Data Mining, Portland, Oregon, USA, 2–8 August.

Hegde, Chiranth, Daigle, Hugh, and Gray, Ken E. 2018. "Performance comparison of algorithms for real-time rate-of-penetration optimization in drilling using data-driven models." *SPE Journal*, SPE-191141-PA. http://dx.doi.org/10.2118/191141-PA

Hegde, Chiranth, Daigle, Hugh, Millwater, Harry, and Gray, Ken. 2017. "Analysis of rate of penetration (ROP) prediction in drilling using physics-based and data-driven models." *Journal of Petroleum Science and Engineering*, 159, 295–306. http://dx.doi.org/10.1016/j.petrol.2017.09.020

Jahanbakhshi, Reza, Keshavarzi, Reza, and Jalili, Sajad. 2014. "Artificial neural network-based prediction and geomechanical analysis of lost circulation in naturally fractured reservoirs: a case study." *European Journal of Environmental and Civil Engineering*, 18(3), 320–335. https://doi.org/10.1080/19648189.2013.860924

Jahanbakhshi, Reza, Keshavarzi, Reza, Shoorehdeli, Mahdi Aliyari, and Emamzadeh, Abolqasem. 2012. "Intelligent prediction of differential pipe sticking by support vector machine compared with conventional artificial neural networks: an example of Iranian offshore oil fields." SPE Drilling and Completion, SEP-16362-PA. https://doi.org/10.2118/163062-PA

Kormaksson, Matthias, Vieira, Marcos R., and Zadrozny, Bianca. 2015. "A data driven method for sweet spot identification in shale plays using well log data." Paper presented at SPE Digital Energy Conference and Exhibition, The Woodlands, Texas, USA, 3–5 March, SPE-173455-MS. https://doi.org/10.2118/173455-MS

Ma, Zheren, Vajargah, Ali Karimi, Lee, Hanna, Kansao, Rami, Darabi, Hamed, and Castineira, David. 2018. "Applications of machine learning and data mining in SpeedWise drilling analytics: a case study." Paper presented at Abu Dhabi International Petroleum Exhibition and Conference, Abu Dhabi, UAE, 12–15 November, SPE-193224-MS. https://doi.org/10.2118/193224-MS

Miri, Rahman, Sampaio, Jorge H.B., Afshar, Mahmoud, and Lourenco, Affonso. 2007. "Development of artificial neural networks to predict differential pipe sticking in Iranian offshore oil fields." Paper presented at the International Oil Conference and Exhibition in Mexico, Veracruz, Mexico, 27–30 June, SPE-108500-MS. https://doi.org/10.2118/108500-MS

Nguyen, D., and Rahman, S.S. 1996. "A three-layer hydraulic program for effective cuttings transport and hole cleaning in highly deviated and horizontal wells." Paper presented at the IADC/SPE Asia Pacific Drilling Technology Conference, Kuala Lumpur, Malaysia, 9–11 September, SPE-36383-MS. https://doi.org/10.2118/36383-MS

Ozbayoglu, Evren M., Miska, Stefan Z., Reed, Troy, and Takach, Nicholas. 2002. "Analysis of bed height in horizontal and highly-inclined wellbores by using artificial neural networks." Paper presented at SPE International Thermal Operations and Heavy Oil Symposium and International Horizontal Well Technology Conference, Calgary, Alberta, Canada, 4–7 November. SPE-78939-MS. https://doi.org/10.2118/78939-MS

Pollock, Jacob, Stoecker-Sylvia, Zachary, Veedu, Vinod, Panchal, Neil, and Elshahawi, Hani. 2018. "Machine learning for improved directional drilling." Paper presented at Offshore Technology Conference, Houston, Texas, USA, 30 April-3 May. OTC-28633-MS. https://doi.org/10.4043/28633-MS

Rooki, Reza. 2015. "Estimation of pressure loss of Herschel-Bulkley drilling fluids during horizontal annulus using artificial neural network." *Journal of Dispersion Science and Technology*, 36, 161–169. https://doi.org/10.1080/01932691.2014.904793

Rooki, Reza, Ardejani, Faramarz Doulati, and Moradzadeh, Ali. 2014. "Hole cleaning prediction in foam drilling using artificial neural network and multiple linear regression." *Geomaterials*, 4, 47–53. https://doi.org/10.4236/gm.2014.41005

Soares, Cesar, and Gray, Kenneth. 2019. "Real-time predictive capabilities of analytical and machine learning rate of penetration (ROP) models." *Journal of Petroleum Science and Engineering*, 172, 934–959. http://dx.doi.org/10.1016/j.petrol.2018.08.083

Woods, Kevin S., Doss, Christopher C., Bowyer, Kevin W., Solka, Jeffrey L., Priebe, Carey E., and Kegelmeyer Jr, W. Philip. 1993. "Comparative evaluation of pattern recognition techniques for detection of microcalcifications." *International Journal of Pattern Recognition and Artificial Intelligence*, 7(6), 1417–1436. https://doi.org/10.1142/S0218001493000698

Zhang, Feifei, Miska, Stefan, Yu, Mengjiao, and Ozbayoglu, Evren. 2018. "A unified transient solid-liquid two-phase flow model for cuttings transport – modelling part." *Journal of Petroleum Science and Engineering*, 166, 146–156. https://doi.org/10.1016/j.petrol.2018.03.027

Zheng, Zhaohui, Wu, Xiaoyun, and Srihari, Rohini. 2004. "Feature selection for text categorization on imbalanced data." *ACM SIGKDD Explorations Newsletter*, 6(1), 80–89. https://doi.org/10.1145/1007730.1007741

Zhong, Ruizhi, Bao, Jinqing, and Fathi, Ebrahim. 2014. "Fully coupled finite element model to study fault reactivation during multiple hydraulic fracturing in heterogeneous tight formations." Paper presented at the SPE Eastern Regional Meeting, Charleston, West Virginia, 21–23 October, SPE-171035-MS. https://doi.org/10.2118/171035-MS

Zhong, Ruizhi, Johnson Jr, Raymond, and Chen, Zhongwei. 2020a. "Generating pseudo density log from drilling and logging-while-drilling data using extreme gradient boosting (XGBoost)." *International Journal of Coal Geology*, 103416. http://dx.doi.org/10.1016/j.coal.2020.103416

Zhong, Ruizhi, Johnson Jr, Raymond L., and Chen, Zhongwei. 2020b. "Using machine learning methods to identify coal pay zones from drilling and logging-while-drilling (LWD) data." *SPE Journal*, 25, 1241–1258. http://dx.doi.org/10.2118/198288-PA

Section IV

Production Data Analysis Applications

7 Machine Learning Assisted Production Data Filtering and Decline Curve Analysis in Unconventional Plays

David Fulford
ARM Energy,
Houston, Texas

CONTENTS

DOI: 10.1201/9781003207009-11

7.1 INTRODUCTION

Production data analysis (PDA) is the analysis of time-rate-pressure data. More specifically, it is the analysis of measured volumes of produced reservoir fluids and pressure readings captured over many days to months. While it includes fields of study such as numerical reservoir simulation, which employs physically rigorous equations to model flow in porous media, in this chapter we seek an alternative to the strict data requirements of such methods. For our purposes in this chapter, PDA refers to methods that operate directly on production data and require minimal inputs for other reservoir, fluid, or petrophysical properties. In other words, we will be focusing on empirical methods for PDA.

PDA, and more specifically rate-time analysis, has several features that make it unusual compared with many other types of data analysis. Primarily, we are dealing with time-series data, but do not use standard methods for time-series analysis. This is because we have a reasonably well understood physical process we are attempting to describe, and nearly a century of research in the field has revealed certain methods to be quite reliable. Despite a well understood physical process, there are many uncertainties regarding the production performance of any individual well. Moreover, PDA can have profound financial impact as economic decisions often rely on the estimated future production volumes.

Production forecasts often carry an obligation for accuracy. Reserves estimates require testament to correctness by a signatory professional engineer. Revenue forecasts based upon estimated future volumes carry the weight of a fiduciary responsibility to investors. For these reasons, empirical decline curve models see widespread use throughout the industry and many applications attempt to regress a decline curve model to predict future production volumes.

However, a growing number of data-driven methods are not based upon any rate-time model. These models may also not focus on long-term ultimate recovery estimates as decline curves have traditionally done but target short-term forecasts days or weeks in the future. For example, they may serve operations as a base case forecast to detect production anomalies. Given the high accuracy standards applied to production forecasts, unsupervised machine learning methods to predict hydrocarbon volumes decades into the future – where little-to-no training data exists for analog well sets – are much less common, but these methods have seen success in outlier filtering.

Production forecasts are used by various disciplines: reservoir engineers concerned with reserves estimates, operations engineers concerned with production status, and finance professionals concerned with asset value. Practical approaches to modeling must be able to be leveraged by various disciplines with reasonable computational time depending on the application requirements. This chapter provides an overview of best practices and current state-of-the-art methods (in the opinion of the author, of course) for empirical production forecasts.

7.1.1 Methodologies

To begin, we outline a framework for classifying various techniques, including empirical, model-based, and hybrid methods. Figure 7.1 presents a matrix for the relationships of various types of machine learning models. Starting with linear regression, we can progress to classification models, kernel models, and/or Bayesian models. We will cover some of these model types in the next few sections. The matrix provides a clean mental model as we stitch additional features onto the simple linear regression model.

As we transition downward in our matrix to kernel methods, we obtain the capability to add nonlinear behavior to linear regression. Applications of kernel regression methods for fitting production data are uncommon but do exist in the literature for prediction of pressure given rate measurements. In these applications, kernel functions encode the functional form of the rate-time relationship for various physical flow regimes. Examples of these flow regimes are transient linear flow, transient radial flow, pseudosteady state or boundary-dominated flow, etc. A matrix of coefficients is determined to act as a weighting for these kernels at each of the observed time values.

Kernel classification methods are more common and tend to be borrowed from more general applications rather than derivations specific to oil and gas. There are well-known kernel classification methods for outlier determination that we will review.

Bayesian methods are of specific interest to us. It is common that analysts have some notion or expectation of future production performance that they wish

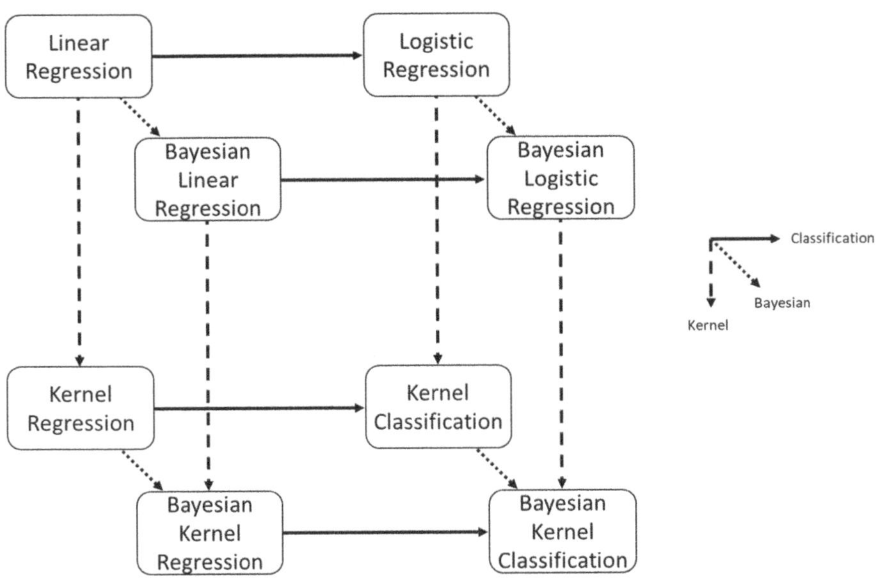

FIGURE 7.1 A matrix describing the relationships of various model types.

a model outcome to reflect in accordance with the fact that we are forecasting a physical process. The inclusion of prior probabilities allows the expression of these expectations in a form that is useful to a regression or classification model. One of the more familiar and powerful Bayesian methods for kernel regression is a Gaussian process. This type of model expresses a prior over functions in one or more dimensions. Simpler Bayesian methods may only express a prior for a specific model parameter such as the intercept, slope, or curvature of a line, but are still useful methods to include information outside of the production dataset. It is a fact to state that the mathematics of these methods are often complex; however, numerical implementations often tend to be rather straightforward.

Moving beyond our matrix of linear regression relationships, we can explore nonlinear regression methods. Here we find applications for regression of the familiar empirical decline curve models: the hyperbolic and exponential decline curve models, and variations of them such as multi-segment models. In certain situations, these models can be linearized, but often it is simpler to apply a nonlinear solver to regress the model directly. Similar linear methods to linear methods, nonlinear methods may also perform classification and be Bayesian in nature.

For PDA, nonlinear methods are of much greater interest to us because they present a means to regress any model we choose. We are also not limited to minimizing the sum of squared residuals (the least squares assumption), which provides a pathway to robustly regress our chosen model while tolerating a greater number and severity of outliers. This includes non-Gaussian outliers, which can cause catastrophic failure of the least squares assumption (Huber 1964).

7.2 DATA FILTERING

Some filtering of production data is nearly always necessary to remove effects that are not captured by empirical rate-time models. Because the underlying assumption of these models is single-phase constant-pressure production, effects such as early-time flowback and rate buildup, periods of downtime, periods of flush production resulting from workovers, etc., are typically removed.

Traditionally, data filtering has been a manual process. When production forecasts are manually adjusted, some filtering process is occurring in the mind of the analyst. Various software has introduced box or lasso selection tools, spray brush tools, etc., to allow a user to select or deselect specific data points prior to performing the regression. A major weakness of workflows that require these tools is that they are by their nature non-reproducible from analyst to analyst. For our discussion, we'll ignore the art of hand-fitting empirical rate-time models and focus solely on machine learning methods.

While it is desirable to have algorithm (or set of algorithms) that can be applied to production data with little to no analyst intervention, in practice this is challenging due to the many aspects of production operations that cannot be determined from production data. Many algorithms have been developed to perform outlier detection, and many of these can also be applied to production data. Alimohammadi and Chen (2022) have provided an excellent overview, comparison, and scoring of various methods. Here we feature selected methods that are also highlighted in their work. Each example operates with a different key mechanism, providing a diverse overview of methods.

7.2.1 Heuristic Filters

We interject to note some filters that do not require any machine learning method in order to apply and can be first applied to production data before any more complex method(s). Because these heuristics are easy to understand and computationally inexpensive, we recommend their application in all cases.

- Elimination of early-time buildup – The peak production data point that occurs within some time period starting from the first production point is identified, and then all data points prior to the peak are removed.
- "High-pass" filter – All values below a specified rate, e.g., 1 barrel per day (BPD), are removed.

7.2.2 Unsupervised Methods

These are methods that do not require any model or notion of how the data are generated, although they may require hyperparameters and hyperparameter tuning for good performance. The algorithms listed below are commonly found in data science packages in various programming languages.

- DBSCAN (Ester et al. 1999) – a density-based clustering method that can identify nonlinearly separable clusters and has a notion of noise.
- Local Outlier Factor (Bruenig et al. 2000) – a *k*-nearest neighbors (KNN) algorithm that compares local densities to identify outliers as those points with the lowest local densities.
- One Class Support Vector Machine (SVM) (Schölkopf et al. 2001) – an SVM algorithm that identifies a decision boundary for outliers. A kernel is often used to generate a nonlinear boundary.
- Isolation Forest (Liu, Ting, and Zhou 2008) – a random forecast algorithm that identifies outliers as those data points with shorter paths in a tree structure.
- Elliptic Envelope (Rousseeuw and Driessen 1999) – identifies outliers as those data points that fall outside a Gaussian ellipse.

A comparison of the results of these models applied to daily data from a Permian Basin oil well is provided in Figure 7.2. While each of these models has its own hyperparameters, we can generally describe the hyperparameter tuning process as a trade-off between false positives and false negatives. False positives are data points that are not outliers but have been classified as such, while false negatives are data points that are outliers but have not been classified to be so. Too many false positives and the filtering is "too strong," whereas with too many false

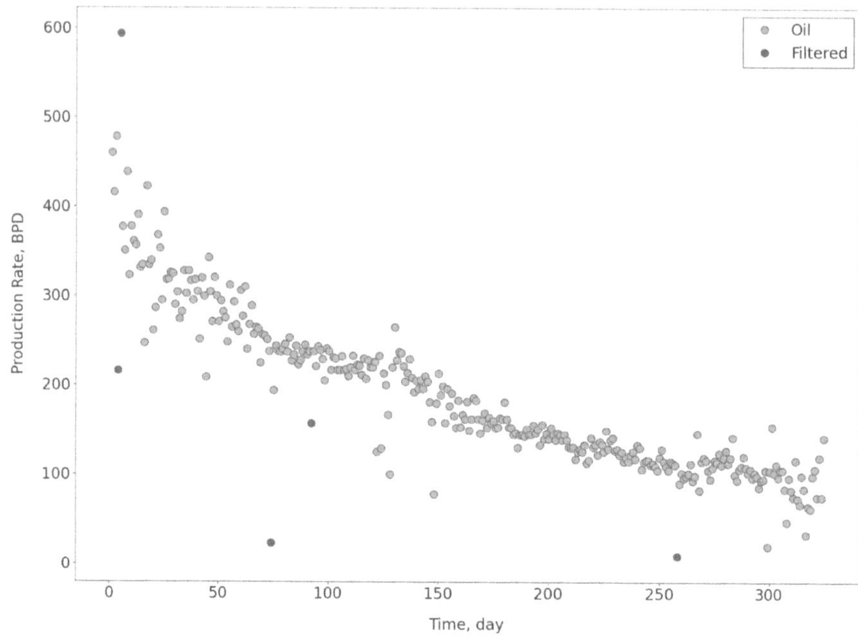

(a)

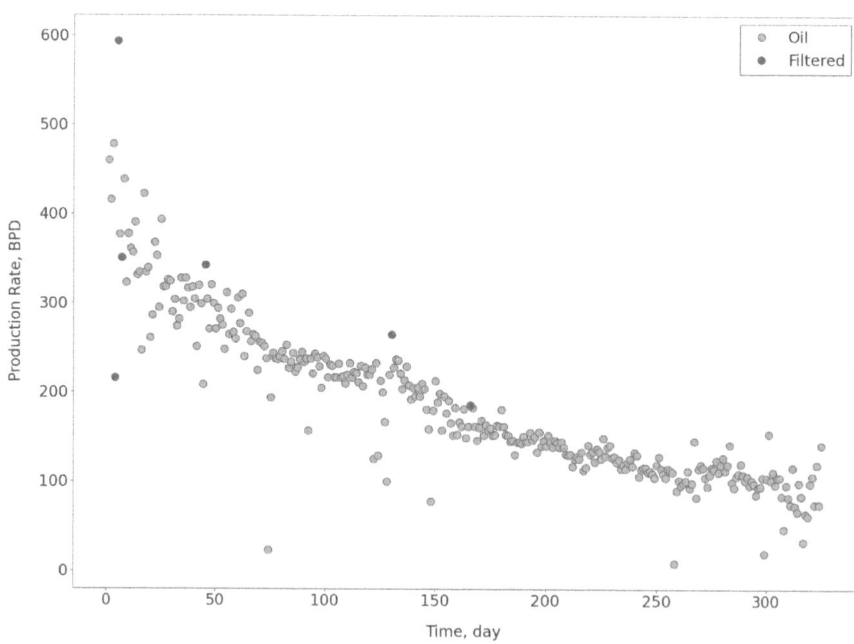

(b)

FIGURE 7.2 Examples of unsupervised outlier classification methods – (a) DBSCAN, (b) Local Outlier Factor, (c) One class SVM, (d) Isolation Forest, (e) Elliptic Envelope, and (f) Ensemble of all approaches. *(Continued)*

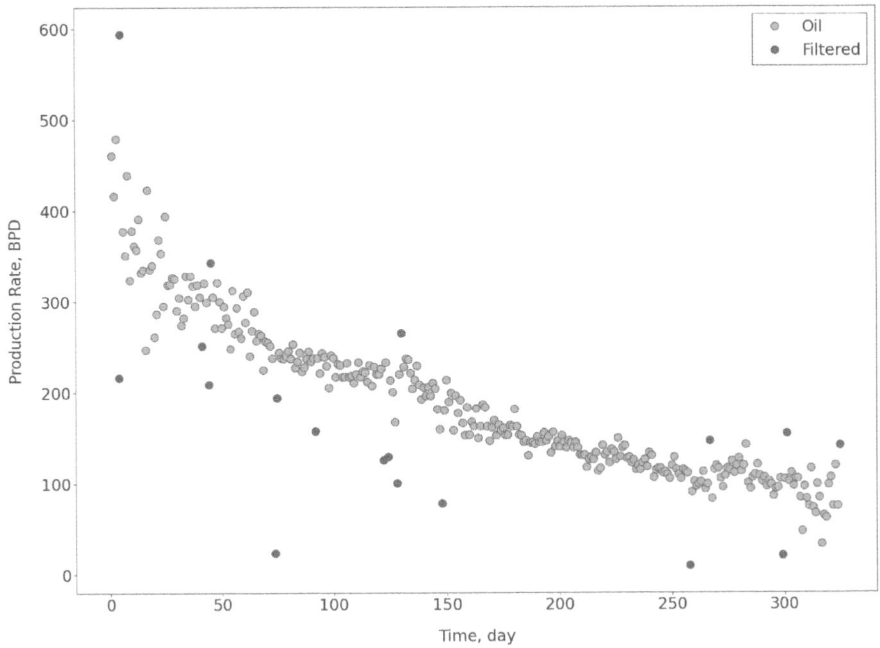

(c)

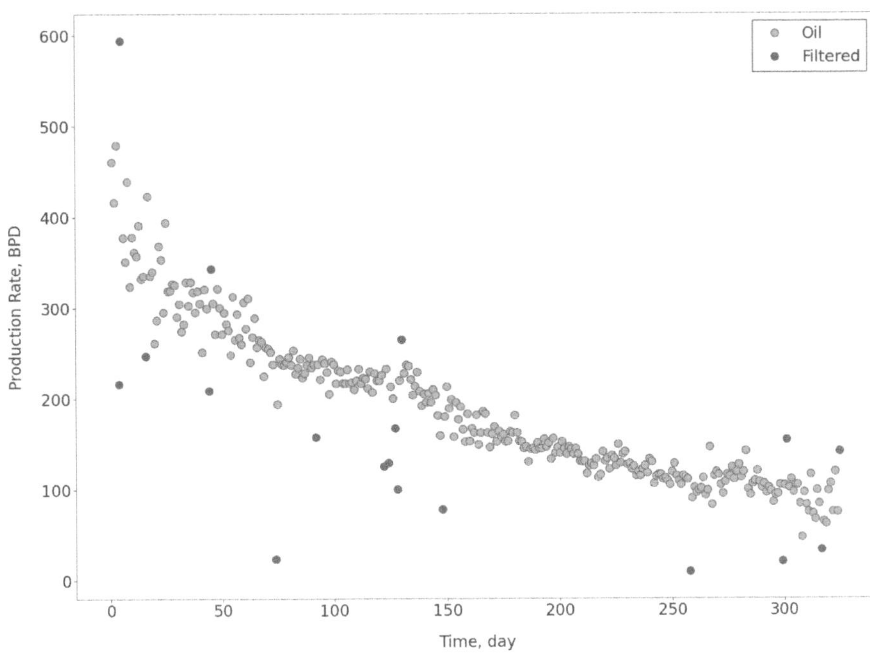

(d)

FIGURE 7.2 *(Continued)*

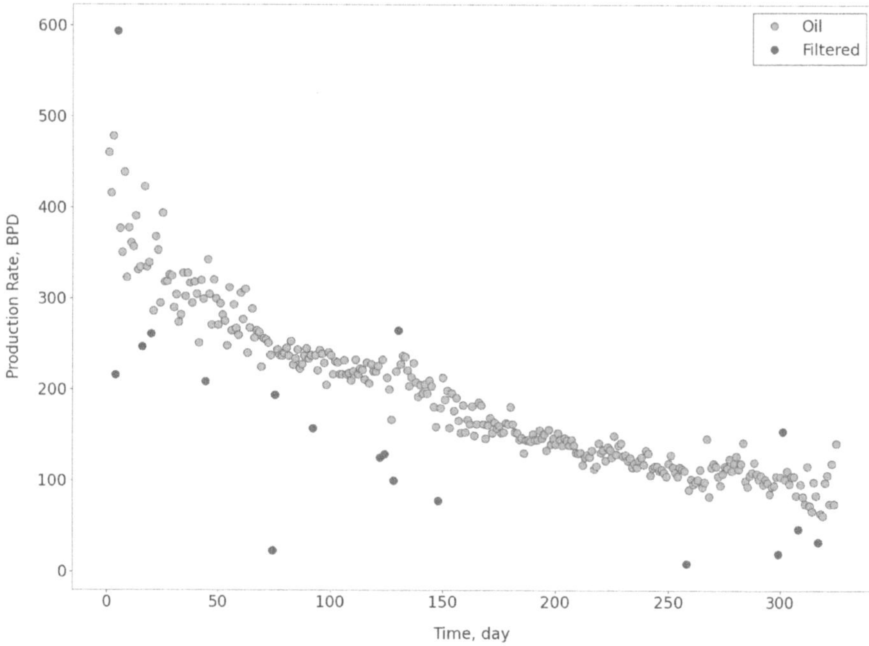

(e)

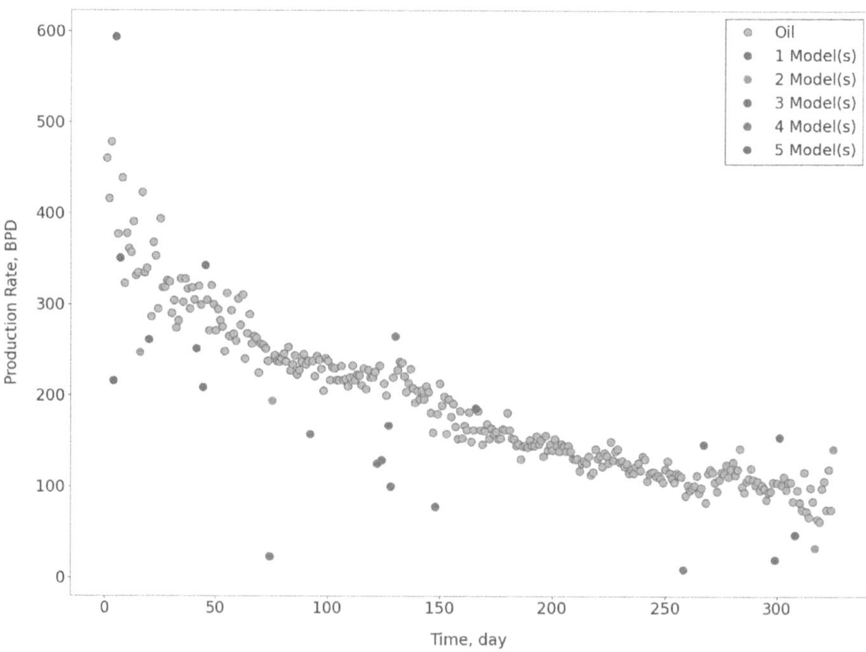

(f)

FIGURE 7.2 *(Continued)*

negatives the filtering is "too weak." Hyperparameter tuning adjusts the hyperparameters until the results "look good" to an observer. Obviously, manual review and calibration of these models, if required for every dataset to which they may be applied, is not ideal.

One approach to address the undesirability of manual review is to combine the models into an ensemble. In Figure 7.2(f), an ensemble approach is employed to combine the results of each algorithm. For each data point, we count the number of algorithms that have classified it as an outlier. With this approach, we may let each algorithm weakly filter the dataset (prefer false negatives) and increase our confidence that any given data point is an outlier by counting the number of algorithms that classified it as such. In the example, there are a few points that all five algorithms classify as an outlier, and the data points that had two or more algorithms classify them as an outlier are those that an analyst would likely manually remove. Those data points, which only one algorithm classified as an outlier, may be seen as having a higher likelihood of being false positives and may or may not need to be removed from the dataset.

7.2.3 Supervised Methods

Supervised methods in this context refer to regression-based approaches. A generating function of the data is chosen to use in the process of determination of outliers. The chosen function can be anything, but for practical purposes should be a function that would have a high goodness-of-fit to the data. Polynomial functions are often used for flexibility, but non-parametric functions such as splines as well as local regression techniques can also be used. Nonlinear regression is also an option, and we again employ the Arps decline model for illustration, although any model could be used.

We classify supervised methods by two general types:

- Regression Filtering – scores data points based upon residuals from a regressed function. A variety of criteria may be used to determine outliers, such as magnitude of residual or fraction of residuals. The residual function may also be chosen, e.g., squared difference, absolute difference, or other.
- Robust Regression Filtering (Fulford and Blasingame 2020) – a stepwise procedure that regresses a chosen function with both a robust ($L2$) and non-robust ($L1$, Soft $L1$, Huber, or other) cost function, and determines an outlier as the data point with the largest residual from the robust fit. The process repeats until the residuals between the two fits is less than the threshold value.

In comparison with regression filtering, The Fulford-Blasingame robust regression filtering method is specifically designed to identify data points that do not conform to an assumption of Gaussian error distribution and will remove all such points until the non-robust fit is no longer influenced by these data points.

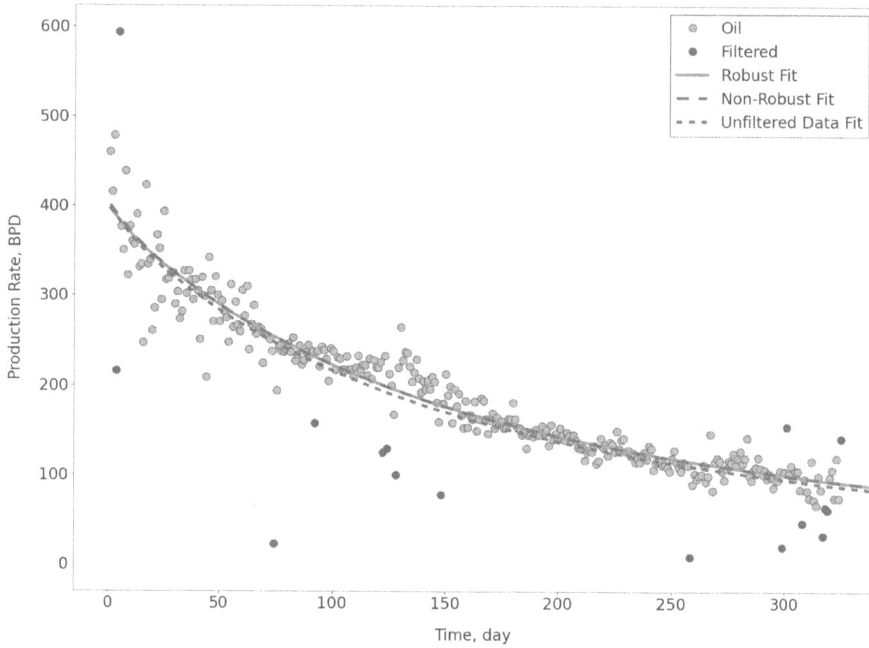

FIGURE 7.3 An example of robust regression filtering to filter production data.

Regression filtering is affected by outliers even as it tries to identify them. This causes it to perform poorly for detection of subtle outliers.

Figure 7.3 shows the result of robust regression filtering applied to the example dataset. In addition to the classification of outliers, we also obtain the final regressions of the robust and non-robust cost functions. These fits are nearly identical and may be used as a forecast for future production rates. However, it would be preferential to utilize a more comprehensive regression technique such as the regularized regression presented in the last section, or the probabilistic regression presented in the next section.

Comparing with the unsupervised techniques, we find that many of the same data points have been classified as outliers. There are a few additional data points at the tail end of the data series that are outliers but were not classified as outliers by any of the unsupervised methods, while there are several points throughout the data series that are not outliers but were classified by several of the unsupervised methods.

It is important to note that the robust regression approach has a unique property. It is less important that any given point is an outlier than it is that, taken together as a dataset, there are no individual points that unduly influence the non-robust fit. In other words, it specifically addresses two of the enumerated challenges with production data: non-Gaussian residuals and leverage points.

7.3 LINEAR METHODS

Classical reservoir engineering approaches have often involved graphical techniques that employ "plotting" functions that transform data to present as a straight line. These "straight- line" analyses involve an assumed functional relationship between rate and time or rate and pressure based upon solutions of the diffusivity equation for specific boundary conditions. These solutions are referred to as "flow regimes" and are named for the geometry of the reservoir used to derive the solution, as well as whether the solution applies to early-time infinite-acting conditions or late-time bounded conditions. They are commonly used in rate-transient analysis (RTA) and pressure-transient analysis (PTA). Their application requires an analyst to first verify that the well exhibits the flow regime that is expected. Then a straight line can be fit to the data in the appropriate region, and from the slope and intercept of the line, reservoir properties are extracted.

7.3.1 Linearization of Production Data

We are more interested in regression of a model to production data as opposed to transformation of production data to fit a model. However, an understanding of flow regimes, and the sequences that evolve over the course of a well's life, are valuable because they provide a physics-based lens for generating hyperparameters and parameter bounds for machine linear methods learning models. While there are many flow regimes, Table 7.1 lists functional forms of the asymptotic solutions for the most common flow regimes we expect to encounter in wells we would analyze with empirical methods. If we have an understanding of which flow regime(s) we expect to see, then we can preprocess our data to remove the need for our algorithm to learn that flow regime.

Inspecting the various functional forms, we observe that all of these regimes can be made less nonlinear by taking the natural logarithm of the rate or pressure data. For the boundary-dominated regime, the transform produces a straight-line function. For the other regimes, we can gain immediate improvement in most algorithms by fitting the logs of rates and pressures.

TABLE 7.1
Functional Forms for Common Flow Regimes

Flow Regime	Pressure Function	Rate Function
Radial	$\Delta p(t) \propto \ln t$	$\Delta q(t) \propto \ln t$
Linear	$\Delta p(t) \propto t^{1/2}$	$\Delta q(t) \propto t^{-1/2}$
Pseudosteady state	$\Delta p(t) \propto t$	—
Boundary-dominated	—	$\Delta q(t) \propto e^{-t}$

7.3.2 Linearization from Empirical Model Basis

Linearization can also be approached from the basis of empirical decline curve models. The Arps (Arps 1945) exponential and hyperbolic decline functions are most commonly used for empirical estimation of future production rates. Recently, other models have been developed to address the long duration transient regimes observed in tight and unconventional shale reservoirs but have not reached the same status of popular use. Accordingly, the Arps functions will be the focus of our study. The Arps family of models are derived from observation that the rate of change of the production rate, which is the definition of the decline rate or D-parameter, can be approximated to have a constant derivative, which is the b-parameter. Eq. (7.1) is the D-parameter

$$D(t) \equiv -\frac{d \ln q}{dt} \equiv -\frac{1}{q}\frac{dq}{dt} \tag{7.1}$$

The b-parameter is defined as the rate of change of the D-parameter:

$$b(t) \equiv \frac{1}{D} t \equiv \text{constant} \tag{7.2}$$

Derivation of the rate function $q(t)$ yields two functions depending on the value of the b-parameter. There is a physical bound of 0 as it would result in a negative value of decline rate. If $b = 0$, then the exponential rate function is the result:

$$q(t) = q_i e^{-D_i t} \tag{7.3}$$

And when $b > 0$, the hyperbolic rate function results:

$$q(t) = \frac{q_i}{(1 + bD_i t)^{1/b}} \tag{7.4}$$

When we express Eq. (7.3) in logarithmic terms, we get the following straight-line equation:

$$\ln q(t) = \ln q_i - D_i t \tag{7.5}$$

When we express Eq. (7.4) in logarithmic terms it will form a logarithmic equation, which when plotted against the log of time will yield a straight-line equation:

$$\ln q(t) = \ln q_i - \frac{1}{b} \ln(1 + bD_i t) \tag{7.6}$$

Considering these results, it is appealing that we reach the same conclusion as when we analyzed the asymptotic solutions for the flow regimes of interest: fitting the natural logarithm of the rate data rather than the untransformed rate data is the optimal choice.

7.3.3 Limitations of Linear Models

It is worth asking why we would not proceed to fit production data using a linear model. There are numerous issues with doing so:

1. Production data is time-series data, and autocorrelation of the residuals is both significant and not constant due to operational effects on production performance.
2. Hyperbolic model parameters are highly correlated.
3. Linearization can be performed for a single flow regime, but the flow regime is not constant. When the flow regime does not match the chosen regime for linearization, then the data will not yield a straight-line.
4. We often expect the flow regime in the future to not be the same as observed in the data. The decline trend of initial production is typically not representative of the decline trend of late-time data.
5. Linear models are not appropriate for extrapolation as data points at the ends of the regression line exert greater leverage than do data points near the middle of the regression line. Because of this fact, the forecast of future production is highly sensitive to outliers at the tail of the production history.

Figure 7.4 presents a set of production data sourced from Enverus for a Delaware Basin 3rd Bone Springs horizontal well. This well was put into production in June 2018 and had been producing for 38 months at the time of analysis. The data display features typical of production data from unconventional shale reservoirs:

- The first month production rates are significantly below the peak rate that occurs in month two due to fracture flowback.
- There are various sections in the production history where downtime is clearly visible.
- There are clear trends occurring over sets of a few months throughout the production history as operational practices or other effects impact the production performance.
- A decline trend is observed over the full production history, but there is considerable noise visible throughout the data.
- There is clear correlation in the production of all three phases.

7.3.4 A Motivating Example

The remainder of this chapter will introduce concepts to address these items. Ultimately, we will reach a set of data filtering and probabilistic regression algorithms that address all of the previous concepts. As evidence to this, we present a

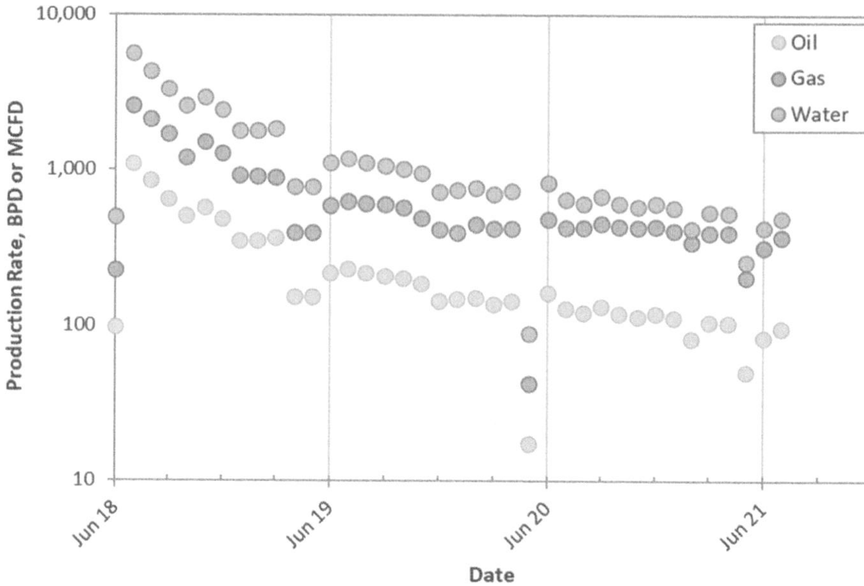

FIGURE 7.4 Production data from a Delaware Basin 3rd Bone Springs Well. BPD, barrels per day; MCFD, thousands of standard cubic feet per day.

probabilistic regression in Figure 7.5. The figure shows samples drawn from a distribution of hyperbolic decline curve models determined from the probabilistic regression. The scatter plots show parameters from the distribution, and Figure 7.5(d) shows a randomly selected subset of models chosen to illustrate the distribution variance. Data in the "filtered" series is not used in the regression.

We will get into the technical details later in this chapter, but for now we focus on a few features that clearly illustrate the limitations we would encounter with linear models. It is obvious that the parameters are not only correlated, but also that the correlation is not linear. These violations of the assumptions of linear regression indicate that attempts to fit linear models to production data—even with transformations of the data and/or model—are unlikely to yield results that meet our expectations of accuracy. Furthermore, they provide no clear path for evaluating the forecast uncertainty we clearly observe in this example.

7.4 NONLINEAR METHODS

Nonlinear regression addresses many of the shortcomings we observe with linear regression. Namely, that we often wish to regress a known empirical decline model to production data. We are confronted immediately with numerous decisions to make as to how we accomplish this. To start, there are many nonlinear solver algorithms to choose from. Some of these require bounds on the model parameters. Some also require a gradient function, whereas others can compute

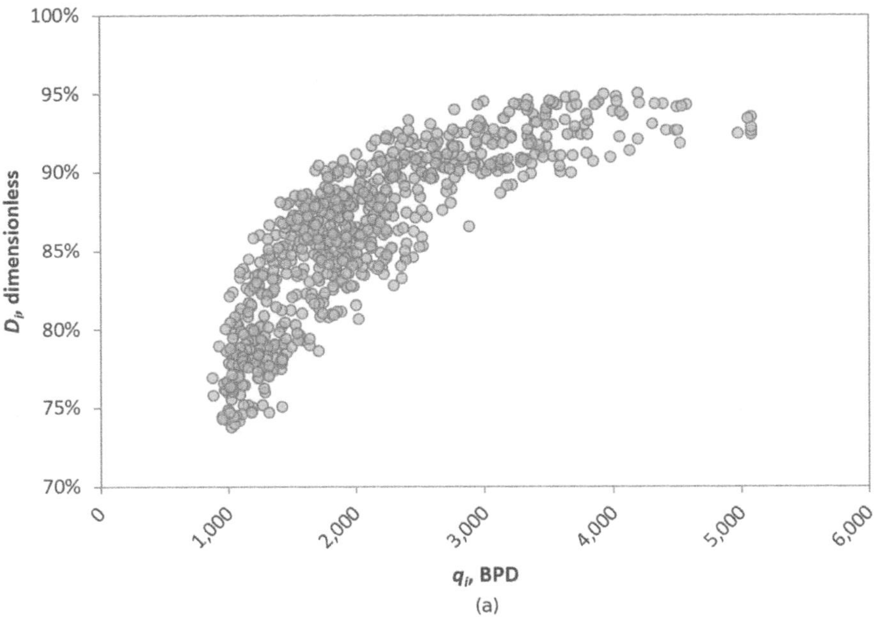

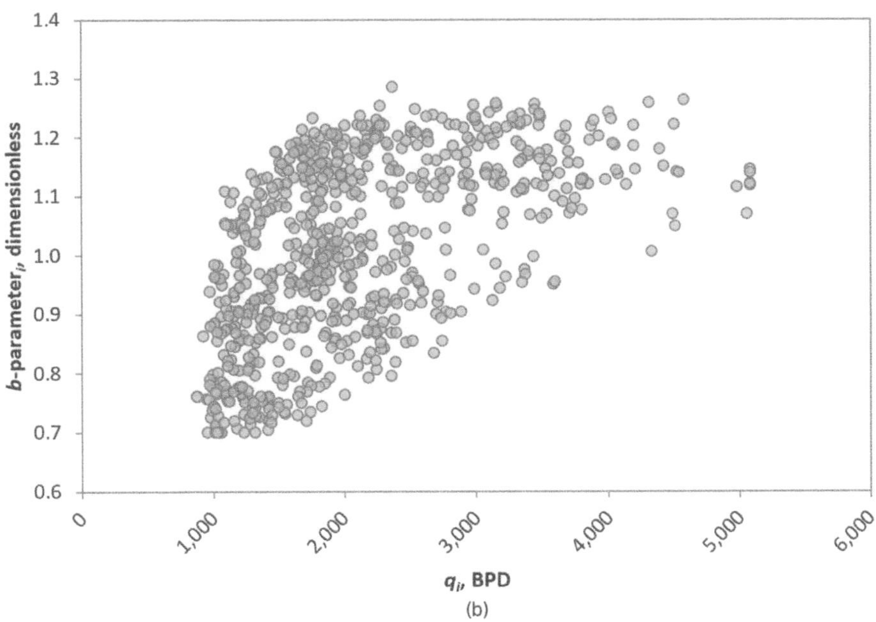

FIGURE 7.5 Arps hyperbolic model parameters sampled from the distribution of models display – (a) D_i versus q_i, (b) b versus q_i, (c) D_i versus b, (d) randomly selected subset of models for rate versus time history. *(Continued)*

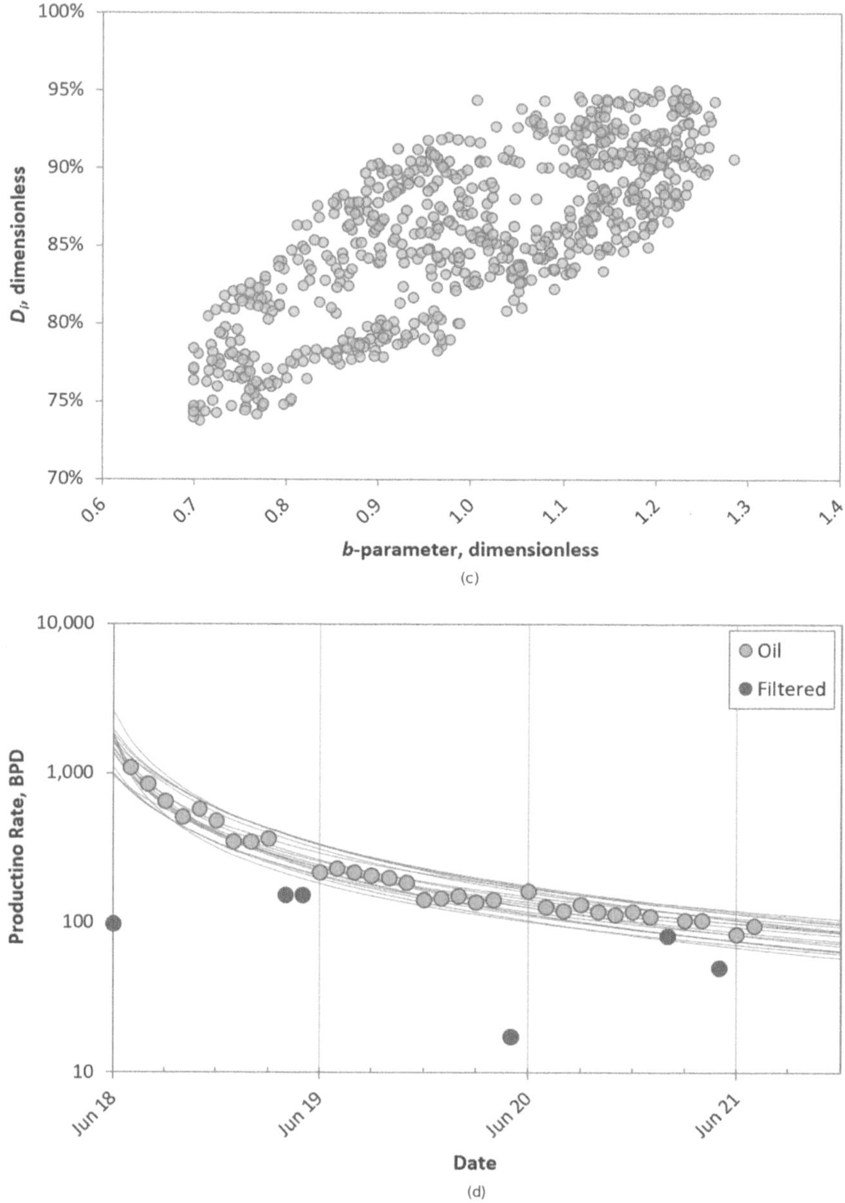

FIGURE 7.5 *(Continued)*

the gradient numerically, and yet still others do not require a gradient at all. We will not review various solver algorithms, as such comparisons have been performed (for example, see Kelley 1999). Rather, we will explore the choice of cost function passed to the solver for minimization and the effect that choice has on the resulting model fit.

7.4.1 Cost Functions and Minimization

A cost function, also known as a residual function or loss function, is a reduce operation that (in most cases) is passed a set of data vectors and model parameters and returns a single scalar. The returned scalar represents a score of the current set of model parameters, and the solver algorithm will iterate the parameters until it satisfies various convergence criteria. The nature of the cost function can greatly affect the resulting model. The most common cost function is the "squared error" function, often implemented as the residual sum of squares (RSS) or the mean squared error (MSE). This is the cost function used in linear regression, due to the fact that it has the property of providing a unique analytic solution. We can generalize the least squares cost function as follows:

$$\left\|\boldsymbol{x}_p\right\| = \left[\sum_{i=1}^{n} |x_i|^p\right]^{1/p} \tag{7.7}$$

This is called the p-norm. Minimization of Eq. (7.7) without the $1/p$ exponent term would have no effect on the result, as a power transform is a monotonic transformation. We can simplify the p-norm to the squared error cost function by substitution of the residuals as the difference of our data, y, and the function to regress, $f(x)$. The result is a general cost function:

$$J_p(x) = \sum_{i=1}^{n} \left|y_i - f(x_i)\right|^p \tag{7.8}$$

Substitution of $p = 2$ into this function provides the familiar 2-norm or $L2$ cost function, i.e., the squared error:

$$J_{L2}(x) = \sum_{i=1}^{n} \left|y_i - f(x_i)\right|^2 \tag{7.9}$$

If we substitute $p = 1$ then we obtain the 1-norm or $L1$ cost function, i.e., the absolute error:

$$J_{L1}(x) = \sum_{i=1}^{n} \left|y_i - f(x_i)\right| \tag{7.10}$$

While any p-norm exists, we will limit our discussion to just these two. Minimization of a cost function is equivalent to solving for the set of parameters that set the gradient $\Delta J_p(x)$ to zero.

$$\min J_p \equiv \Delta J_p(x) \rightarrow 0 \tag{7.11}$$

We find that the value that satisfies Eq. (7.11) for the $L2$ cost in Eq. (7.9) is the mean. For the $L1$ cost, the value that satisfies Eq. (7.10) is the median. For applications to production data, we have cause for concern. First, as we have seen in our production data in Figure 7.4, we expect that our data contains severe outliers, which are often correlated. Second, because the physical equations that describe the production rate involve the multiplication of many properties, we expect a right-skewed distribution of production rates.

Correlated outliers present problems for the $L2$ cost function because they are severely penalized due to the squaring of residuals. The right-skewed distribution of production exacerbates the problem of outliers, and presents a situation in which the probability of achieving the mean is less than even. Both of these problems are solved by using the $L1$ cost function.

In the case where the production data errors are symmetrically distributed, the same result would be achieved as the $L2$. However, the advantage of the $L1$ is that it is much more robust to outliers. Even if we take the natural logarithm of the production data as we have recommended, the $L1$ and the $L2$ provide the same result in the cleanest of datasets, while the $L1$ is much less affected by outliers in the dirtiest of datasets.

The $L1$ does have drawbacks. It cannot be solved by analytic means, and it may have multiple local minima. The first is not a problem for us when we would be using a numerical solver regardless, but the second may present as an issue — especially when regressing on daily data.

7.4.2 Maximum Likelihood Regression

Alternatively, to cost function minimization, we can instead maximize likelihood. While similar for simple cases such as evaluation of squared residuals, the use of a full probability distribution offers additional flexibility to include the probability of additional terms, such as the probability of a parameter having a specific value. We might wish to do so to include regularization in the regression problem. While regularization can also be performed for cost function minimization, by casting it in terms of probability we can then treat it as a data-based process wherein the probability distribution can be determined from data sources external to the rate-time data, as opposed to a hyperparameter tuning exercise that tends to be subjective.

Cost function minimization can be viewed as a reduction of maximum likelihood estimation. Starting with the density function for the series product of a Gaussian distribution, where we are estimating the production rate $\hat{q}_i$ at n total various intervals i of time t given measurements of production rate q_i:

$$f\left(\hat{q}_i | q(t=0),\, D(t=0), b\right) = \prod_{i=1}^{n} \frac{1}{\sqrt{2\pi\sigma^2}} \exp\left[\frac{-\left(\hat{q}_i - q_i\right)^2}{2\sigma^2}\right] \tag{7.12}$$

If we take the negative logarithm, we obtain the "log-likelihood" (actually the negative log-likelihood, but we often leave out the "negative" as implied). We refer to this as "logp":

$$\text{logp}f\left(\hat{q}_i\right) = n \ln \sqrt{2\pi\sigma^2} + \frac{1}{2\sigma^2} \sum_{i=1}^{n} \left(\hat{q}_i - q_i\right)^2 \tag{7.13}$$

which is the form of the $L2$ cost function. While the first term is a constant and can be disregarded if only the log-likelihood of the data is considered, in practice we often compose other terms to obtain a total likelihood. To do so is as simple as adding additional log-likelihoods. For this reason, it is convenient to carry around the constant.

7.4.3 Regularization

One of the most common log-likelihoods to include is a regularization term. In general terms, regularization is a penalty for complexity. In the maximum likelihood framework, it can be conceptualized as the probability of deviation from an expected value. The expected value can be almost anything – a model parameter or a model calculation. Typical lasso or ridge regression in linear regression places a penalty on coefficients. In nonlinear regression, we have the analog of placing a penalty on a specific model parameter. For example, the value of the initial decline rate D_i or b-parameter in the Arps rate-time model. If we have an analog dataset to inform the value of these parameters, then we can choose any probability distribution we wish to model the parameter likelihood.

Less common but also useful would be to regularize a model outcome. For example, we may wish to fit a decline curve to production data for a well that we believe is likely to achieve a specific ultimate recovery of oil or gas. The probability distribution used to estimate reserves can come from an analog dataset, more rigorous analytic or numerical modeling, or simply expert estimation. Determination of decline curve parameters from the current producing rate and ultimate recovery is extremely common for conventional reservoirs as well as early in the producing life of unconventional wells when a "type well profile" may be used as a basis for ultimate recovery with some adjustment to the well's actual production data.

No matter which parameter we wish to model, we simply create a probability distribution for it. For example, to model the b-parameter taking a certain value, such as based upon an analog type well profile, we would write:

$$\text{logp}f(b) = \ln\sqrt{2\pi\sigma_b^2} + \frac{1}{2\sigma_b^2}\sum_{i=1}^{n}\left(\hat{b} - \mu_b\right)^2 \tag{7.14}$$

We would then simply add Eq. (7.14) to Eq. (7.13) to calculate the total log-likelihood. It is of interest to point out if we define $\lambda = \frac{1}{2\sigma_b^2}$ as the strength of the regularization, then we have:

$$\text{logp}f(b) = \frac{1}{2}\ln\pi - \frac{1}{2}\ln\lambda + \lambda\sum_{i=1}^{n}\left(\hat{b} - \mu_b\right)^2 \tag{7.15}$$

And this would reduce to a typical $L2$ regularization in the cost function minimization framework.

7.4.4 Example of Maximum Likelihood Regression

Using the outlined principles of maximum likelihood regression, we can regress the Arps hyperbolic model with regularization of the *b*-parameter. We use the data of oil production rates from the same Delaware Basin 3rd Bone Springs Well as before. The result is shown in Figure 7.6, and model parameters are shown in Table 7.2.

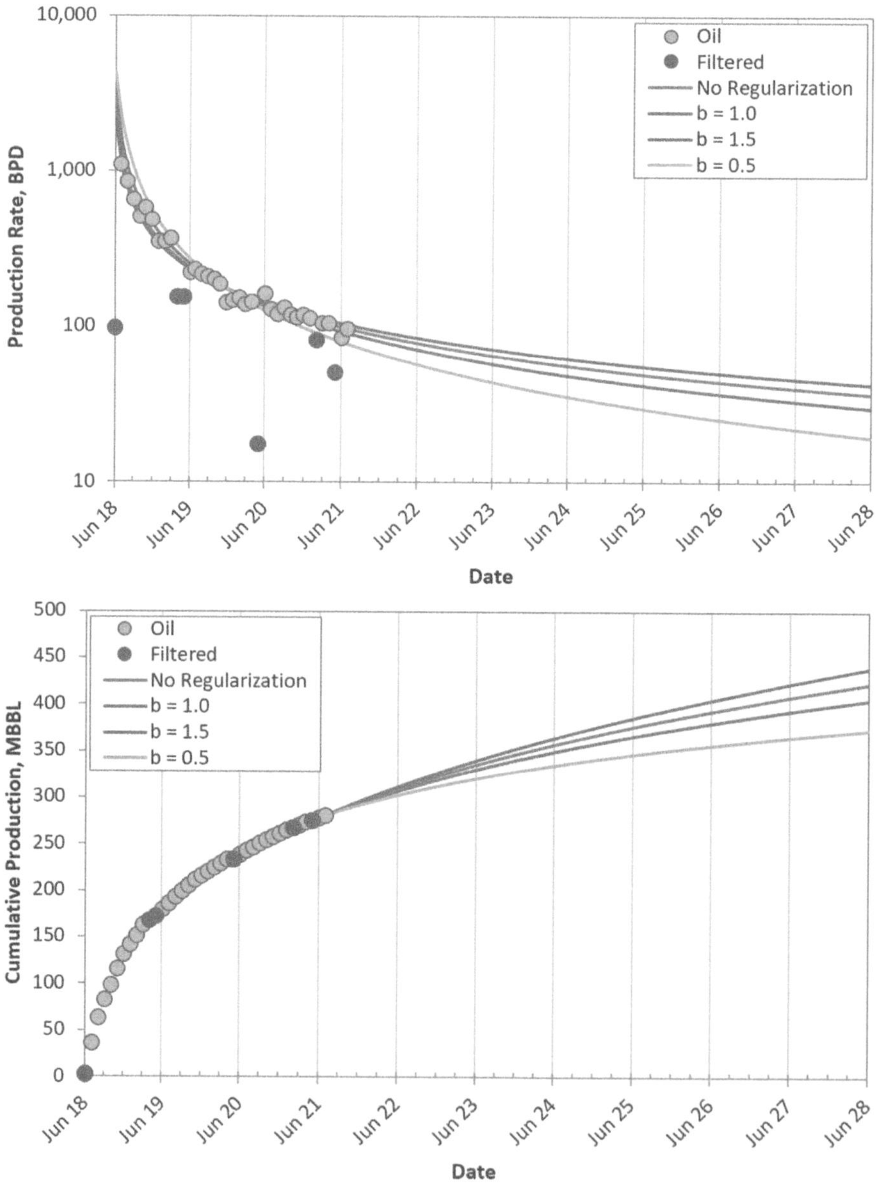

FIGURE 7.6 Various applications of regularization to a Delaware Basin 3rd Bone Springs Well. MBBL, thousand barrels.

TABLE 7.2
Maximum Likelihood Regression Model Parameters

Case	q_i, BPD	D_i, %	b-parameter, *Dimen.*	30 Yr EUR, MBbl
No regularization	7855.2	91.7	1.21	584.7
$b = 1.0$	7855.2	89.2	1.05	527.9
$b = 1.5$	7855.2	93.2	1.33	635.8
$b = 0.5$	7855.2	85.0	0.84	443.1

Each case lists the regularized value (or expected value) for the b-parameter as the case name, with the result from the regression. We observe that the value of the b-parameter for the regressed model is not equal to the regularized value. Instead, we see that the case with no regularization as a b-parameter of 1.21, and each regularized case has the regressed value fall somewhere between this 1.21 and the regularized value. Put another way, the data expresses a probability distribution for the b-parameter, and the regularization is also a probability distribution for the b-parameter, with the actual regressed value resulting from the interaction of the two.

Because empirical PDA is often unable to incorporate additional information an analyst may have—such as nearby offset wells, type well profiles, analytic or numerical models, etc.—regularization gives a way to express this information in a form that can be incorporated into the empirical rate-time model. The data itself may also contain noise, as we observe in this example, which makes determination of the "right" model a difficult proposition. Lastly, limited production history may not allow proper determination of the b-parameter. It typically takes over a year of production—and often longer—to resolve a reasonable value. In these cases, without any regularization, the result may be nonsensical. In a sense, regularization provides a way to "force" the regression to be more physically realistic.

We note that this is actually a discrete Bayesian regression. Eq. (7.13) is the likelihood function, and Eq. (7.15) is a prior probability placed upon the b-parameter. Table 7.2 lists the maximum likelihood of the posterior distribution of models for each case. These examples, which result in discrete outcomes, are helpful to understand the concepts at work. In the following section we'll discuss how to sample from the full distribution of models.

7.5 BAYESIAN REGRESSION METHODS

Given the noise inherent in production data, and the uncertainty in interpretation of decline trends, it is often advantageous to quantify uncertainty by a full distribution of potential model fits. Typically, this takes the form of a 90th percentile, 50th percentile, 10th percentile, and mean forecast. In oil field nomenclature, we call these the P90, P50, P10, and PMean, respectively. Additionally, percentiles are defined as exceedance probability rather than cumulative probability; in other words, the P10 is greater than the P90.

To evaluate a given percentile from a distribution of model fits, we need a way to represent the distribution. However, there is no analytic representation of this distribution! This means that we cannot simply generate discrete percentile models on demand. Thankfully, there are ways around this. Markov Chain Monte Carlo (MCMC) is one way we can directly sample from the distribution of model fits. But we cannot sample any specific percentile we wish. Instead, we must sample thousands of times, and then evaluate these samples to determine which set corresponds to a desired percentile.

7.5.1 Introduction to Markov Chain Monte Carlo

In the Bayesian context, a probability distribution that reflects our belief about the value of a model parameter is called a prior, and it affects the output in a manner similar to regularization. A probability distribution that represents the model error is called a likelihood. We can view Bayesian methods as an update of prior(s) given likelihood(s), but the conceptualization of regularizing likelihood(s) given prior(s) is mathematically equivalent. In either case, we would arrive at the same posterior distribution. That we do so is one of the key advantages of Bayesian methods; they are not order dependent. This gives us a nice guarantee of convergence to the same posterior distribution no matter in what order we include our prior(s) and likelihood(s). Eq. (7.16) is Bayes' theorem for continuous probability distributions. We have $\pi(\theta)$ as the prior, $f(y)$ as the data, $f(y \mid \theta)$ as the conditional probability of the data given the prior, and $\pi(\theta \mid y)$ as the conditional probability of the prior given the data — or more simply, the posterior distribution.

$$\pi(\theta \mid y) = \frac{f(y \mid \theta)\pi(\theta)}{f(y)} \tag{7.16}$$

When the data is fixed, such as our case, then the posterior probability is proportional to the product of the prior and likelihood. In this case, we can simplify Bayes' theorem to Bayes' rule:

$$\pi(\theta \mid y) = \frac{f(y \mid \theta)\pi(\theta)}{\int f(y \mid \theta)\pi(\theta)} \tag{7.17}$$

The integral in the denominator is intractable, but it is also a constant. Instead of evaluation of the integral, we can instead simply calculate a ratio of two randomly generated models, which would lead to cancellation of the integral. If this ratio is greater than unity, than the model in the numerator is the more likely model, and the converse is true if the ratio is less than unity. This algorithm is called the Metropolis algorithm (Metropolis et al. 1953) and is defined by Eq. (7.18), where n is the current proposed model, and $n-1$ is the prior model.

$$\alpha = \frac{f(y \mid \theta_n)\pi(\theta_n)}{f(y \mid \theta_n - 1)\,\pi(\theta_n - 1)} \tag{7.18}$$

How do we remove randomly generated model proposals that are not actually part of the posterior distribution? This is solved by rejecting unlikely models using α as the probability of acceptance. As we generate model proposals, if α is greater than unity, then we know that the proposed model is more likely than the prior model. Thus, it is always accepted as part of the posterior distribution. However, if α is less than unity, then α is compared with a random draw from the uniform distribution $\mathcal{U}\{0, 1\}$ and the model is accepted if α is greater.

It is important to ensure the randomly generated models are independent of one another. If we sample parameters by a stochastic Markov process, then we achieve a chain of parameter values that has the Markov property of being memoryless. Thus, we can generate independent model proposals to feed into the Metropolis algorithm.

How do we determine which function to fill the role of the likelihood distribution? The Metropolis algorithm is our friend here. Because we removed the integral in Eq. (7.17), we do not need our function to be the true probability distribution or even a valid probability distribution—defined as having an integral equal to unity. Instead, it must only be proportional to the true probability distribution. In theory, we can use any function we desire. In practice, a function that behaves similarly to a density function is ideal.

While Gaussian and Laplace distributions are convenient choices, they tend to not work well due to non-Gaussian residuals and having long-tails, respectively. In this author's opinion, they do not create a distribution that most analysts would expect (they do not look correct upon observation), and overly smooth production data can lead to a collapse in variance of the distribution they sample. Attempts to correct this latter problem have led researchers to add a white noise term to the likelihood function. An alternative approach to solve this led Fulford et al. (2016) to introduce a likelihood function that allows independent tuning of bias and variance as separate hyperparameters.

Lastly, during startup of the MCMC it is likely that the Markov chain(s) has not yet converged to the space of the posterior distribution. This period is call "burn-in", and it is important to discard the set of iterations during which converges occurs.. A few details remain: discussion of jumping distributions to sample each parameter, how many burn-in iterations to run, and a method of discretization from the posterior distribution to discrete percentiles. We refer to Fulford et al. (2016) for the details of these items.

The steps of the MCMC simulation are as follows:

1. Estimate prior probabilities and generate starting points of Markov chains for each model parameter.
2. Calculate the product of prior(s) and likelihood.
3. Generate a proposed model by use of jumping distributions for each parameter.
4. Calculate the product of prior(s) and likelihood for the proposed model.
5. Calculate α and compare with $\mathcal{U}\{0, 1\}$ to accept or reject the proposed model.
6. Repeat steps 3 through 5 to sample the posterior distribution.

After sampling of the posterior, analysis can be performed to identify parameter distributions, identification of forecast corresponding to specific percentiles, etc.

7.5.2 Case Example

Putting our MCMC algorithm into practice, Figure 7.7 shows the resulting probabilistic regression on the Delaware Basin 3rd Bone Springs well. Table 7.3 lists the parameters of the percentile forecasts. Inspecting the figure, we can observe the bounds that have been placed around each parameter in each respective plot of the Markov chain. The bounds are placed to improve numerical efficiency of the algorithm; that is, so that less iterations are needed, and better "mixing" of the chain occurs. Mixing, i.e., the rapid movement of the chain between the upper and lower bounds is a good result. We also observe that the epsilons reached their minimum values prior to the end of the burn-in period, which is the first 10,000 iterations.

We observe a rather narrow distribution for *b*-parameter. This is partly due to the fact that a prior was expressed for the *b*-parameter to have a value of 1.0, but more significantly that the data exhibits a clear range of potential values of the *b*-parameter. In contrast with the maximum likelihood regression, we do not represent the most likely model from the posterior distribution. This causes a significant difference with the No Regularization case and highlights a weakness of discrete regression. We are much more interested in a decline model with a 50%

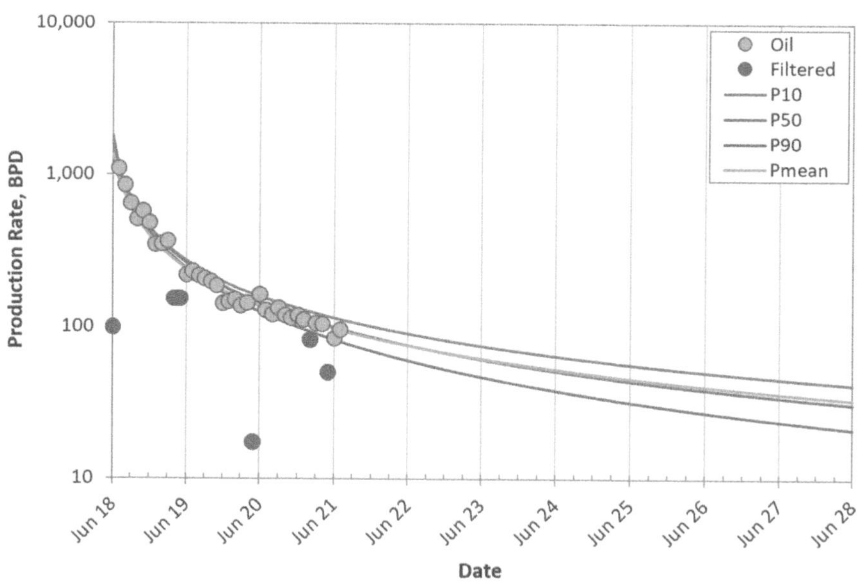

(a)

FIGURE 7.7 Example of MCMC to perform probabilistic regression to a Delaware Basin 3rd Bone Springs Well – (a, b) probabilistic regression and forecasts for production rate and cumulative production, respectively, (c–e) progression of MCMC iterations for various model parameters, and (f) full ensemble of probabilistic regression fits. *(Continued)*

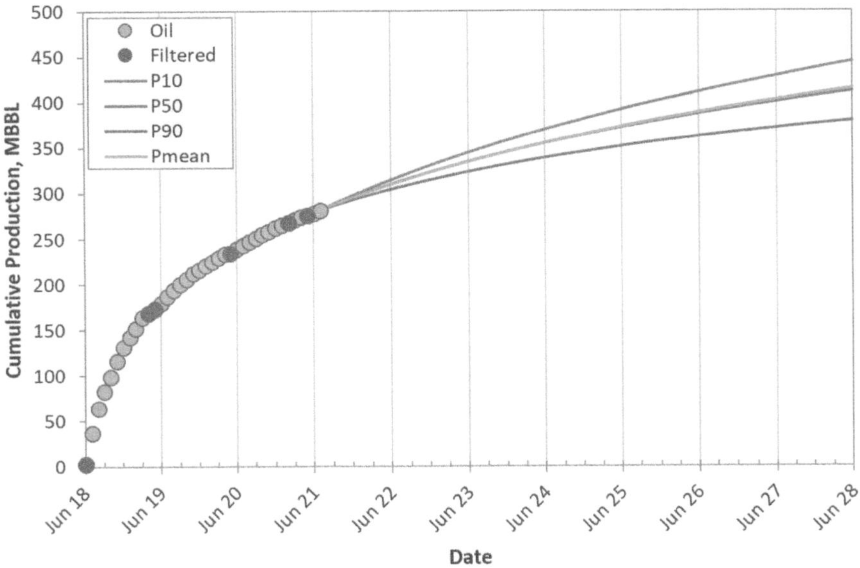

(b)

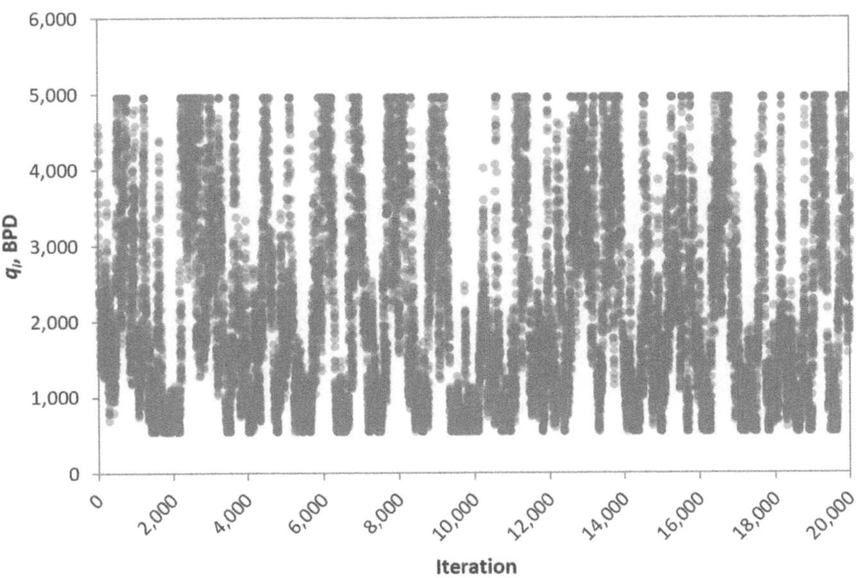

(c)

FIGURE 7.7 *(Continued)*

chance of exceedance, or an expected value decline model, than we are with the most likely model. We do not know the shape of the posterior distribution and where the mode lies; it may be that the distribution has multiple local maximum likelihoods.

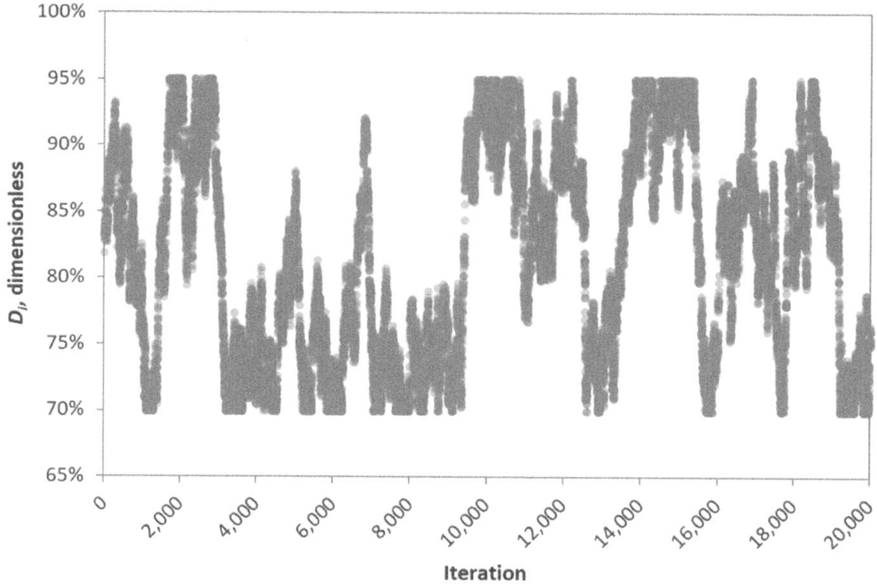

(d)

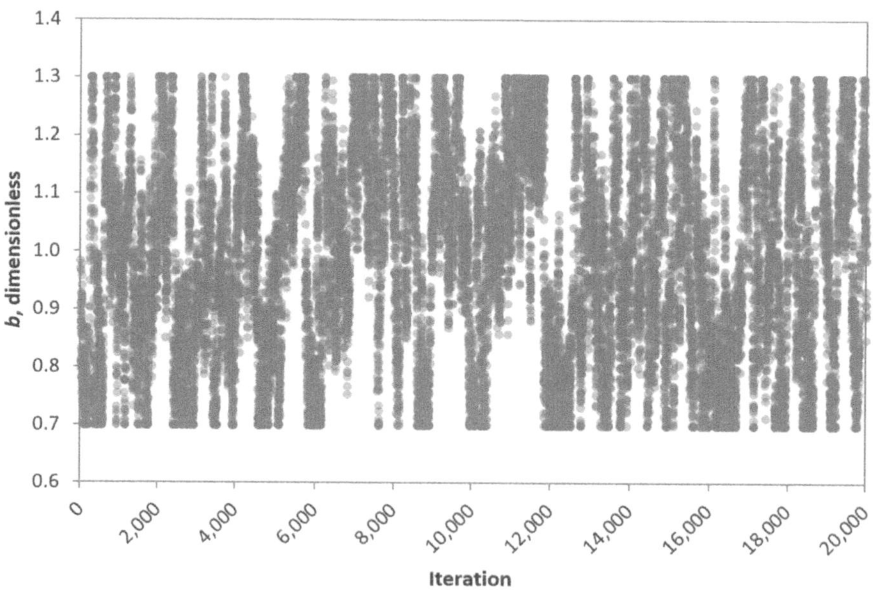

(e)

FIGURE 7.7 *(Continued)*

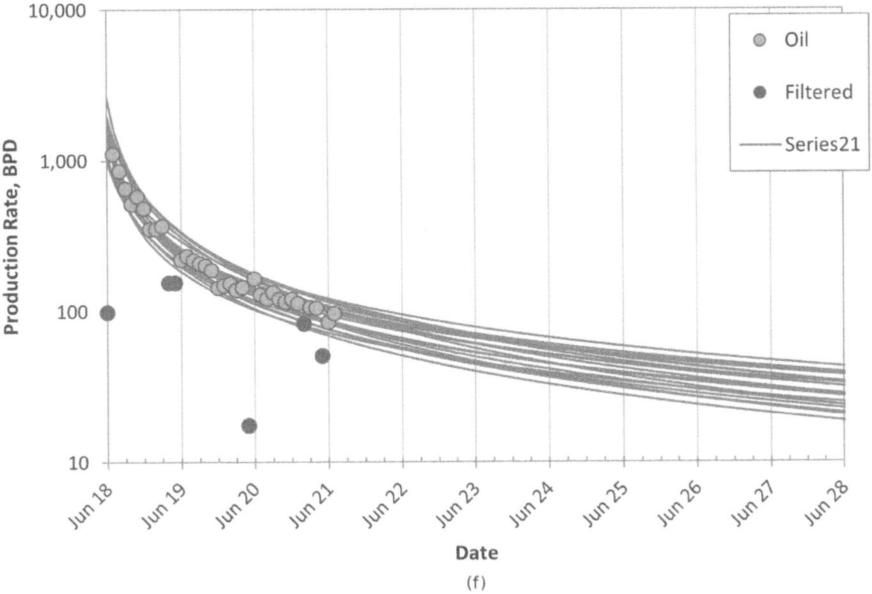

FIGURE 7.7 *(Continued)*

TABLE 7.3
MCMC Probabilistic Regression Model Parameters

Case	q_i, BPD	D_i, %	b-parameter, *dimensionless*	30 Yr EUR, MBbl
No regularization	2635.6	89.3	1.17	319.8
$b = 1.0$	2007.9	86.4	1.00	532.0
$b = 1.5$	1401.9	82.0	0.86	457.4
$b = 0.5$	1752.5	85.9	1.09	551.6

Instead, the probabilistic regression allows us to determine the median and mean models given all possible models that are reasonable interpretations of the production data. The median and mean are much more useful measures of central tendency than the mode.

7.6 CONCLUDING REMARKS

7.6.1 Current Challenges

- There are tens of thousands of multi-fractured horizontal wells (MFHW) in the United States, and millions of wells if we include vertical and directional wells. There is no one-size-fits-all model and associated hyperparameters for production decline forecasts. Regularized regression and

Bayesian regression present opportunities to capture petrotechnical knowledge about well performance in the various landing zones across a basin. These techniques can be applied to ensure consistency of forecasts with more rigorous models and analog datasets.

- The forecasting of production data is an attempt to predict something that has never been observed, which inhibits many unsupervised approaches. Simulated datasets offer the potential to improve the training dataset. However, better calibration and predictive capability are needed to generate these datasets. Currently, there is no ideal approach for generation of simulated production data to improve long-term production forecasts, mostly due to the fact that there are no real data points for the type of late-life unconventional well production we would desire to simulate.
- Varying pressure and multiphase behavior introduce severe nonlinearities that simple models are unable to explain. Data limitations, e.g., no available pressure or fluid composition data in public datasets, further complicate attempts to model these affects. Often, this leaves accountment of these effects as a process of inference by evaluators.

7.6.2 Near-term State-of-Practice

- Expect wide deployment of autofiltering and autoforecasting as well as probabilistic analysis capabilities in leading commercial software.
- Commercial tools should soon recognize the importance of storage and retrieval of hyperparameters, creating a pathway to hyperparameter optimization workflows.
- Active research continues to search for "correction factors" for well spacing and parent-child issues, which will eventually yield practical workflows to model these effects.

ACKNOWLEDGMENTS

The author would like to thank Yannis Katsaros of ARM Energy for assistance in developing the code base for the unsupervised data filtering algorithms.

REFERENCES

Alimohammadi, H. and Chen, S. N. 2022. "Performance Evaluation of Outlier Detection Techniques in Production Timeseries: A Systematic Review and Meta-analysis." *Expert Systems with Applications* **191** 116371. https://doi.org/10.1016/j.eswa.2021.116371

Arps, J. J. 1945. "Analysis of Decline Curves." *Transaction of AIME* **160** (1) 228–247.

Bruenig, M. M., Kriegel, H., Ng, R. T., and Sander, J. 2000. "LOF: Identifying Density-Based Local Outliers." *Proc. ACM SIGMOD* **29** (2) 93–104.

Ester, M., Kriegel, H., Sander, J., and Xu, X. 1999. "A Density-Based Algorithm for Discovering Clusters in Large Spatial Databases with Noise." *Proceedings of 2nd International Conference on Knowledge Discovery and Data Mining*. AAAI Press, 226–231.

Fulford, D. S. and Blasingame, T. A. 2020. "Optimization Methods for Time-Rate-Pressure Production Data Analysis using Automatic Outlier Filtering and Bayesian Derivative Calculations." *SPE Annual Technical Conference & Exhibition*. SPE 201404-MS. Denver, Colorado. https://doi.org/10.2118/201404-MS

Fulford, D. S., Bowie, B., Berry, M. E., Bowen, B., and Turk, D. W. 2016. "Machine Learning as a Reliable Technology for Evaluating Time/Rate Performance of Unconventional Wells." *SPE Economics and Management* **8** (1) 23–39. https://doi.org/10.2118/174784-PA

Huber, P. J. 1964. "Robust Estimation of a Location Parameter." *The Annals of Mathematical Statistics* **35** (1) 73–101. https://doi.org/10.1214/aoms/1177703732

Kelley, C. T. 1999. "Iterative Methods for Optimization." Philadelphia, Society for Industry and Applied Mathematics.

Liu, F. T., Ting, K. M., and Zhou, Z. 2008. "Isolation Forest." 2008 Eighth IEEE International Conference on Data Mining. Pisa, Italy.

Metropolis, N., Rosenbluth, A. W., Rosenbluth, M. N., Teller, A. H., and Teller, E. 1953. "Eq. of State Calculations by Fast Computing Machines." *The Journal of Chemical Physics* **21** (6) 1087–1092. https://doi.org/10.1063/1.1699114

Rousseeuw, P. J. and Driessen, K. V. 1999. "A Fast Algorithm for the Minimum Covariance Determinant Estimator." *Technometrics* **41** (3) 212–223.

Schölkopf, B. Plat, J. C., Shawe-Taylor, J., and Smola, A. J. 2001. "Estimating the Support of a High-Dimensional Distribution." *Neural Computation* **13** 1443–1471.

8 Hybrid Data-Driven and Physics-Informed Reservoir Modeling for Unconventional Reservoirs

Sathish Sankaran and Hardik Zalavadia
Xecta Digital Labs,
Houston, Texas

CONTENTS

8.1 INTRODUCTION

The development of unconventional reservoirs has triggered a reimagination of reservoir engineering models and methodologies. It is important to properly understand well performance across the whole life of the wells to optimally develop

DOI: 10.1201/9781003207009-12

unconventional reservoirs and extract maximum economic value. Identification of the key drivers and production behaviors is critical to generate robust production forecasting and enable proper field development planning and production optimization through the most appropriate operational strategy. In this work, we focus on shale and tight reservoirs with very low matrix permeability, where standard reservoir models describing fluid flow in porous media are not adequate.

Traditional well performance analysis and modeling methodologies, which have been developed over the last century through accumulated learnings from conventional field development, originally proved insufficient to properly represent the production behavior in unconventional wells. These are characterized by rapid production decline and long periods of transient flow generated by a complex fractured system in very low permeability. For the past two decades, several researchers and practitioners have focused on trying to understand rate- and pressure-transient behaviors of unconventional reservoirs under different production strategies.

The challenges in unconventional reservoirs include (1) unavailability of required data for traditional full-physics reservoir modeling for every well in the field (e.g., PVT, downhole pressure, logs, etc.), (2) lack of understanding of relevant physics to be modeled (e.g., hydraulic fracture growth and closure, multiscale flow, etc.), and (3) inability of models to be continuously calibrated and run multiple simulations to keep up with the pace of operations for thousands of wells.

8.2 CURRENT METHODS

Decline curve analysis (DCA) and its variants such as Arps (Arps 1945), modified hyperbolic, Valko (Valko 2009), power law (Ilk et al. 2008), Duong (Duong 2010), logistic growth (Clark et al. 2011), etc., have been adopted as one of the most popular production forecasting methods (Lee and Sidle 2010). DCA methods are very popular because they are fast, repeatable, and easy to implement and maintain, and they remain a favorite method for reserves booking and production planning. Analog-based type curves are another common method to match early production trends and are used for forecasting. However, by only considering rate vs time trends and ignoring the pressure, they implicitly assume constant operating conditions, and cannot fully capture changes in the well production such as choke changes, artificial lift installations, and change in the artificial lift operational parameters. Furthermore, many wells are surface constrained, and often do not exhibit a discernable production decline. In such wells, the DCA method cannot be applied, and surrogate type curves are often used, with little fidelity to the actual production data. Finally, the decline models are single-phase and empirical, and do not capture fundamental physics, such as the expected gas-oil ratio (GOR) increase on oil systems as the solution gas is liberated when pressure declines below saturation pressure.

Analytical (or semi-analytical) methods, such as rate-transient analysis (RTA), address changing well operating conditions by incorporating pressure and capturing some physics, including rock and fluid properties and the transient flow regimes, and generate a more representative production forecast. The focus is on identifying and modeling key flow regimes with analytical or semi-analytical models (Ozkan et al. 2011, Nobakht and Clarkson 2012, Chen and Raghavan 2013, Uzun et al. 2016), often assuming homogeneous and isotropic reservoirs. However, in practice these models are difficult to generalize for all wells, as they require knowledge of rock properties (such as porosity and permeability in both matrix and "enhanced" stimulated zones, net pay, initial saturation) and completion attributes (number of fractures, fracture conductivity, fracture half-length, fracture height, etc.) that cannot be accurately defined and measured, introducing significant input uncertainty in the models. A hypothesis needs to be made on the RTA model mechanisms (e.g., linear flow, bounded stimulated reservoir volume (SRV), trilinear zones, etc.) and on fracture characterization (e.g., planar fractures, fracture symmetry, uniform properties, number of contributing fractures, etc.). With so many modeling knobs, solutions are non-unique and difficult to constrain. Often, a pragmatic model implementation involves assuming many of these input parameters through rules of thumb and engineering judgment, which adds significant bias to the models as many of these inputs cannot be empirically validated. Under special circumstances, when these field data are collected, unconstrained modeling with large degrees of freedom can lead to significant variations in forecasts. Further, the final work product is interpretive and not conducive to automation. Therefore, implementing RTA requires significant effort for routine work and, consequently, is not often practiced regularly at most companies due to the previously mentioned limitations.

Numerical simulation models allow inclusion of more complex physics, such as irregular fracture geometries, heterogeneous reservoir properties, stress-dependent rock properties (geomechanical effects), compositional PVT, adsorption, Knudsen diffusion, PVT suppression, etc. While numerical models may provide mechanistic understanding of relevant phenomena, there is no industry consensus on which are the "right" physics to properly model unconventional reservoirs in a particular setting for routine business processes such as forecasting and production optimization. Each company and researcher may use a different set of assumptions and modeling principles, resulting in various modeling philosophies such as using planar fractures in a homogeneous field, dual porosity systems (Li et al. 2011), discrete fracture networks (Mi et al. 2016, Li et al. 2020), dynamic stimulated rock volumes (Sen et al. 2018), etc. Each model represents a different "belief system," embedding some physics that are thought to fully represent the modeled phenomena. In practice, diverse modeling approaches produce a wide and often diverging set of answers. Also, numerical models are computationally intensive, taking time to build and history match, which requires expensive and extensive data collection, often not available for most wells. In real field applications with hundreds or thousands of wells, simulation models are unable to keep up with the pace of development and cannot scale to cover every well. Typically, they are applied in few representative wells for special studies and the results are extrapolated, often loosely, to other wells in the pad or area. Thus, any conclusion will be affected by the sample bias depending on the selected history-matched wells, while not all the data is being leveraged. Further, numerical models suffer the same disadvantages as RTA such as non-uniqueness, manual interpretation, and lack of well coverage for thousands of wells.

Data-driven models (such as statistical learning techniques and machine learning) have been successfully implemented in some cases as alternatives to the traditional domain-specific physics-based reservoir models. Given sufficient training data, these models can capture complex relationships and predict hydrocarbon production from a variety of data inputs, including rock-fluid properties, completion design parameters, and operational factors (Wicker et al. 2017, Burton et al. 2019, Mohaghegh 2019). Invariably, many operating companies use some form of data analytics in their internal evaluation and practical decision-making processes. These models can be very powerful for various applications in descriptive, predictive, and prescriptive analytics, but have their own inherent drawbacks. However, sufficient training data is required to properly capture meaningful correlations, both in quantity (large number of wells) and quality (representing variability in the input ranges). Data-driven models are only valid within the training data features and ranges, so they tend to be very specific to a given field, basin, or sector and not fully generalizable. Significantly, most training data available does not include pressure and, inherently, the dataset is biased toward the specific operating mode of the well, which is often ignored. They are not guaranteed to provide a robust extrapolation for more general conditions as they do not incorporate any physics acting as a constraint. For most pure machine learning models, there is also the problem of interpretability, as the models are often perceived as a black box, and difficult to explain and audit, which may limit their adoption in a business setting. Advances in application of machine learning and deep learning models can address some of the limitations of pure data-driven approaches such as physics-driven proxy modeling (Park et al. 2020) to generate training data from physics-based models, such as the physics-informed neural network (PINN) (Raissi et al. 2019) that ensures that the underlying governing physical laws are satisfied provided they are known a priori and neural operators (Kovachki et al. 2020) that learn mapping between infinite dimensional functional spaces.

It is desirable to have fast, robust, repeatable, and scalable methods to model well performance in unconventional reservoirs that are able to capture every well, and combine the strengths of data-driven methods (namely, speed, data centricity and flexibility) and physics-based models (namely, robust extrapolation, generalization, and interpretability). A few approaches that embrace hybrid models include diffusive diagnostic function (Zhang et al. 2018), physics-guided deep learning (Razak et al. 2021) and physics-assisted transfer learning (Cornelio et al. 2021), and transient well performance (TWP) (Molinari et al. 2019a, 2019b, Molinari and Sankaran 2021).

8.3 HYBRID RESERVOIR MODELS FOR TRANSIENT WELL PERFORMANCE

While pure data-driven models are searching for better architectures and assume data availability, can reliable models be built with "small data," where the local prediction for each well is more accurate? Can physics-informed data-driven models (hybrid models) improve the prediction in unconventional reservoirs using routinely collected field data?

In the following sections, we present a comprehensive, hybrid reservoir modeling methodology for TWP. The method is based on a novel formulation that combines diffusive time of flight (DTOF), succession of pseudosteady state (SPSS) material balance, and transient productivity index (PI) concepts for performing production diagnostics and forecasting (Molinari et al. 2019a, 2019b, Molinari and Sankaran 2021). The approach has been successfully implemented in thousands of wells in several shale plays and tight reservoirs in North America. The key idea behind TWP is that the drainage volume is increasing continuously with time, but the exact geometry is unknown. The DTOF approach to calculate drainage volume ensures a no-flow outer boundary, and the reservoir withdrawal is known through cumulative volume produced. Under the SPSS assumption, material balance can be performed at each timestep at a high resolution (e.g., daily). This allows the calculation of pressure depletion (i.e., average reservoir pressure) in the contacted drainage volume at any timestep.

The TWP methodology (Figure 8.1) uses routinely available production rates, combined with the PVT and flowing bottomhole pressure (BHP). It also handles variable rate and BHP, pressure depletion, variable compressibility, nonlinear pressure-dependent PVT properties, and represents both liquid and gas systems producing under any production method (e.g., natural flow, gas lift, electrical submersible pump (ESP), sucker rod pumps, etc.). The results of the TWP method enable field-scale automation of a variety of critical applications such as artificial lift optimization, well performance analysis, PI-based forecasting (PIBF), and well interference analysis.

8.3.1 Diffusive Time of Flight

The DTOF is physically associated with the peak propagation of pressure for an impulse source. The 3D diffusivity equation can be reduced to a 1D diffusivity equation (King et al. 2016, Zhang et al. 2018), if we assume that the pressure gradients are aligned with the "time of flight" gradients. Without using a well and

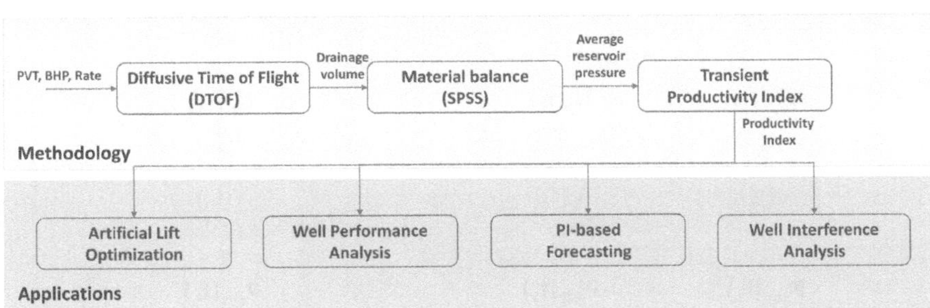

FIGURE 8.1 Transient well performance (TWP) methodology and selected applications.

reservoir model, the drainage volume is calculated using the pressure and rate data as follows (Xue et al. 2018, Molinari et al. 2019b):

$$V_d \approx \frac{1}{c_t \dfrac{d}{dt_e}(RNP)} \tag{8.1}$$

The calculated drainage volume in Eq. (8.1) represents the contacted reservoir pore volume due to the propagation of the pressure front at any given timestep in the reservoir. Analogous to the concept of investigation radius in homogeneous fields, it tracks the DTOF contour of an irregular geometry due to the draining of a lumped fracture system, stimulated and unstimulated matrix. The rate-normalized pressure (RNP) formulation represents the production behavior that would be observed if the well were produced at a constant reference rate.

8.3.2 Material Balance

Material balance is strictly valid only during steady-state or pseudosteady-state conditions, when the well drains a fixed volume and all reservoir boundaries have been contacted. However, we seldom reach true boundary-dominated conditions in unconventional wells. Even when the well contacts all hydraulically fractured regions (loosely referred to as the stimulated rock volume or pseudo-boundary dominated flow), the lower permeability matrix and natural fractures are still being incrementally contacted.

We solve this problem by using an *expanding control volume* approach, where the outer boundary expands at the speed of our pressure contours. Basically, assume that transient flow (where the drainage volume is dynamic and continuously increasing) can be approximated as "snapshots" or SPSS, as shown in Figure 8.2. The drainage volume, already calculated in Eq. (8.1), defines the size of the "container" in each timestep. By our definition of drainage volume based on DTOF, there's no flow into the container at each timestep. However, we know how much volume is taken out of the system through production at the well. By applying material balance, we can estimate the loss of mass in the contacted drainage volume, manifested as reservoir depletion. Cumulative production is

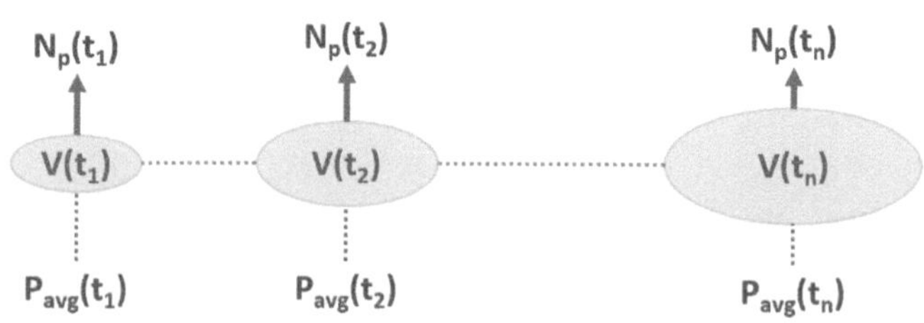

FIGURE 8.2 Material balance applied to expanding drainage volume with SPSS conditions.

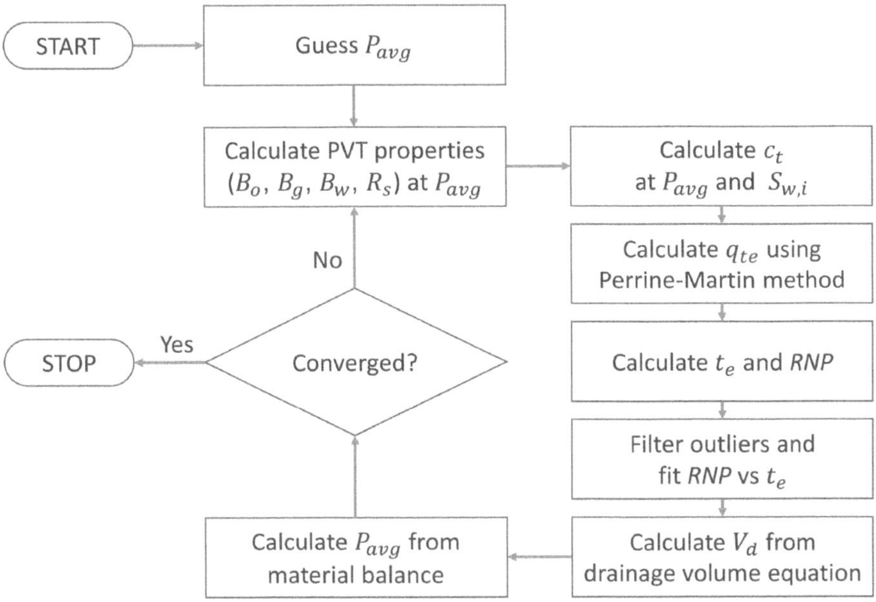

FIGURE 8.3 Algorithm to calculate drainage volume and average reservoir pressure depletion.

tracked daily in the field and these small timesteps also help in improving the accuracy of our material balance.

The overall algorithm is outlined in Figure 8.3 and incorporates pressure-dependent PVT properties (e.g., formation volume factor, compressibility, and solution GOR), which is posed as an optimization problem.

The average reservoir pressure represents the volumetric-averaged pressure in the contacted drainage volume at any instantaneous time, and it approximates the depletion ($\Delta p = P_i - P_{avg}$) in the reservoir due to production. Figure 8.4 illustrates typical drainage volume and average reservoir pressure trends for several wells from a major liquid-rich U.S. shale play.

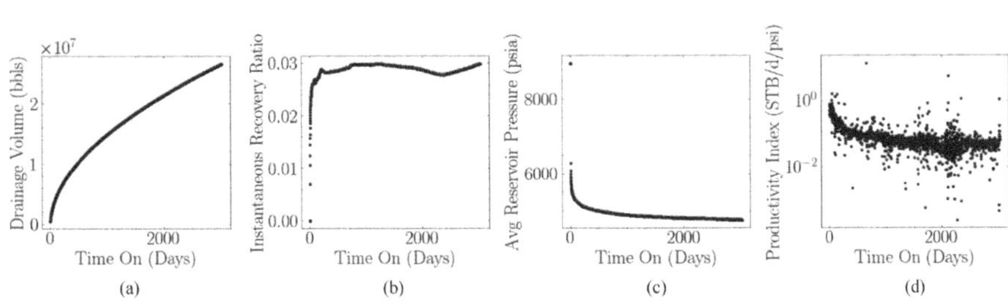

FIGURE 8.4 (a) Drainage volume, (b) instantaneous recovery factor, (c) average reservoir pressure, and (d) transient PI for a well from a liquid-rich U.S. shale play.

The drainage volume (V_d) for a liquid system can be solved as:

$$V_d = \frac{N_p\left[B_o + \left(R_p - R_s\right)B_g\right] + W_pB_w}{\frac{\left(B_o - B_{oi}\right) + \left(R_{si} - R_s\right)B_g}{\left(1+\omega\right)B_{oi}} + \left(\frac{\omega}{1+\omega}\right)\frac{\left(B_w - B_{wi}\right)}{B_{wi}} + \left(\frac{S_{wc}c_w + S_{or}c_o + c_F}{1 - S_{wc} - S_{or}}\right)\Delta p} \tag{8.2}$$

8.3.3 Transient Productivity Index

As discussed in Molinari et al. (2019a, 2019b), the PI defines a representative well performance metric and the true reservoir inflow potential. It normalizes production volumes by the flowing pressures and incorporates the impact of pressure depletion and PVT. It is an effective diagnostic metric as it allows consistent comparison of wells experiencing different operating conditions, such as different choke openings or artificial lift, mitigating biases often encountered in rate-time based DCA and type curves. Also, by representing the relationship between rates and pressure drawdown, PI can be used for production optimization purposes to predict performance under various operational strategies.

The PI depends on the producing rates, flowing BHP and average reservoir pressure. For unconventional reservoirs, we need to consider PI as a transient quantity (Figure 8.4d) that is updated at each timestep.

$$q_l(t) = PI(t) \times \left(p_{avg}(t) - p_{wf}(t)\right) \tag{8.3}$$

8.4 PRODUCTION FORECASTING

Traditional DCA does not incorporate pressure data or fluid phase behavior and only strictly works when the operating conditions are constant (e.g., constant BHP or choke setting, with limited production interruptions). In some scenarios, DCA models cannot even be applied as the well may not exhibit any discernable production decline when production is curtailed by facility constraints. We propose using PI as the forecasting variable, as it allows normalizing both surface effects and considers phase behavior, thus reducing noise. It produces cleaner trends that are easier to fit with less data, resulting in more accurate models and generates better predictions. This PI decline behavior has also been empirically validated on several wells across multiple shale plays and tight reservoirs with constrained rate production. In comparison, note that pressure normalized rate-based decline methods, though similar, do not handle PVT effects or pressure depletion.

A PI-based production forecasting workflow was introduced in Molinari et al. (2019b) and further developed in Molinari and Sankaran (2021) that can be used above and below saturation pressure for a variety of flow conditions. The liquid PI trend may be fitted with time-series-based forecasting models such as

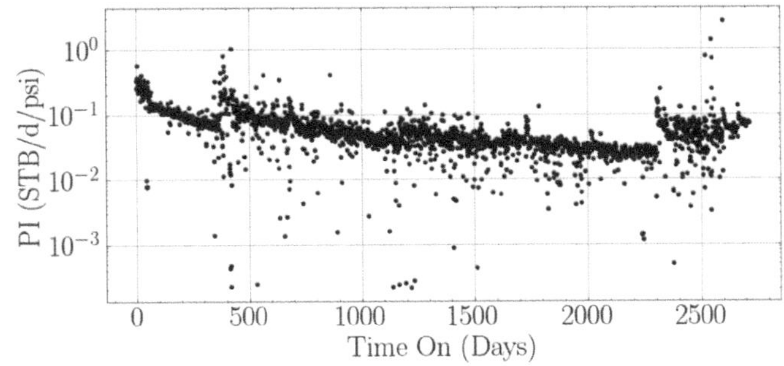

FIGURE 8.5 Example of well interference identified by PI after 400 days and 2,300 days due to child well completion and production.

modified hyperbolic, ARIMA (Box et al. 2015), recurrent neural network (RNN) (Jozefowicz et al. 2015), etc., based on past values of PI and cumulative volumes. The PI decline model allows extrapolating the PI at future timesteps. The key assumption is that the well will remain on primary depletion without sudden changes in PI, due to events such as a refrac or offset frac hit. On the other hand, well interference can also be automatically identified using PI trends more clearly than just using rate or pressure data (Rahman et al. 2019). Figure 8.5 highlights a breakpoint in the PI trend after 400 days and 2,300 days, when a nearby child well was completed and there was subsequent change in well performance. In such cases, multiple PI segments can be fitted after each valid interference event to create a composite PI decline model. When parent-child interference needs to be included, PI forecast will depend on cumulative volumes of both parent and child wells.

The PI forecast is converted into a liquid rate profile by combining it with reservoir pressure and BHP forecasts (Figure 8.6). The average reservoir pressure is extrapolated as a function of cumulative liquid, following the profile defined through material balance in the prior step. The BHP profile is completely controllable by the operator, as it represents the expected operating conditions under the planned production strategy for each well (e.g., choke schedule, artificial lift installs, and operational set points). Once the liquid rate forecast is generated as a single-phase forecast, it is necessary to derive a multiphase forecast, obtaining the corresponding oil, gas, and water rate profiles. This is achieved in two steps, first modeling the water cut, and subsequently modeling the GOR. Both the water cut and GOR models are independent and modular, and various mathematical representations can be used as part of the workflow, tailored to a given field specific conditions.

The factors that affect GOR trends include flowing BHP, saturation pressure, PVT property suppression, pressure-dependent permeability, changes in critical gas saturation, fracture geometry, pore size distributions, and gas-oil relative permeability (Pradhan 2020). A data-driven approach for GOR forecasting uses a two-segment power law by plotting cumulative oil and cumulative gas in a log-log

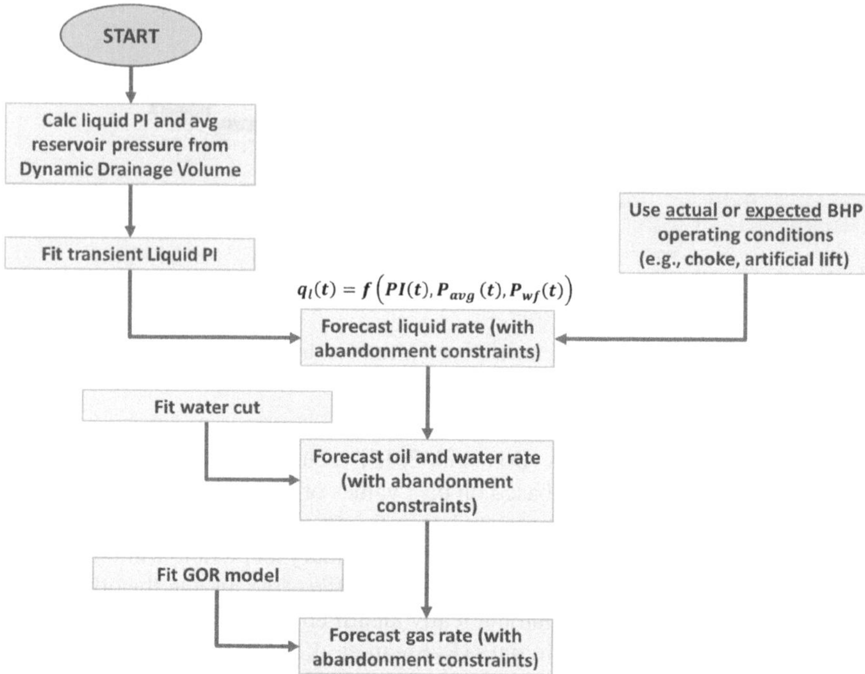

FIGURE 8.6 PI-based forecasting methodology.

plot and matching two straight-line models, as illustrated for a sample well in Figure 8.7.

Using a method that combines physics and data-driven trend, the second GOR segment is estimated by matching the observed GOR increase after an inflection point has been detected. For young wells, when such transition has not been

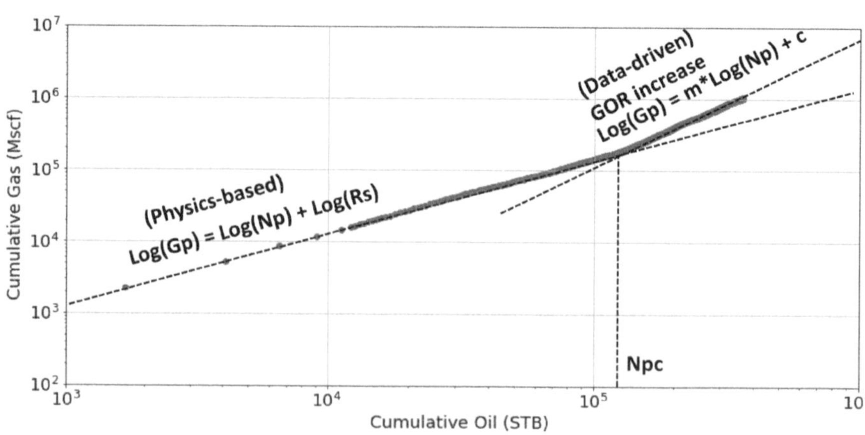

FIGURE 8.7 Gas-oil ratio (GOR) model from two-segment power law.

observed, offset wells can be used as analogs to define expected future GOR trends, provided they are producing under similar drawdown conditions. The two-segment trend has been observed in thousands of wells across various basins, and it is generally adequate to represent most situations. However, when a different GOR model is needed to adapt to specific field trends, such expressions can be implemented without replacing the other parts of the forecasting workflow.

The proposed PIBF methodology is a combination of the following:

- Data-driven methods, which are used to extrapolate PI, water cut, and GOR trends, and
- Physics-based methods, which are PVT-aware, capture pressure depletion, represent PI reduction at low pressures due to multiphase effects, and capture production sensitivity at different pressure drawdown strategies.

8.5 CASE STUDIES

The proposed approach has been applied on several wells (over 5,000 wells) in multiple unconventional fields and tight reservoirs across North America through an automated production surveillance and optimization platform. The application covered wells with slightly compressible fluids (black oil) and variable compressibility fluids (volatile oil, gas condensate, and dry gas) effectively handle production above and below saturation pressure.

The methodology has been successfully implemented with the following objectives.

8.5.1 Derive Value Insights through Normalized Well Performance Drivers

By using only routinely captured field data, without making geometrical assumptions about the well or requiring reservoir properties that are difficult to capture for most wells, the model can be applied to every well in the field continuously, generating consistent performance metrics (such as PI, drainage volume, and instantaneous recovery ratio [IRR]) to characterize well performance at field scale. Multi-well trends in drainage volume, IRR, and PI have been effectively used to quantify differences in completion effectiveness and reservoir quality.

With field-wide drainage volume and PI metrics, data-driven unsupervised clustering techniques can be used to find wells with similar characteristics. Figure 8.8 shows three drainage volume clusters for a sample dataset with about 600 wells in an unconventional field. Distance-based multidimensional scaling (Mead 1992) and *k*-means clustering (MacQueen 1967) were used with drainage volume trends over the first year of production to identify the distinct clusters. Since drainage volume represents dynamic well response, such groupings can inform the key performance drivers describing well performance.

For example, in the case study shown in Figure 8.8, an increasing drainage volume is correlated with a higher PI. However, this does not always translate into a higher produced volume, as seen in cluster 3 (group "High"), which also

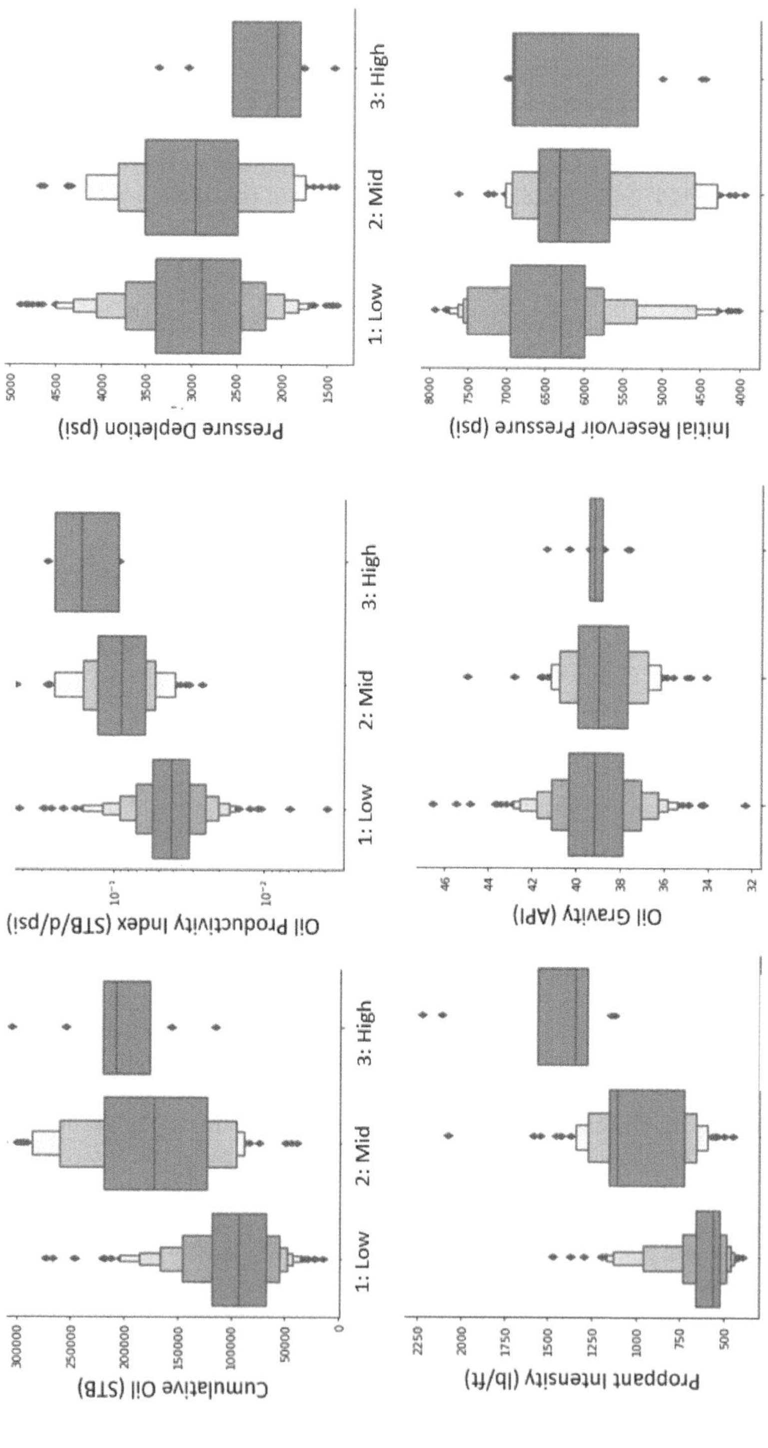

FIGURE 8.8 Key performance drivers based on drainage volume clusters in Field A.

experiences a smaller amount of pressure depletion. Wells with larger drainage volume might be constrained (e.g., choke management, surface offtake capacity limits, line back pressure, lack of artificial lift, etc.) and are potential candidates to increase production by systematically removing these constraints. Similarly, the drainage volume can be correlated with completion design (as evidenced with the strong correlation with proppant intensity). In this example, we do not notice any significant correlation with rock and fluid properties, such as initial reservoir pressure and oil gravity.

8.5.2 Improve Forecasting Accuracy Compared with Rate-Based Methods

By normalizing production for surface effects and changing operating conditions, and quantifying the depletion effect, PIBF provides a more robust way to predict future production, with favorable model accuracy when compared with DCA models. The proposed approach provides a better definition of the long-term gas forecasting and natively captures the PI decline due to depletion and multiphase effects, resulting in an analytical production forecast that is "pressure- and PVT-aware".

Figure 8.9 shows the forecasting results comparison for a single well in Field C between DCA and PIBF methods, where the PIBF method can better forecast production under severe constraints, as the well does not experience a clear rate decline trend. The comparison of the cumulative oil prediction shows nonphysical values with DCA compared with the PIBF method. Further, to validate the PIBF method against traditional DCA, a hindcasting exercise was carried out in an unconventional field with more than 1,000 wells. In this example, each well was forecasted with various training data windows up to a full year (e.g., 30, 60, 90, days, etc.) and used the remaining "known" data as a blind hold-out sample. As expected, for any forecasting model, the prediction errors can be very large when using a very limited dataset but systematically reduce as more training data is included. The PIBF method consistently performs better relative to DCA particularly at early time (Molinari and Sankaran 2021), when production declines are not yet very well defined, achieving a median well-level error reduction between 1.4% and 18.3%.

8.5.3 Continuously Generate Production Optimization Opportunities

Besides predicting production, the PIBF method has other applications, allowing the evaluation of production optimization opportunities (Sankaran et al. 2022). Various BHP scenarios are generated to reflect different operational scenarios (for example, constrained vs unconstrained facilities, or various conservative vs aggressive artificial lift strategies), and the PI-based forecast can predict the expected production for each scenario, determining the most economically optimal case. The availability of key diagnostics such as average reservoir pressure and PI (for historical, current, and future time periods) enable several production

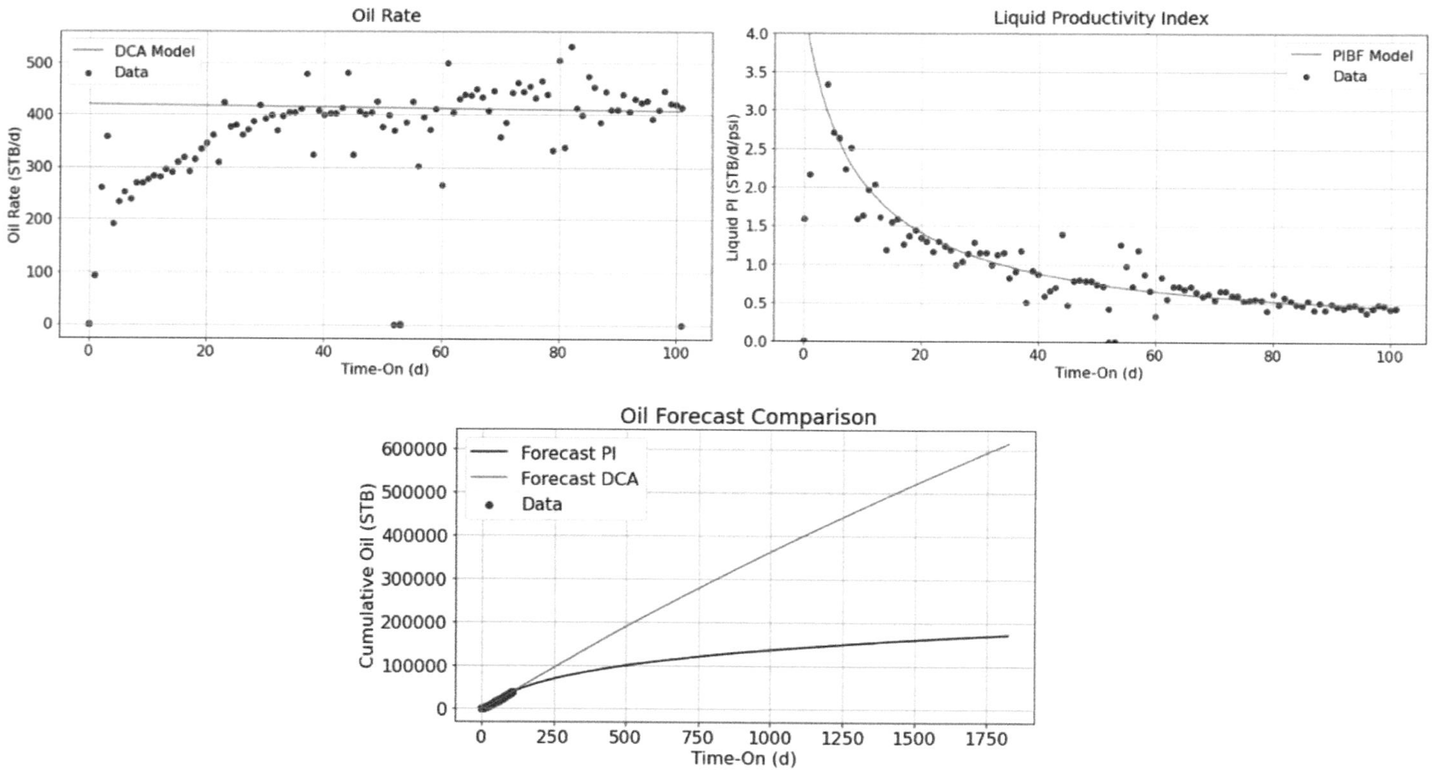

FIGURE 8.9 Forecasting results for single well with production constraints in Field C.

optimization opportunities (Zalavadia et al. 2022) for drawdown management, liquid loading, artificial lift selection and design, and adjustment of lift parameters (e.g., gas lift rate, ESP frequency, etc.).

Figure 8.10 illustrates a case study from Field D, where the base case generated a production forecast with the actual flowing conditions, extrapolated with a constant BHP profile. Two additional production sensitivity cases were evaluated – Case 1 (aggressive drawdown) and Case 2 (conservative drawdown). In this case, by reducing the BHP, there is a noticeable increase in oil production, especially at early time while still honoring material balance (since the profiles converge to similar ultimate recovery volumes at late time), which influences the well economics. These types of sensitivities allow the evaluation of what-if scenarios, quantifying the production impact of various field constraints and artificial strategies and computing the impact on economic value.

The well performance analysis has also been very useful evaluating pilot tests and field trials. When comparing various completion designs, often pilot and control wells are not produced with the same operating conditions, generating biases when the analysis is carried out purely with produced rates. By using PI and drainage volume as key analysis variables, all wells are effectively normalized, to extract their true reservoir potential, and estimate the best economic design.

8.6 SUMMARY

The main contribution of this work is to establish a field-scale, PI-based method to understand well performance and forecast multiphase production in unconventional reservoirs. A hybrid data-driven and physics-based workflow has been developed to characterize production performance through computation of the transient PI by combining DTOF and material balance concepts. PI represents the TWP and is a key indicator of the well potential, thus normalizing operational impacts. The proposed method is very practical, uses only routinely available field data, and does not make any assumptions about reservoir homogeneity, flow regimes, or production mechanisms. The computed features (namely, PI, drainage volume, IRR, average reservoir pressure) are used to normalize and compare well performance. Additionally, average reservoir pressure and PI are used to represent well deliverability (as a dynamic inflow performance relationship (IPR) curve) at various field conditions (past and future) that are used for drawdown management, artificial lift planning, and optimization (Vogel, 1968).

A new PIBF method has been developed to provide a robust and sustainable approach to incorporate pressure into multiphase rate forecasting. The method is applicable to a wide range of fluid systems (incompressible, slightly compressible, and compressible), while considering pressure depletion below the saturation pressure. The workflow has been successfully implemented in thousands of wells in major unconventional basins in North America, using scalable and automated algorithms to capture well performance and derive value through fit-for-purpose production optimization and improved field development planning.

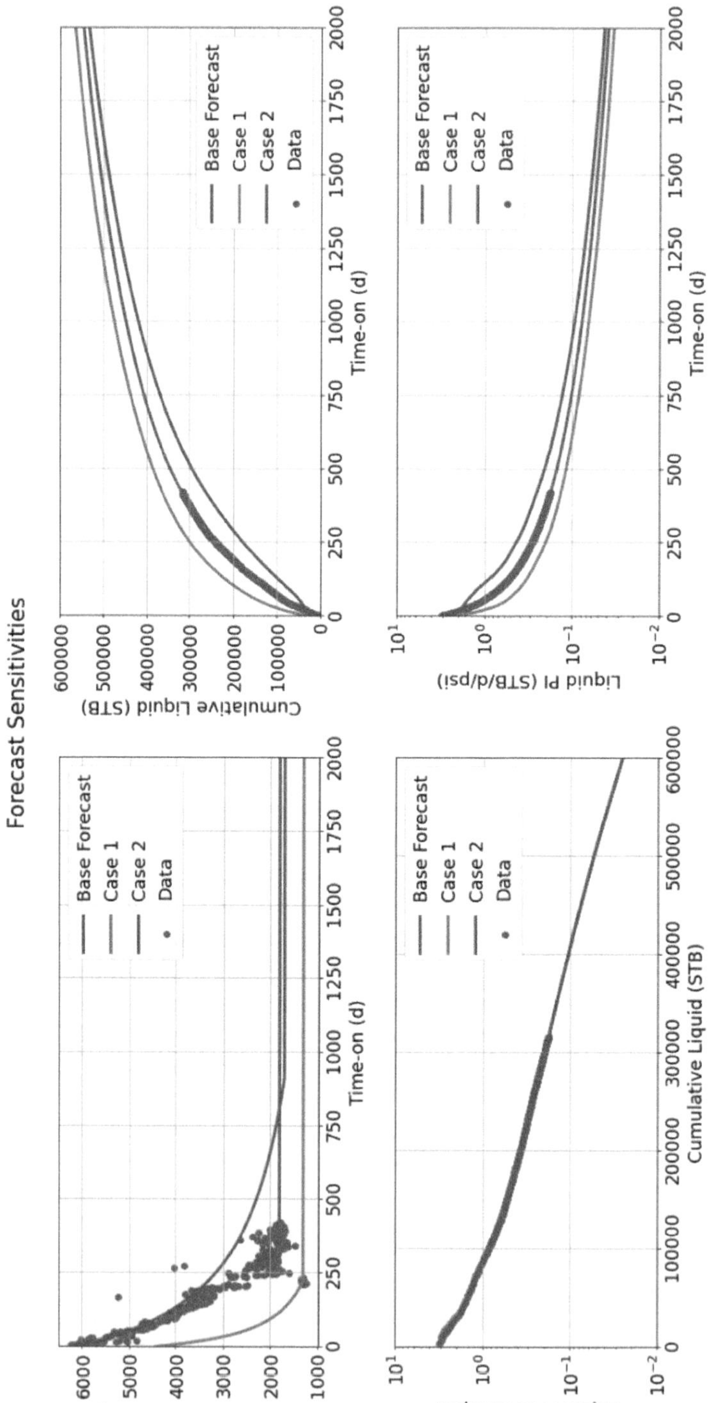

FIGURE 8.10 Production sensitivities for PI-based forecasting on various BHP profiles in Field D.

8.7 FUTURE TRENDS

In several studies conducted by the authors and as shown in a few previously mentioned case studies, pressure-based forecasts have been shown to be consistently better than rate-based forecasts. Where BHP is not measured, it can be estimated automatically from surface pressures using nodal analysis. Several operators are investing in BHP estimation to aid such analysis. In such cases, with more widespread awareness and adoption, it is expected that PIBF would replace rate-based forecasts for routine business processes such as operational planning, field development, production optimization, and reserves.

The TWP method avoids the common limitations of RTA and can be applied pretty much from first production, instead of waiting for significant flow regimes (linear, boundary-dominated, etc.) to develop for manual interpretation. Where subsurface performance is critical (see the previous case studies), it is expected that the TWP method would be more suitable than existing RTA methods to provide automated and consistent results without making overbearing assumptions. Availability of several key diagnostics such as drainage volume, average reservoir pressure, PI, IRR, etc., provides a rich set of engineering features that can further improve the data analytics models to support field development planning. This can address the current limitations with data-driven models that correlate geological and completions parameters with well performance, typically characterized by cumulative production but do not consider pressure impacts.

Several liquids-rich reservoirs have experienced long production declines and are approaching or producing below saturation pressure. Predicting GOR increase and forecasting future gas rates is becoming increasingly critical in such reservoirs. While the authors have empirically observed characteristic trends for GOR above and below saturation pressure, it is desirable to correlate the empirical GOR forecasts with physical mechanisms that can better inform the hybrid models. This can further improve GOR forecasts as lift mechanisms change over the life of the well.

Computational advances in full physics simulations such as GPU-enabled simulations and physics-informed machine learning (PIML) and its variants, along with better surrogate modeling methods, may address the challenge of computational limitations. However, other advances in data-driven discovery of physics using routinely collected, low-dimensional data are needed to address the practical limitations and scale challenges in unconventional reservoir modeling.

8.8 NOMENCLATURE

B_g gas formation volume factor at pressure of interest
B_o oil formation volume factor at pressure of interest
B_{oi} oil formation volume factor at initial reservoir pressure
B_w water formation volume factor at pressure of interest
B_{wi} water formation volume factor at initial reservoir pressure
c intercept of second segment

c_f	formation compressibility
c_o	oil compressibility at pressure of interest
c_t	total compressibility at pressure of interest
c_w	water compressibility at pressure of interest
GOR	production gas-oil ratio (instantaneous ratio of gas rate/oil condensate rate)
G_p	cumulative gas production
IRR	instantaneous recovery ratio at evaluated timestep
N_p	cumulative oil production
p_{avg}	average reservoir pressure
p_i	initial reservoir pressure
p_{wf}	flowing bottomhole pressure
Δp	reduction in reservoir pressure due to depletion
PI	liquid productivity index
q	instantaneous production rate
q_l	liquid rate
R_p	cumulative gas-oil ratio (calculated as cumulative gas divided by cumulative oil)
R_s	solution gas-oil ratio at pressure of interest
R_{si}	initial solution gas-oil ratio
RNP	rate-normalized pressure
S_{or}	residual oil saturation
S_{wc}	connate water saturation
S_{wi}	initial water saturation
t	time
t_e	material balance time
V_d	drainage volume
W_p	cumulative water production

REFERENCES

Arps, J. J. 1945. "Analysis of Decline Curves. Transactions of the American Institute of Mining." Metallurgical and Petroleum Engineers AIME 160: 228–247.

Box, G. E. P., Jenkins, G. M., Reinsel, G. C., and Ljung, G. M. 2015. "Time Series Analysis: Forecasting and Control," 5th Edition, John Wiley and Sons Inc., Hoboken, New Jersey:712.

Burton, M., Matringe, S., Atchison, T. et al. 2019. "A Data-Driven Modeling Methodology to Support Unconventional Reservoir Development Decisions: Application to the STACK Play in Oklahoma". URTEC-2019-573-MS, SPE/AAPG/SEG Unconventional Resources Technology Conference, Denver, CO. https://doi.org/10.15530/urtec-2019-573

Chen, C., and R. Raghavan. 2013. "On the Liquid-Flow Analog to Evaluate Gas Wells Producing in Shales." SPE Reservoir Evaluation and Engineering 16: 209–215. https://doi.org/10.2118/165580-PA

Clark, A. J. 2011. "Decline Curve Analysis in Unconventional Resource Plays using the Logistic Growth Model." MS Thesis, Department of Petroleum and Geosystems Engineering, The University of Texas at Austin.

Cornelio, J., Razak, S. M., Jahandideh, A., Cho, Y., Liu, H-H., Vaidya, R., and Jafarpour, B. 2021. "Physics-Assisted Transfer Learning for Production Prediction in Unconventional Reservoirs." URTEC-2021-5688-MS, SPE/AAPG/SEG Unconventional Resources Technology Conference, Houston, Texas, USA. https://doi.org/10.15530/urtec-2021-5688

Duong, A. N. 2010. "An Unconventional Rate Decline Approach for Tight and Fracture-Dominated Gas Wells." SPE-137748-MS, Canadian Unconventional Resources and International Petroleum Conference, Calgary, Alberta, Canada: 1–15. https://doi.org/10.2118/137748-MS

Ilk, D., Rushing, J. A., Perego, D. A., and Blasingame, T. A. 2008. "Exponential vs. Hyperbolic Decline in Tight Gas Sands: Understanding the Origin and Implications for Reserve Estimates Using Arps' Decline Curves." SPE-116731-MS, SPE Annual Technical Conference and Exhibition, Denver, Colorado, USA: 1–23. https://doi.org/10.2118/116731-MS

Jozefowicz, R., Zaremba, W., and Sutskever, I. 2015. "An Empirical Exploration of Recurrent Network Architectures." In International Conference on Machine Learning, PMLR: 2342–2350.

King, M. J., Wang, Z., and Datta-Gupta, A. 2016. "Asymptotic Solutions of the Diffusivity Equation and Their Applications." SPE-180149-MS, SPE EUROPEC featured at 78th EAGE Conference and Exhibition, Vienna, Austria: 1–37. https://doi.org/10.2118/180149-MS

Kovachki, N., Li, Z., Liu, B., Azizzadenesheli, K., Bhattacharya, K., Stuart, A., and Anandkumar, A. 2020. "Neural Operator: Learning Maps Between Function Spaces." 1–89. https://doi.org/10.48550/arxiv.2108.08481

Lee, J., and Sidle, R. 2010. "Gas-Reserves Estimation in Resource Plays." SPE-130102-PA, SPE Economics & Management 2: 86–91. https://doi.org/10.2118/130102-PA

Li, J., Du, C. M., and Zhang, X. 2011. "Critical Evaluation of Shale Gas Reservoir Simulation Approaches: Single-Porosity and Dual-Porosity Modeling." SPE-141756-MS, SPE Middle East Unconventional Gas Conference and Exhibition, Muscat, Oman: 1–15. https://doi.org/10.2118/141756-MS

Li, W. F., Frash, L. P., Welch, N. J., Carey, J. M., and Meng, M. 2020. "A Simple Transient DFN Model with Stress-Dependent Fracture Permeability for Shale Gas Production." ARMA-2020-1723, 54th U.S. Rock Mechanics/Geomechanics Symposium.

MacQueen, J. 1967. "Some methods for classification and analysis of multivariate observations". In Proceedings of the fifth Berkeley Symposium on Mathematical Statistics and Probability 1, no. 14: 281–297.

Mead, A. 1992. "Review of the Development of Multidimensional Scaling Methods." Journal of the Royal Statistical Society. Series D (The Statistician) 41, no. 1: 27–39. https://doi.org/10.2307/2348634.

Mi, L., Jiang, H., Tang, L., Liu, C., Li, J., Wang, Y., Cao, Y., and Killough, J. 2016. "A Utility Discrete Fracture Network Model for Field-Scale Simulation of Naturally Fractured Shale Reservoirs." SPE-180968-MS, SPE Argentina Exploration and Production of Unconventional Resources Symposium, Buenos Aires, Argentina: 1–19. https://doi.org/10.2118/180968-MS

Mohaghegh, S. D. 2019. "Shale Descriptive Analytics: Which Parameters are Controlling Production in Shale." SPE-196226-MS, SPE Annual Technical Conference and Exhibition, Calgary, Alberta, Canada: 1–25. https://doi.org/10.2118/196226-MS

Molinari, D., and Sankaran, S. 2021. "A Reduced Physics Modeling Approach to Understand Multiphase Well Production Performance for Unconventional Reservoirs." URTEC-2021-5023-MS, SPE/AAPG/SEG Unconventional Resources Technology Conference, Houston, Texas, USA. https://doi.org/10.15530/urtec-2021-5023

Molinari, D., Sankaran, S., Symmons, D., and Perrotte, M. 2019a. "A Hybrid Data and Physics Modeling Approach Towards Unconventional Well Performance Analysis." SPE-196122-MS, SPE Annual Technical Conference and Exhibition, Calgary, Alberta, Canada: 1–27. https://doi.org/10.2118/196122-MS

Molinari, D., Sankaran, S., Symmons, D., Perrotte, M., Wolfram, E., Krane, I., Han, J., and Bansal, N. 2019b. "Implementing an Integrated Production Surveillance and Optimization System in an Unconventional Field." URTEC-2019-41-MS, SPE/AAPG/SEG Unconventional Resources Technology Conference, Denver, Colorado, USA: 1–19. https://doi.org/10.15530/urtec-2019-41

Nobakht, Morteza, and Clarkson, C. R. 2012. "A New Analytical Method for Analyzing Linear Flow in Tight/Shale Gas Reservoirs: Constant-Flowing-Pressure Boundary Condition." SPE Reservoir Evaluation and Engineering 15: 370–384. https://doi.org/10.2118/143989-PA

Ozkan, E., Brown, M, Raghavan, R., and Kazemi, H. 2011. "Comparison of Fractured-Horizontal-Well Performance in Tight Sand and Shale Reservoirs." SPE Reservoir Evaluation and Engineering 14: 248–259. https://doi.org/10.2118/121290-PA

Park, J., Datta-Gupta, A., Singh, A., and Sankaran, S. 2020. "Hybrid Physics and Data-Driven Modeling for Unconventional Field Development – Onshore US Basin Case Study." URTEC-2020-2573-MS, SPE/AAPG/SEG Unconventional Resources Technology Conference, Virtual. https://doi.org/10.15530/urtec-2020-2573

Pradhan, Y. 2020. "Observed Gas-Oil Ratio Trends in Liquids Rich Shale Reservoirs." URTEC-2020-3229-MS, SPE/AAPG/SEG Unconventional Resources Technology Conference, Virtual: 1–15. https://doi.org/10.15530/urtec-2020-3229

Rahman, M., Gioria, G., Sankaran, S., and Molinari, D. 2019. "Automatic Well Interference Identification and Characterization: A Data-Driven approach to Improve Field Operation." SPE-195813-MS, SPE Annual Technical Conference and Exhibition, Calgary, Alberta, Canada: 1–21. https://doi.org/10.2118/195813-MS

Raissi, M., Perdikaris, P., and Karniadakis, G. E. 2019. "Physics-informed neural networks: A deep learning framework for solving forward and inverse problems involving nonlinear partial differential equations." Journal of Computational Physics 378: 686–707. ISSN 0021-9991, https://doi.org/10.1016/j.jcp.2018.10.045

Razak, S. M., Cornelio, J., Cho, Y., Liu, H-H., Vaidya, R., and Jafarpour, B. 2021. "A Physics-Guided Deep Learning Predictive Model for Robust Production Forecasting and Diagnostics in Unconventional Wells." URTEC-2021-5059-MS., SPE/AAPG/SEG Unconventional Resources Technology Conference, Houston, Texas, USA. https://doi.org/10.15530/urtec-2021-5059

Sankaran, S., Molinari, D., Zalavadia, H., Stoddard, T., Sun, W., Singh, G., and James, C. 2022. "Unlocking Unconventional Production Optimization Opportunities Using Reduced Physics Models for Well Performance Analysis – Case Study." 22IPTC-22493-MS, International Petroleum Technology Conference, Dhahran, Saudi Arabia. https://doi.org/10.2523/IPTC-22493-MS

Sen, V., Min, K. S., Ji, L., and Sullivan, R. 2018. "Completions and Well Spacing Optimization by Dynamic SRV Modeling for Multi-Stage Hydraulic Fracturing." SPE-191571-MS, SPE Annual Technical Conference and Exhibition, Dallas, Texas, USA: 1–11. https://doi.org/10.2118/191571-MS

Uzun, I., Kurtoglu, B., and Kazemi, H. 2016. "Multiphase Rate-Transient Analysis in Unconventional Reservoirs: Theory and Application." SPE Reservoir Evaluation and Engineering 19: 553–566. https://doi.org/10.2118/171657-PA

Valko, P. P. 2009. "Assigning value to stimulation in the Barnett Shale: a simultaneous analysis of 7000 plus production histories and well completion records." SPE-119639-MS, SPE Hydraulic Fracturing Technology Conference, The Woodlands, Texas: 1–19. https://doi.org/10.2118/119369-MS

Vogel, J. V. 1968. "Inflow Performance Relationships for Solution-Gas Drive Wells." Journal of Petroleum Technology 20: 83–92. https://doi.org/10.2118/1476-PA

Wicker, J., Courtier, J., Gray, D., Jeffers, T., and Trowbridge, S. 2017. "Improving Well Designs and Completion Strategies Utilizing Multivariate Analysis." URTEC-2693211-MS, SPE/AAPG/SEG Unconventional Resources Technology Conference, Austin, TX.https://doi.org/10.15530/URTEC-2017-2693211

Xue X., Yang C., Park J., Sharma V. K., Datta-Gupta, A., and King, M. J. 2018. "Reservoir and Fracture-Flow Characterization Using Novel Diagnostic Plots." SPE Journal 24: 1248–1269. https://doi.org/10.2118/194017-PA

Zalavadia, H., Stoddard, T., Sinha, U., Corman, A., and Sankaran, S. 2022. "Using Hybrid Models for Unconventional Production Opportunities and Value Generation – Case Studies." URTEC-2022-3724117, SPE/AAPG/SEG Unconventional Resources Technology Conference, Houston, Texas, USA. https://doi.org/10.15530/urtec-2022-3724117

Zhang, Y., He, J., Yang, C., Xie, J., Fitzmorris, R., and Wen, X. 2018. "A Physics-Based Data-Driven Model for History Matching, Prediction, and Characterization of Unconventional Reservoirs." SPE-191126-PA, SPE Journal 23: 1105–1125. https://doi.org/10.2118/191126-PA

9 Role of Analytics in Extracting Data-Driven Models from Reservoir Surveillance

Raj Banerjee
Schlumberger,
Houston, Texas

CONTENTS

9.1 INTRODUCTION

Reservoir surveillance is the continuous process of acquiring and analyzing data with the aim of improving reservoir and well performance. Surveillance methods have been used in the oil industry for a long time; however, in the very early days, the use of these methods was sporadic. Since then, advancements in hardware,

DOI: 10.1201/9781003207009-13

software, and science have made surveillance an integral part of reservoir management. In the operation of oil and gas reservoirs, particularly for mature fields, huge amounts of data are generated because of reservoir surveillance programs. These data provide both valuable insight into the behavior of the reservoir and help to optimize performance.

The industry has endeavored to meet challenges imposed by an ever-increasing flow of data from a variety of sources such as permanent downhole gauges, multiphase flow meters, and smart completions. A reservoir surveillance system should have the following characteristics: (1) predict the reservoir performance and recovery mechanisms, which would most likely deviate from the the initial development plan, (2) identify any discrepancy in the reservoir performance early, (3) understand the root cause of these deviations, and (4) support on-time decision-making with all available data (Zangl and Hannerer 2003).

Traditionally, considerable focus has been on graphical and analytical techniques that are fast and flexible. However, many of the underlying assumptions often render the solutions inaccurate. On the other hand, complex numerical models that capture the full physics of the problem have been too slow for applications such as reservoir surveillance. Operators have striven to bridge the gap through automated "living asset" models.

Digital transformations within the oil industry are having considerable effects on operations including reservoir surveillance programs. Operators are adopting cloud-based ecosystems for data management, workflow integration, and heavy number crunching.

One of the consequences of digital transformation is the widespread application of data analytics, including machine learning (ML). Simply put, data analytics is the science of analyzing datasets to find trends, answer questions, and draw conclusions. With this broad definition, a holistic approach to data analytics includes classical production data analysis techniques such as decline curve analysis; statistical analysis involving univariate, bivariate, and multivariate data; and ML.

ML is the process of applying algorithmic models to data so that the computer discovers hidden patterns that can be used to make predictions. ML is a subset of artificial intelligence (AI). It is aimed at teaching computers to learn from data and to improve with experience. ML has wide and ever-expanding applications, including those in the oil industry.

An integrated and iterative data analysis workflow must be implemented for successful application in reservoir surveillance. The four components of a data analysis cycle, which aligns well with reservoir surveillance, are described (Mishra and Datta-Gupta 2018):

- Data Collection and Management: This step involves the acquisition, aggregation, quality control, and storage of both structured and unstructured data.
- Exploratory Data Analysis: The aim here is to develop a preliminary understanding of the data such as the characteristics of the variables

and the relationships among various variables. Relevant tools include clustering and dimension reduction techniques, and graphical and non-graphical analysis of univariate and multivariate data.

- Predictive Modeling: The goal here is to make predictions about future outcomes using historical data combined with statistical modeling, data mining techniques, and ML.
- Visualization and Reporting: This step involves the presentation of what has been learned in the previous steps in the form of visual summaries, reports, and insights for decision-making.

This chapter discusses how data analytics is used to build models from data acquired during reservoir surveillance. These models could be the outcome of traditional multivariate data analysis techniques or a result of a more contemporary ML method. The models are characterized by predictive capability and can be used in workflows for optimization of operational processes in oil and gas fields.

9.2 NOVEL ENABLING TECHNOLOGIES

In the last few years, there have been rapid developments in technologies that have affected oil field operations, including reservoir surveillance. As these developments progress and mature, there are not only huge efficiency gains but also new workflows. The key technologies are covered in this section.

9.2.1 Internet of Things (IoT)

The Internet of Things (IoT) aims to extend the power of the internet beyond smartphones and computers to numerous other things like pumps and sensors. In other words, it means connecting any device to the internet and to other connected devices. When a device with a built-in sensor is connected to the internet, it can send or receive information, or both. This means that the device can take advantage of massive data storage and supercomputing capability elsewhere. Being connected to an IoT platform allows integration of data from the different devices and application of analytics to share relevant information with applications built to address specific needs.

Edge computing is an emerging area of significance in digital transformation. Edge computing is a distributed computing paradigm that brings computation and data storage closer to the location where it is needed to improve response times and save bandwidth. Broadly, edge computing operates on real-time data, whereas cloud computing operates on big data. Both have significance in reservoir surveillance and oil field operations. Edge computing is suitable for fast-loop workflows that require quick decision-making such as changing operating conditions of an electrical submersible or rod pump, while cloud computing is ideal for optimizing slow-loop workflows related to changes in reservoir behavior.

Edge computing has facilitated the application of connected and intelligent hardware in the oil field. Intelligence is enabled by models that describe the operational behavior of the equipment. The models are data-driven models and learn as more data streams in. These models depend not only on operational data but also life-cycle data such as manufacturing and maintenance, and contextualization data such as location and environment. Generating these models requires knowledge of multiple domains including data science, physical sciences, and engineering (Pafitis 2020).

Platforms incorporating edge computing and Industrial IoT (IIoT) solutions that derive intelligence from data generated during oil field operations and provide the ability to implement changes in real-time are now becoming commercial. Openness is a key attribute of such a platform. This not only means the ability to interface with field equipment from any manufacturer, but also the flexibility to connect to a wide variety of data ecosystems. Security is another key consideration that ensures the security of data and system resources against unauthorized access, disclosure, or corruption – both intentional and unintentional. Additionally, scalability to manage the complexity of oil field operations is critical. It involves agility in independently managing multiple operational use cases as well as dealing with ever-increasing streams of real-time data (Kumar 2020).

Figure 9.1 is a schematic diagram of an oil field surveillance platform showing the relationship between cloud and edge. The cloud offers a centralized data center for insights and analytics. Edge computing automates decisions local to the source. The surveillance system addresses both slow-loop production optimization and fast-loop operations workflows.

The industry is still in the early stages of development in the application of IIoT and edge computing for oil field operations and reservoir surveillance. With the increase in quality and quantity of data coupled with increased understanding of reservoir physics, more sophisticated systems will evolve.

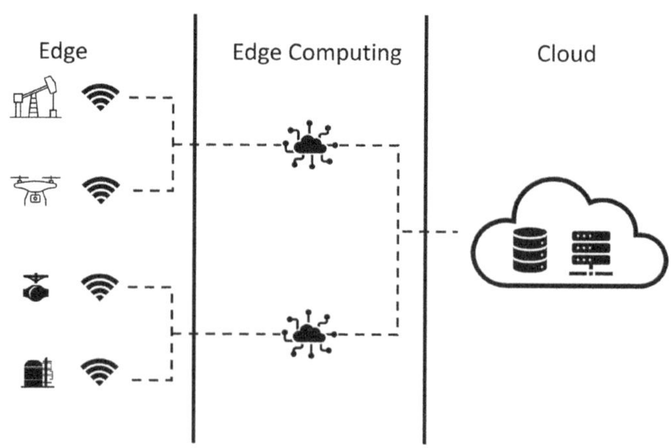

FIGURE 9.1 Edge and cloud concepts in oil field operations.

9.2.2 Big Data and Cloud Computing

Solving the data challenge has always been a major issue in the oil industry. The total amount of data that passes through an oil field surveillance system can be potentially petabyte-sized or larger per day and can be termed as "big data." Big data is the ability to manage a massive volume of disparate data fast and within a reasonable time frame to allow real-time analysis and actions. Big data is typically broken down into six characteristics: volume signifying the quantity of data; velocity signifying the speed at which the data is generated and processed; variety signifying different types and variety of the data including structured and unstructured data; veracity indicating how reliable the data is; variability denoting how the source, structure, and format of the data is changing; and value signifying the worth of the processed information.

Cloud computing is closely associated with big data because it involves the processing of any data, including big data, in a set of high-powered servers in the cloud. In addition to these, cloud-based analytics engines can manage this highly distributed and complex data and provide optimized results to business problems.

The complexity of the data used in a reservoir surveillance process is presented in Table 9.1. This data creates many challenges to manage, process, and interpret in a timely manner for actionable decisions because it takes into consideration all the wells in a reservoir, field, or a business unit. In this context, the adoption of a

TABLE 9.1
Data Types Used in Reservoir Surveillance

Data Type and Source	Data Acquisition Process/Instrument
Logs	
Mud logging	Cutting description, gas analysis
Openhole wireline logs	Compensated neutron log/formation density log, gamma ray, resistivity. Magnetic resonance image (MRI) logging. Modular formation dynamic tester (MDT). Formation microimage
Logging-while-drilling (LWD)	Neutron/density/resistivity/gamma ray/sonic logs
Cased hole logs	Production logging tool (PLT)
	Circumferential acoustic scanning tool/cement bond log (CBL)
	Tubing caliper logs
Fluid Rates and Composition	
Production	Production tests, fluid sampling, production logging tool, multiphase flowmeters
Injection	Gas metering, gas sampling, production logging tool
Pressure and Temperature	
Reservoir, wellbore, and surface pressures and temperatures	Well tests, permanent downhole gauge, production logging tool, wireline formation tester slickline pressure surveys

Source: Kikani (2013).

big data and cloud computing environment, along with IIoT, have the potential of delivering huge benefits.

Intimately linked with big data is the data platform. A cross-industry collaboration is addressing the challenge of storing, organizing, migrating, and accessing subsurface data, by developing the OSDU™ Data Platform[1] (OSDU 2021). The aim is to have a single data platform that liberates data from traditional data silos and makes all data discoverable and usable. OSDU Data Platform enables new ways of performing integrated exploration and production workflows. While the OSDU Data Platform is applicable for the cloud and on-premise, as well as the Edge, it particularly takes advantage of cloud technology.

DELFI (Schlumberger 2022) is one example of an AI-enabled cloud ecosystem that is compatible with the OSDU Data Platform. It is a secure, cloud-based environment that harnesses data, scientific knowledge, and domain expertise to provide integrated multidomain workflows. Data and models are available for number crunching by full physics engines, as well as by AI and ML applications to provide powerful insights.

9.2.3 Machine Learning Algorithms

There are many types of ML algorithms, and based on the application, an algorithm type must be selected for the best result. Furthermore, there are many algorithms that are suitable for each ML algorithm type, and selection must be made based on the accuracy of each algorithm for solving the problem at hand. An easy-to-understand description of ML algorithms was given by Pierson (2021). The key algorithms used in the examples in this chapter are as follows:

- Artificial neural network (ANN): Inspired by the human brain, ANNs are computational algorithms that can be used in several ways including to classify information, predict outcomes, and cluster data. An ANN has hundreds to millions of artificial neurons or nodes arranged in a series of layers – an input layer, one or more hidden layers, and an output layer. A neural network that consists of more than three layers is generally considered a deep learning algorithm.
- Convolutional neural network (CNN): A CNN is a type of deep learning algorithm used in image recognition and processing that is specifically designed to process pixel data.
- Kernel ridge regression (KRR): KRR models are nonparametric regression models that are capable of modeling linear and nonlinear relationships between independent variables and outcomes. KRR is a supervised learning method particularly useful for continuous variables.
    - *K*-means clustering: *K*-means clustering separates objects into clusters that share similarities and are dissimilar to the objects belonging to another cluster. *K*-means clustering is an unsupervised learning algorithm and is used for application such as anomaly detection and pattern recognition.

- Random forest: A random forest is an ML technique that is used to solve regression and classification problems. It builds decision trees on different samples and takes their majority vote for classification problems and average in case of regression problems.
- Support vector machine (SVM): SVMs are a set of supervised learning methods used for both classification and regression problems. However, they are more popular for classification.

9.3 DATA ANALYTICS WORKFLOWS

The scope of reservoir surveillance is extensive, and it is beyond the scope of this chapter to describe fully all the methods in which data analytics can be applied. The selection cases presented here relate to the use of dynamic data to extract models that are ultimately used to optimize production. One or more of the technologies described in the previous section have enabled these cases.

9.3.1 Artificial Lift Surveillance

Artificial lift is a method applied to a well to lower the producing bottomhole pressure on the formation to obtain a higher production rate from the well. It can be done with a positive displacement pump such as a rod pump, a downhole centrifugal pump such as an electrical submersible pump (ESP), or by reducing the density of the fluid in the tubing using gas lift systems. The ability to analyze data coming from these pieces of equipment in real time to take immediate remedial action is seen as a key benefit of artificial lift surveillance. The lack of adequate surveillance of pumps can lead to low efficiency, high lifting costs, and frequent repair and replacements. This in turn has adverse effects on reservoir performance. The application of data analytics and ML is becoming increasingly popular for artificial lift surveillance. This is described in the following two cases: one for rod pump surveillance and another for ESP.

Sharma et al. (2020) presented a solution for automatic analysis of a rod pump dynacard to identify anomalies in pump behavior and notify the operator. A rod pump is a positive displacement pump used widely for pumping out fluids from oil wells. A dynamometer is an instrument used to record the variation between the rod load and the rod displacement. A dynacard is a plot of rod load against displacement over a pump cycle. The shape of the dynacard plot gives the operator an indication about the performance of the pump (Figure 9.2). Traditionally, this is a manual process and requires special skills to interpret the patterns. The authors used *k*-means clustering for pattern recognition to automate identification of critical pump signatures as they happened. Subsequently, a CNN was applied. ML libraries were deployed at the edge for analysis in real time.

For an ESP, the overall aim is to minimize production deferment and costly interventions caused by equipment failure by proactive intervention planning. Surveillance involves analyzing trends in pump characteristics such as pump intake pressure, discharge pressure, amperage, motor temperature, and frequency

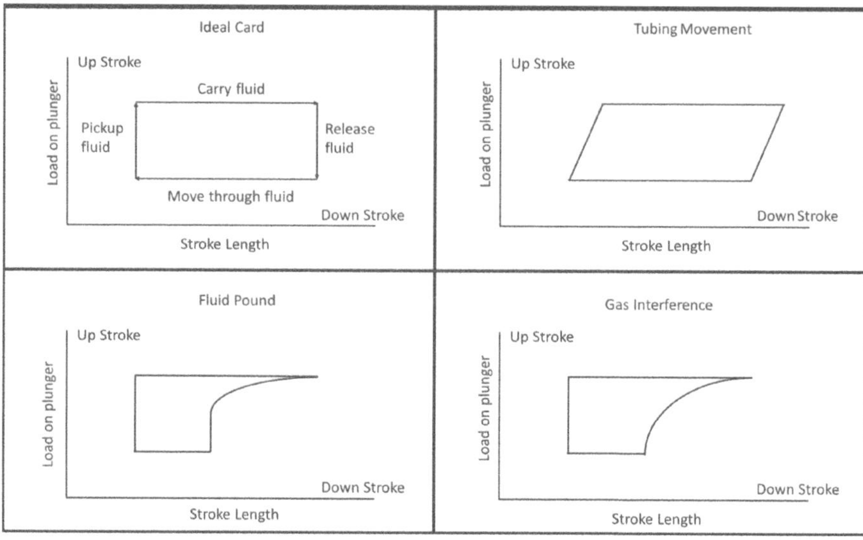

FIGURE 9.2 Schematic diagrams of a few dynacard shapes and associated problems.

data, as well as taking corrective actions. A solution for ESP surveillance was presented by Sinha et al. (2021). Three workflows utilizing ML were discussed.

- Anomaly detection: This workflow was used to detect anomalies in ESP operational data. An important part of the workflow identified health signals using inputs from domain experts, and exploratory statistical analysis followed by feature engineering and selection. A deep neural network model was built and tested with each of the health signals during healthy operating conditions, and then the trained model predicts the health signals using the incoming new data. The magnitude of the differences between the predicted values from the normal model vs the new data defined the level of abnormality observed.
- Event detection: This workflow specifically focused on the detection of emulsion formation and excessive gas production. Both these events can cause a decrease in fluid production as well as pump health-related issues. The solution involved analyzing ESP data for signatures of the two events. The sensor data used for this included pump intake pressure, pump discharge pressure, motor temperature, motor current, and vibration. Historical events were labeled and key features in the data were identified to establish a relationship between the parameters. Tree-based models like random forest were used for modeling.
- ESP failure analysis: The objective of this workflow was to analyze historical failure patterns and provide insight for taking preventive measures in the future. Historical records of failure-related data can be both structured and unstructured. This required the need for natural language

processing techniques to convert the textual data into numerical features. For the ML part, deep learning algorithms were used to train with the data and make predictions on the root cause of failure and failed components.

9.3.2 Automated Interpretation of Pressure and Rate Data

Dynamic data comprising of pressure and fluid flow rates provide information on reservoir properties and well productivity. Dynamic data can be acquired during sporadic well tests designed with specific test objectives. For high-value wells, dynamic data can be continuously acquired. The latter gives many opportunities for interpretation of pressure buildup data due to both planned and unplanned shutdowns. Additionally, where shutting-in the well is not feasible, deconvolution or rate transient methods may be applied to rate and pressure data for interpretation.

Traditionally, the workflow for pressure-transient data is graphical and the engineer's judgment must be provided during the process. The books by Kuchuk et al. (2010) and Horne (1995) contain detailed explanations of this process. After the data is acquired and stored, the workflow generally proceeds as per the following steps:

1. Data preparation step involving various data manipulation and quality checks, which includes sampling, denoising, outlier removal, time-shifting, and transient identification.
2. Diagnosis step where different types of plots, such as derivative plots and specialized plots, are used to visually obtain some insights on the dynamic model.
3. Model recognition step involving identification of the appropriate flow regimes and selection of the appropriate model often through visual inspection of the plots.
4. Parameter estimation step where the pressure and rate data are matched against a corresponding model by fine-tuning model parameters.
5. Validation step involving determination of confidence in the estimated parameters, and consistency of the interpretation with other sources of static and dynamic information.

The process can be onerous and depends, to some extent, on the skills of the engineer. When we have streaming data from permanent downhole gauges in wells, large amounts of data are collected every day. For such surveillance operations to be successful, automation and inherent intelligence in the system are highly desirable. We have seen steady progress in this area over more than 10 years and, recently, the application of ML is becoming increasingly more common.

Keeping well performance surveillance in mind, Chang et al. (2019) described an integrated workflow that addressed many of the efficiency issues in traditional well-test interpretation with the application of ML methods. Their workflow began with data preparation for outlier removal, denoising, and sampling using the wavelet technique (Athichanagorn et al. 1999). These authors used *k*-means clustering to automatically identify pressure buildups and maximum slope point on the pressure data to identify the start of the buildups. Another use of ML was to identify the well-test model automatically from data corresponding to each of the buildups. For this, the authors used SVMs to train and validate the ML model. The training set consisted of synthetic response generated with analytical models, as well as field datasets, which had been analyzed by experts and the corresponding well-test models identified. Once the well-test model had been identified, the usual nonlinear regression technique was used with the selected analytical model to estimate the reservoir and well parameters.

Chang et al. (2019) applied their integrated workflow in a deepwater oil field in the Gulf of Mexico for continuous surveillance, obtaining important insights on well productivity changes, well interference, and permeability alterations.

In another case, Sankaran et al. (2017) presented an integrated production surveillance and optimization platform to automate the analysis of pressure transient behavior of wells with permanent downhole gauges in a deepwater asset in the Gulf of Mexico. Their workflow included algorithms to automatically detect flow periods, filter data, and perform transient analysis. Their application of ML for transient analysis was based on algorithms developed by Tian and Horne (2015). Most notably, a KRR method was shown to successfully recover the complete reservoir model successfully from data without recourse to physics-based models, thus facilitating automation.

9.3.3 Workover Candidate Recognition

One of the main purposes of information obtained for reservoir surveillance programs is to identify, rank, and plan workover and intervention opportunities with the ultimate objective of sustaining and increasing production. Often termed as "opportunity maturation process" (OMP), it involves screening, planning, and executing well workover and intervention opportunities in a field.

In mature fields, the process can be challenging because wells are prone to have more problems. While huge amounts of data generated can be used to analyze the problems and provide solutions, the data is typically disorganized and in different formats. This process can take weeks or even months and must be carried out frequently during the life of the field. The challenge is to achieve efficiency gain by automating and speeding up the workflow.

Cio Cio et al. (2020) presented a hybrid method that used ML with petroleum engineering analysis methods and business rules to autonomously identify well performance signatures and workover opportunities. The method was applied in the Ujung Pangkah asset – an offshore oil and gas field in Indonesia.

The analysis techniques used in this asset were further elaborated by Sinha et al. (2020). Two interesting applications of data analytics are highlighted below:

9.3.3.1 Water Control Diagnostic Plot

Chan (1995) plots are widely used for identifying produced water sources based on the shapes of the water-oil ratio and derivatives of water-oil ratio plots on a log-log plot. This allows for the diagnosis of problems such as water coning and channeling. Traditionally, the engineer visually inspects the shape of the plot to interpret the problem. This manual process was automated using the logistical regression method (Sinha et al. 2020). During this process, change point identification was used to identify discontinuities in the plot. Thereafter, nonparametric trend evaluation was used to capture the trend of the identified segments. Sets of models were trained with these features and were ultimately used for interpretation.

9.3.3.2 Behind-Casing Opportunities

Behind-casing opportunity is a cost-effective production enhancement solution that uncovers additional production opportunities in a well. One opportunity is to open unperforated layers in a well. However, these layers could have been opened in other wells in the reservoir. With the objective of determining the behind-casing production potential, a set of production, petrophysical, and depletion features were selected for modeling. The features used include porosity, permeability, pay thickness, water saturation, distance-weighted average water-cut of neighboring wells, and decline rate of neighboring wells. The behind-casing potential target variable for the model was selected to be the best 6-month moving average oil production rate declined using the distance-weighted decline rate of neighboring wells in the given reservoir. Finally, the trained ANN model was used to cross-validate with the producing completions data and to predict production of nonperforated layers (Sinha et al. 2020).

One of the challenges in planning workover in mature fields is the ability to extract meaningful information from a large amount of diverse data sources. This may not always be structured like time-series production history. Workover information, particularly in older wells, may simply be typed or even handwritten paper reports. Karnik et al. (2021) presented a method that used natural language processing and ML to extract workover-related information from the reports and then correlate the events with historical production and cost information. The solution provided a good understanding of the relevance and chance of success for new workovers in each field.

9.3.4 Waterflood Surveillance

An essential component of a successful waterflood project is an efficient surveillance program. While each field has its own characteristics, there are some common ingredients that need to be considered. These include reservoir conditions, well conditions, facilities, and operating conditions (Talash 1988). A schematic view of the key monitoring focal points is presented in Figure 9.3.

Managing waterflood assets requires an integrated approach that is multidisciplinary in nature. An excellent treatise on the subject was presented by Thakur

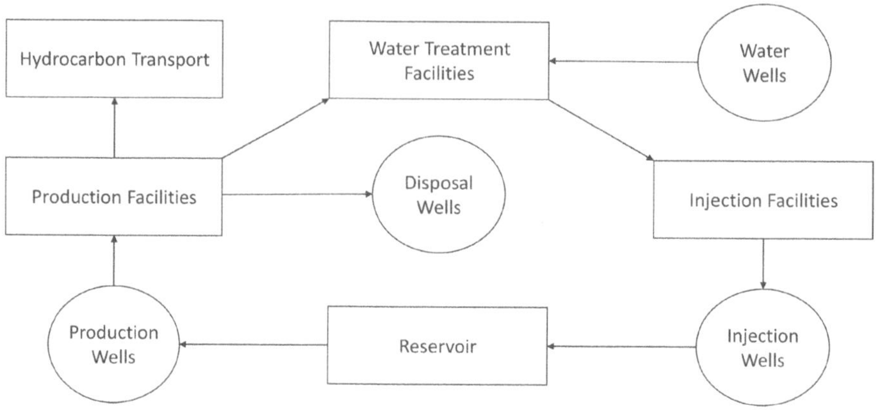

FIGURE 9.3 Waterflood cycle. (After Talash 1988.)

and Satter (1998). This section focuses on the production performance of waterflood reservoirs.

9.3.4.1 Waterflood Production and Injection Data Analysis

Traditionally, specialized plots of production and injection data have been extensively used for waterflood surveillance. Composite plots at field or pattern level of injection/production rates, water-cut, gas-oil ratio, cumulative volumes, and number of active wells are useful for identifying large-scale trends. They are the first steps in understanding historic performance and for planning accordingly.

The voidage replacement ratio (VRR) is an important consideration in a waterflood reservoir. VRR is the ratio of reservoir barrels of injected fluid to reservoir barrels of produced fluid. It gives a quantitative idea on replacing the volume of oil, gas, and water produced from the reservoir by injected fluids. In an ideal scenario, VRR should be near to 1 and the reservoir pressure should be constant. This is a balanced steady-state flood in a finite reservoir. If VRR is considerably greater than 1 and the reservoir pressure is constant, it signals possible injection loss or thieving. On the other hand, if VRR is less than 1 and the reservoir pressure declines, it could indicate insufficient injection. However, if VRR is less than 1 and the reservoir pressure is constant, it could be due to fluid influx from an aquifer. Figure 9.4 shows cumulative and instantaneous VRR plots for a group of wells. Grouping can be done considering pattern, a sector, or the reservoir.

In conformance plot, the recovery factor (RF) is plotted against cumulative drainable hydrocarbon pore volume injected (DHCPVI). It is a good indicator of the performance of the waterflood and can be used to predict future recovery if operating conditions do not significantly change. A linear trend between RF and cumulative DHCPVI indicates that an overall breakthrough has not yet occurred, and the short-term future recovery can be easily extrapolated. When the slope of the curve starts constantly decreasing, water channeling or breakthrough is indicated. Conformance plots for historical data and forecast is

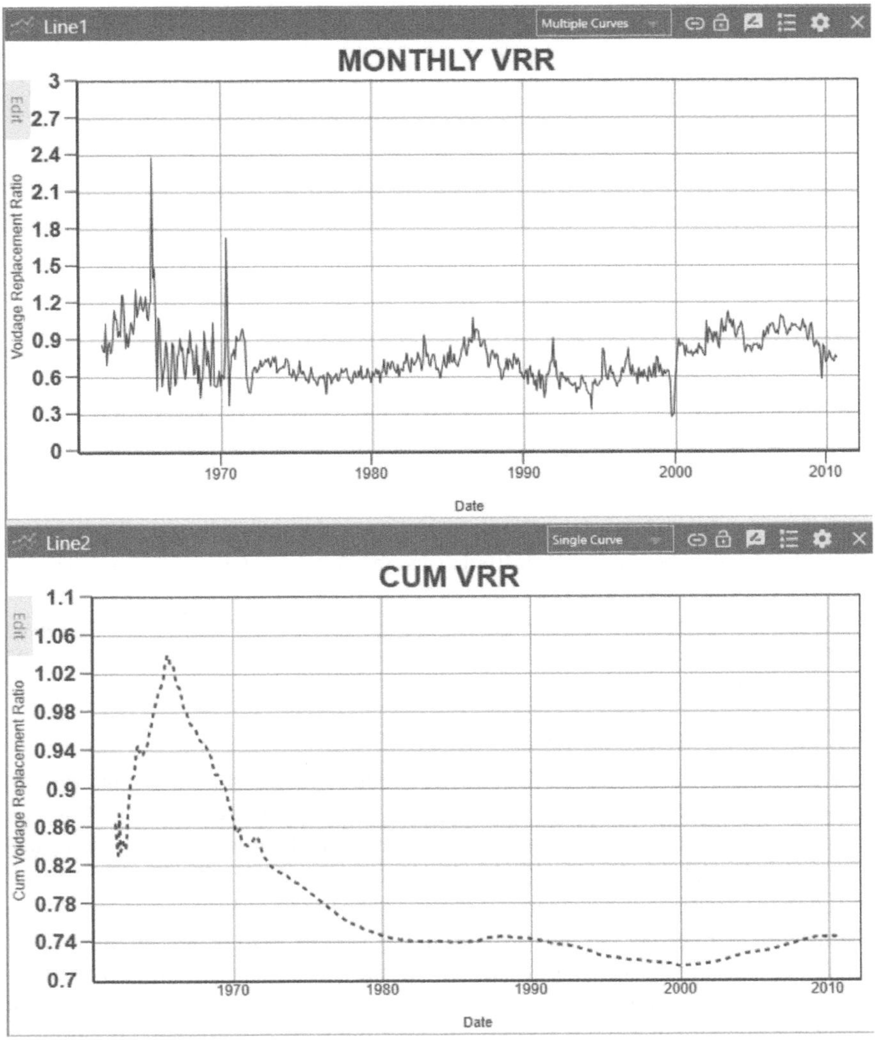

FIGURE 9.4 VRR plots using monthly and cumulative production and injection data. (OFM software – Sensia.)

presented in Figure 9.5. The decrease in slope in the latter part of the curve is due to an increase in water-cut. The curve will level off toward the end of the life of the field arriving at the maximum RF.

There are other diagnostic plots used for surveillance – some examples include the Hall, Chan, and Cobb plots. Practical examples of these plots for a carbonate reservoir under waterflood were presented by Konwar et al. (2019).

Undoubtedly, the plots described have brought significant value to numerous waterflood surveillance programs. However, they require visual inspection

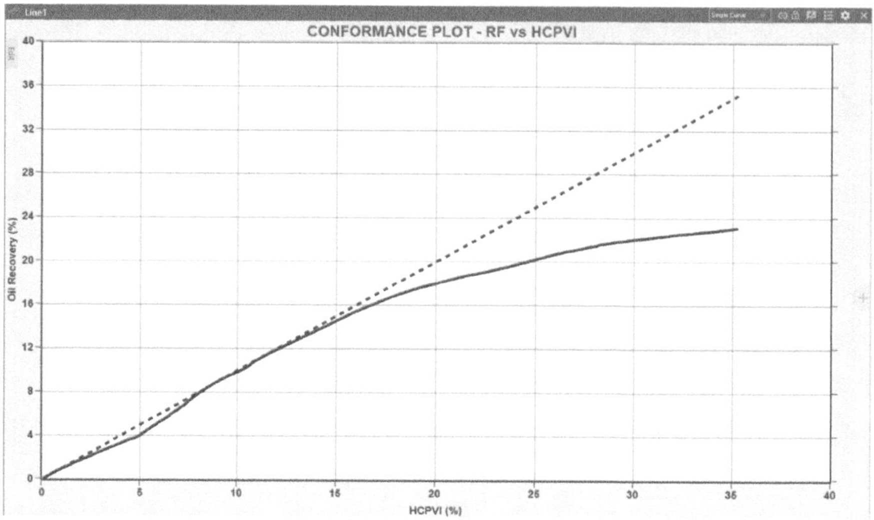

FIGURE 9.5 Conformance plots historical and forecast data. (OFM software – Sensia.)

to recognize the shape and to interpret any change in trend. This is being alleviated as the process is automated by using change point and slope estimation algorithms.

9.3.4.2 Capacitance Resistance Model (CRM)

The capacitance resistance model (CRM) is one of the most popular data-driven techniques for characterizing a waterflood reservoir using only production and injection history from wells. Typically, no geological information or numerical simulation model is required. CRM for waterflood application was first introduced in the 1940s with an electrical device consisting of capacitors and resistors (Bruce 1947). In today's context, the term CRM is used to denote mathematical modeling where the governing equations are like those used by Bruce for his electrical circuit. These models not only honor material balance but also account for interference between wells. CRM models are important for reservoir surveillance because they can be rapidly and expediently history-matched and used as a quick predictive tool for reservoir performance.

In the last decade, this area has been extensively explored and a large amount of research is available covering various aspects of the topic. A thorough review of the available literature was done by Holanda et al. (2018).

There are a few variations of the CRM model. One variation is the CRMT model. CRMT considers a single tank with all the producers represented by a single producer and all the injectors represented by a single injector (Sayarpour et al. 2009). CRMT helps estimate overall injection losses. On the other hand, CRMP is a producer-based representation where the assumption is that every injector influences a producer (Liang et al. 2007). This method works best when the reservoir is relatively homogeneous, and the wells are in a uniform pattern. CRMIP is an injector-producer pair-based representation that characterizes each injector-producer pair (Yousef et al. 2006) and can mitigate some of the limitations of CRMP.

9.3.4.3 Interwell Numerical Simulation Model (INSIM)

The interwell numerical simulation model (INSIM) is a physics-based data-driven model for history-matching and prediction of waterflood reservoirs (Zhao et al. 2015). In INSIM, the reservoir is described as a coarse model consisting of interwell control units. The parameters characterizing these control units are transmissibility and control pore volume. Interwell fluid rates and saturations are computed by solving the mass material balance and front tracking equations for the control units. This in turn allows for the calculation of well oil and water production rates. INSIM allows injector-producer connections as well as injector-injector and producer-producer connections. INSIM is followed by the front tracking version, INSIM-FT (2018), which addresses some of the initial shortcomings. INSIM-FT has more robust physics in front tracking for calculation of saturation. INSIM-FT is restricted to two-dimensional (2D) problems in which all wells are completely penetrating and vertical. When extended to three-dimensional (3D) problems, it is referred to as INSIM-FT-3D (Guo and Reynolds 2019, Zhao et al. 2021).

The data-driven models described do not need detailed geological and petrophysical description and depend on the production and injection data. The model building, history-matching, and prediction steps are simpler compared with numerical simulation models and therefore require less expertise. Furthermore, the simulation run times of the data-driven models are much less than those of numerical simulation. These models can be efficiently used for production optimization and are, therefore, ideal for waterflood surveillance. However, they are of limited value in complex reservoirs.

9.3.5 Reservoir Simulation in Surveillance

Closed-loop reservoir management, otherwise known as "smart reservoir management" (SRM), aims to maximize reservoir performance in terms of recovery or financial measures (Jansen et al. 2009). Reservoir management becomes a continuous loop incorporating model-based optimization. In such a system, a robust reservoir surveillance program is of paramount importance. Consequently, a reservoir simulation model needs to be frequently updated with dynamic data and history-matched to enable a "living model." In such a workflow, multiple simulation runs of the model need to be performed for optimization, uncertainty assessment, history-matching, and understanding "what-if" scenarios. Complex numerical models may be too cumbersome and slow to execute for such a system in many situations. Therefore, engineers often consider alternative methodologies such as reduced-physics models, reduced-order models, top-down models, and proxy or surrogate reservoir models (Mohaghegh 2009, 2011).

A proxy model is an accurate replica of a complex reservoir-simulation model that may include tens or hundreds of wells. Proxy runs provide similar outputs as a numerical simulator, such as well pressure and production profiles, as well as time variant pressure and saturation distribution throughout the reservoir but run many times faster. Proxy models incorporating physics-aware ML techniques are being used, which include physical laws, along with the purely data-driven methods, to improve reliability

Embed-to-control (E2C) methods have generated considerable interest in the last couple of years. Originally presented by Watter et al. (2015) for model learning and control of a nonlinear dynamical system using raw pixel images as input, these methods use a CNN encoder-decoder coupled with a transition layer to predict the time evolution of the system state. The encoder transforms the state in the original dimension space to a reduced dimension space. The transition layer predicts the time evolution of the state in the reduced dimension space. The decoder transforms the state from the reduced dimension space to the original dimension space. Subsequently, Jin et al. (2020) modified this system to build a surrogate for a numerical reservoir simulator by learning from multiple simulation runs. A major difference between both systems is the introduction of a physical loss function, which honors the physics of flow. The surrogate reservoir model can predict the state variables, i.e., pressure and saturation values of the simulation grid blocks. The well fluid rates are calculated from the block pressure and saturation values through appropriate well models. Coutinho et al. (2021) extended this by using an additional neural network component to predict well outputs directly without any dependency on the simulator well model. This is known as the embed-to-control and observe (E2CO) method.

Both works are limited to only 2D reservoir models. Atadeger et al. (2022) expanded the E2C and E2CO methods to handle subsurface flow applications in 3D reservoir systems by implementing the use of 3D convolutional layers and modifying the physical loss functions that address the 3D flux conditions.

In predictive mode, the proxy models are many orders of magnitude faster than the full physics reservoir simulation models they emulate. In reservoir surveillance operations, the proxy models can be used to investigate what-if scenarios by altering well controls. Moreover, when implemented along with optimization methods, the proxy models can be used in production optimization workflows such as waterflood optimization.

9.4 FUTURE OUTLOOK

Reservoir surveillance will play a greater role in the exploitation of mature reservoirs to improve recovery and reduce costs. The oil industry's ongoing digital transformation helps meet the challenges of the current commercial environment. There are many aspects of this transformation that are critical to reservoir surveillance, such as connected intelligent hardware for real-time measurement and control, and ML integrated with domain science in petroleum engineering workflows.

Currently, there are several challenges that must be addressed. First, there is the issue of data integration; there is still some way to go before all relevant data from disparate sources can be brought together to provide a unified view. The data should be more easily available and simpler to consume and process by engineers and algorithms. Moreover, a reservoir surveillance program involves the use of numerous data-driven and physics-based models working in isolation. This results in considerable effort in managing the models, for instance, keeping the models updated and training with new data.

Ideally, a reservoir surveillance program should encompass "sense, compute, and act." Actionable decisions from reservoir surveillance systems should drive the field's automation. Considerable automation is necessary to make the process fully integrated and seamless.

In the future, more attention will be paid to the deployment and maintenance of data-driven models in the oil field operational environment. Practices, like ML operations (MLOps), which focus the operationalization of ML models, will be of importance. Another area of significance for reservoir surveillance is automated ML (AutoML), which involves automating the task one must perform to build and train the ML models. AutoML can assist engineers to build predictive models without in-depth ML expertise so their attention can be focused on optimizing oil field operations.

Additionally, reinforcement learning (RL) also holds substantial promise for the future. It is an aspect of ML in which software agents learn to behave in an environment by performing certain actions and observing the consequential rewards and results. Among the interesting possibilities are waterflood optimization and ESP optimization.

Lastly, because modern surveillance systems almost always need to be connected to the internet, there is a pressing need for them to be more secure and deal with increasingly greater threats. In cyber security, ML models learn by analyzing internet data to better detect these threats and anomalies.

NOTE

1. OSDU™ and the OSDU logo are trademarks of The Open Group.

REFERENCES

Alemán, Dante Orta, and Roland Horne. 2020. "Improved Robustness in Long-term Pressure Data Analysis Using Wavelets and Deep Learning." *SPE Annual Technical Conference and Exhibition*. Denver (online): Society of Petroleum Engineers (SPE). doi:10.2118/201597-MS.

Atadeger, Aykut, Gerard Vera, Soham Sheth, Raj Banerjee, and Mustafa Onur. 2022. "Deep Learning-Based Proxy Models to Simulate Subsurface Flow of Three-Dimensional Reservoir Systems." *ECMOR* 2022: 1–32.

Athichanagorn, Suwat, Roland N. Horne, and Jitendra Kikani. 1999. "Processing and Interpretation of Long-Term Data from Permanent Downhole Pressure Gauges." *SPE Annual Technical Conference and Exhibition*. Houston: Society of Petroleum Engineers (SPE). doi:10.2118/56419-MS.

Bruce, W.A. 1947. "An Electrical Device for Analyzing Oil-Reservoir Behavior." *Transactions of the AIME* 151(1) 112–124.

Chan, K.S. 1995. "Water Control Diagnostic Plots." *SPE Annual Technical Conference & Exhibition*. Dallas: Society of Petroleum Engineers.

Chang, O., Y. Pan, A. Dastan, D. Teague, and F. Descant. 2019. "Application of Machine Learning in Transient Surveillance in a Deep-Water." *SPE Western Regional Meeting*. San Jose, CA, USA: Society of Petroleum Engineers.

Cio Cio, Mario, Rini Saputra, Sera Saifuddin, Triyono Triyono, Sabrina Metra, Praprut Songchitruksa, Rajiv Ranjan Sinha, and Karthick Vadivel. 2020. "Developing a Smart Workover and Intervention Strategy, Using Automation and Advanced Data Analytics." *World Oil* June: 29–32.

Coutinho, Emilio J.R., Marcelo J. Aqua, and Eduardo Gildin. 2021. "Physics-Aware Deep-Learning-Based Proxy Reservoir Simulation Model Equipped with State and Well Output Prediction." *SPE Reservoir Simulation Conference*. Society of Petroleum Engineers.

Guo, Zhenyu, and Albert C. Reynolds. 2019. "INSIM-FT-3D: A Three-Dimensional Data-Driven Model for History Matching and Waterflooding Optimization." *SPE Reservoir Simulation Conference*. Galveston: Society of Petroleum Engineers.

Holanda, Rafael Wanderley de, Eduardo Gildin, Jerry L. Jensen, Larry W. Lake, and C. Shah Kabir. 2018. "A State-of-the-Art Literature Review on Capacitance Resistance Models for Reservoir Characterization and Performance Forecasting." *Energies* 11: 3368.

Horne, Roland N. 1995. *Modern Well Test Analysis*. 2nd ed. Palo Alto: Petroway, Inc.

Jansen, J.D., S.D. Douma, D.R. Brouwer, P.M.J. Van den Hof, O.H. Bosgra, and A.W. Heemink. 2009. "Closed-Loop Reservoir Management." *SPE Reservoir Simulation Symposium*. Woodlands, TX, USA: Society of Petroleum Engineers.

Jin, Zhaoyang Larry, Yimin Liu, and Louis J. Durlofsky. 2020. "Deep-Learning-Based Surrogate Model for Reservoir Simulation with Time-Varying Well Controls." *Journal of Petroleum Science and Engineering* 192: 107273.

Karnik, S., S. Gupta, J. Baihly, and D. Saier. 2021. "Data Driven Solution for Enhancing Workover Intervention Activities." *SPE Middle East Oil & Gas Show and Conference*. Manama, Bahrain: Society of Petroleum Engineers.

Kikani, J. 2013. *Reservoir Surveillance*. Richardson, TX, USA: Society of Petroleum Engineers.

Konwar, Lakshi, Ebrahim AlOwainati, Isa AlAnaisis, Nadia Nemmawi, and Dowen Michael. 2019. "Understanding Mauddud Waterflood Performance in a Heterogeneous Carbonate Reservoir with Surveillance Data and Ensemble of Analytical Tools." *SPE Middle East Oil and Gas Show and Conference*. Manama, Bahrain: Society of Petroleum Engineers.

Kuchuk, F.J., M. Onur, and F. Hollander. 2010. *Pressure Transient Formation and Well Testing*. Elsevier.

Kumar, Sujit. 2020. "On the Edge, the Next Wave of Digital Transformation." *World Oil* 35–37.

Liang, X., D.B. Weber, T.F. Edgar, L.W. Lake, M. Sayarpour, and A. Al-Yousef. 2007. "Optimization of Oil Production Based on a Capacitance Model of Production and Injection Rates." *SPE Hydrocarbon Economics and Evaluation Symposium*. Dallas, TX, USA: Society of Petroleum Engineers.

Mishra, S., and A. Datta-Gupta. 2018. *Applied Statistical Modeling and Data Analytics – A Practical Guide for the Petroleum Geosciences*. Elsevier.

Mohaghegh, Shahab D. 2009. "Artificial Intelligence and Data Mining: Enabling Technology for Smart Fields." *The Way Ahead* 15: 14–19.

Mohaghegh, Shahab D. 2011. "Reservoir Simulation and Modeling Based on Artificial Intelligence and Datamining (AI&DM)." *Journal of Natural Gas Science and Engineering* 3: 697–705.

Pafitis, Demos. 2020. "Digitally Transforming Oil Fields." *Journal of Petroleum Technology* 72: 12–13.

Pierson, L. 2021. *Data Science for Dummies*. 3rd ed. Wiley.

Sankaran, Sathish, David Wright, Huan Gamblin, and Dhilip Kumar. 2017. "Creating Value by Implementing an Integrated Production Surveillance and Optimization System – An Operator's Perspective." *Annual Technical Conference and Exhibition*. San Antonio: Society of Petroleum Engineers.

Sayarpour, M., E. Zuluaga, C.S. Kabir, and L.W. Lake. 2009. "The Use of Capacitance–Resistance Models for Rapid Estimation of Waterflood Performance and Optimization." *Journal of Petroleum Science and Engineering* 69: 227–238.

Schlumberger. 2022. *DELFI Portal*. Accessed January 31, 2022. https://delfi.slb.com/.

Sharma, Abhishek, Prince Samuel, Gian-Marcio Gey, and Sujit Kumar. 2020. "Edge Computing: Continuous Surveillance and Management of Production Operations in a Cost Effective Manner." *SPE Annual Technical Conference & Exhibition*. Denver: Society of Petroleum Engineers.

Sinha, R.R., S. Gupta, P. Songchitruksa, S. Karnik, and A. Ambade. 2021. "Electrical Submersible Pump Prognostics and Health Monitoring Using Machine Learning and Natural Language Processing." *SPE Symposium: Artificial Intelligence – Towards a Resilient and Efficient Energy Industry*. Society of Petroleum Engineers.

Sinha, R.R., Praprut Songchitruksa, Ralf Holy, Karthick Vadivel, Sreekrishnan Ramachandran, and Henry Martinez. 2020. "Well Portfolio Optimization: Rapid Screening of Production Enhancement." *Abu Dhabi International Petroleum Exhibition & Conference*. Abu Dhabi: Society of Petroleum Engineers.

Talash, A.W. 1988. "An Overview of Waterflood Surveillance and Monitoring." *Journal of Petroleum Technology* (Society of Petroleum Engineers) 40: 1539–1543.

Thakur, Ganesh C., and Abdus Satter. 1998. *Integrated Waterflood Asset Management*. PennWell.

Tian, Chuan, and Roland N. Horne. 2015. "Applying Machine Learning Techniques to Interpret Flow Rate, Pressure and Temperature Data from Permanent Downhole Gauges." *SPE Western Regional Meeting*. Garden Grove, CA, USA: Society of Petroleum Engineers.

Watter, Manuel, Jost Tobias Springenberg, Joschka Boedecker, and Martin Riedmiller. 2015. "Embed to Control: A Locally Linear Latent Dynamics Model for Control from Raw Images." *Advances in Neural Information Processing Systems*.

Yousef, Ali A., Pablo Gentil, Jerry L. Jensen, and Larry W. Lake. 2006. "A Capacitance Model to Infer Interwell Connectivity from Production- and Injection-Rate Fluctuations." *SPE Reservoir Evaluation & Engineering* 9: 630–646.

Zangl, Georg, and Josef Hannerer. 2003. *Data Mining: Applications in the Petroleum Industry*. Round Oak Publishing.

Zhao, Hui, Zhijiang Kang, Xiansong Zhang, Haitao Sun, Lin Cao, and Albert C. Reynolds. 2015. "INSIM: A Data-Driven Model for History Matching and Prediction for Waterflooding Monitoring and Management with a Field Application." *SPE Reservoir Simulation Symposium*. Houston: Society of Petroleum Engineers.

Zhao, Hui, Wei Liu, Xiang Rao, Guanglong Sheng, Huazhou Andy Li, Zhenyu Guo, Deng Liu, and Lin Cao. 2021. "INSIM-FPT-3D: A Data-Driven Model for History Matching, Water-Breakthrough Prediction and Well-Connectivity Characterization in Three-Dimensional Reservoirs." *SPE Reservoir Simulation Conference*. Society of Petroleum Engineers.

10 Machine Learning Assisted Forecasting of Reservoir Performance

Emre Artun
Istanbul Technical University,
Istanbul, Turkey

CONTENTS

10.1 INTRODUCTION AND SCOPE

Forecasting reservoir performance is an integral part of economic evaluation studies for a hydrocarbon asset as the process quantifies the recoverable volume of hydrocarbons over a period of time. By combining economic conditions and operational constraints with the forecast, the actual reserves portion of the hydrocarbon resource-base is determined. These forecasts are critical for proper management of the reservoir, which requires making the best possible decision at any given time during the life of the asset. Dealing with uncertainties of the subsurface, challenges related to reservoir characterization and uncertainties in hydrocarbon markets altogether make this decision-making process a challenging one. While reliable forecasts would always help to make better decisions, poorly made forecasts may heavily affect the economic success of the recovery process and all related projects (Gupta et al. 2016). Traditionally, petroleum engineers have been

DOI: 10.1201/9781003207009-14

estimating flow rates incorporating approximations on reservoir characteristics, drainage architecture, and recovery methods to the analytical methods such as material balance and decline curve analysis (Sankaran et al. 2020). Decline curve analysis (Arps 1945) is a data-driven approach where observed production history is matched by changing a set of parameters that characterize a mathematical decline-curve function. Decline curves were enhanced in terms of practicality by a set of graphical representations known as type curves (Agarwal et al. 1970, Fetkovich 1980, Carter 1985, Palacio and Blasingame 1993). These approaches are effective and practical for known and typical operating conditions. However, since it is not possible to extend these forecasts to varying conditions, different well locations, or different recovery mechanisms, they are not effective for optimizing operational conditions. Over the years, as computational capabilities have improved drastically, numerical reservoir simulation has become the standard approach for modeling and forecasting. However, numerical reservoir simulation requires a relatively good understanding of the reservoir, accurate representation of fluid-flow dynamics, reasonably good amount of data that can be used for model building, and sufficient amount of resources in terms of manpower, finances, computational capabilities, and time. Whenever any of these conditions is not satisfied, typically the simulation study does not result in a reliable model that fully represents the reservoir under consideration. For example, due to the challenges related to natural-fracture characterization and its representation in reservoir models, it may get difficult to build a robust flow-simulation model for fractured systems. Especially in cases with long histories of water or gas flooding, matching production rates and phase ratios become difficult. This results in a poor forecast and it becomes challenging to optimize development plans based on these models (Artun et al. 2016). Another good example is the complex features of unconventional reservoir systems. Existing forecasting methods such as decline-curve analysis and material balance fail in some cases due to their unique characteristics. While extensions of these formulations to these systems have been studied, development of accurate representations remains an active area of development. Mechanisms, such as adsorption/desorption, turbulent flow in nanoscale pores, flow dynamics in induced and natural fractures, and geomechanical effects due to fracturing, make it rather challenging to construct accurate numerical reservoir models (Al-Alwani et al. 2019). Developments in data collection, processing, and data-driven model building in recent years have been a source of motivation to try to improve the modeling efforts in such problems. Therefore, an increasing number of publications that present fine applications of machine learning algorithms to develop forecasting models for hydrocarbon resources have been published.

The petroleum industry has also been faced with other types of challenges recently such as environmental concerns and climate change, as well as economic concerns due to fluctuating oil/gas prices. These challenges call for extreme efficiency in all aspects of the operations. Data management plays an important role in understanding complex phenomena, diagnosing problems, and developing solutions. The recent global pandemic also has made it clear that we must be prepared for the unexpected by having the ability to make quick and solid decisions under changing circumstances. Data mining, visualization, and exploratory data analysis can help to better manage all operations and can create value by improving both economic and environmental efficiencies. Machine learning-based forecasting models can help to make better decisions in this regard. In this chapter, we focus on performance forecasting of oil and gas reservoirs using machine learning models. A literature review is followed by a general methodology to develop data-driven forecasting models and case studies. The summary and a future outlook are discussed in the final section of this chapter.

10.2 LITERATURE REVIEW

Machine learning-based performance forecasting models for hydrocarbon reservoirs can be broadly divided into two groups: (1) proxy models that accelerate numerical-simulation-based forecasting and (2) models that are purely based on real (field) observations. Existing work and recent developments on these two approaches are summarized next.

Numerical reservoir-simulation models are built by incorporating both static and dynamic characteristics of a hydrocarbon reservoir. For green (undeveloped) fields, the model is built by considering early data from appraisal wells and exploration surveys. For brown (developed) fields, historical pressure and production measurements are used for calibrating the model by adjusting uncertain reservoir characteristics (history-matching). In either case, the aim is to have a model that represents the reservoir as much as possible, so that it can be used as a reliable forecasting tool for reservoir performance. To study a wide range of possibilities in terms of operating decisions, a large number of simulation runs (scenarios) should be analyzed. Due to the computational load of numerical solutions, studying a large number of scenarios is not always practical. Therefore, running a limited number of cases and studying the interactions between uncertain/unknown parameters (reservoir and operational) and one or more performance-related parameters, the so-called proxy modeling approach, has been adopted. Such an analysis results in a proxy model, which approximates the response of a system by a mathematical or statistical function (Sankaran et al. 2020). Experimental design procedures can be followed to determine the number of cases to study such that interaction among parameters is understood better by maximizing the information gained from a limited number of runs. Machine learning algorithms have been frequently used to develop proxy models, which allow us to study a large number of cases in a very short period of time when compared with numerical models. This speed of execution allows us to solve associated optimization problems such as determining optimum operating conditions for reservoir management and field development (Yeten et al. 2003, Mohaghegh et al. 2006, Zangl et al. 2006, Kalantari-Dahaghi et al. 2015, Sen et al. 2021), uncertainty analysis (Chen and Sen 2022), or history-matching (Cullick et al. 2006, Silva et al. 2007, Zhao et al. 2015) in a more computationally efficient manner.

Another useful application area was the development of enhanced oil recovery (EOR) forecasting and screening models either for multiple methods simultaneously (Parada and Ertekin 2012), or for specific EOR methods such as chemical flooding (Ebaga-Ololo and Chon 2017, Alghazal and Ertekin 2020, Zarepakzad et al. 2021), thermal methods (Durzman et al. 2013, Yalgin et al. 2018, Ersahin and Ertekin 2020, Sun and Ertekin 2020), and cyclic pressure pulsing (Artun et al. 2010, Artun 2020). Assuming that numerical simulators can adequately predict the behavior of a given EOR process, these data-driven models, once trained and validated, can be used for quick screening and evaluation of the EOR process in a given reservoir. This helps the modeler to analyze a large number of scenarios, even in a probabilistic manner, without expending resources that are needed for a numerical-simulation study. A combination of these models can result in a comprehensive toolbox that would help to investigate the possibility of different EOR methods in a given reservoir and optimize the design of the EOR application (Ertekin 2021). Applications of neural-network-based forecasting models for tight gas sands (Kulga et al. 2017), long short-term memory neural network as a time-series-based forecasting approach (Song et al. 2020), and random-forest regression (Xue et al. 2021) are all aimed to accelerate conceptual simulation models for unconventional reservoirs that focus on a limited section of the reservoir such as the drainage area of a single well. In some studies, data-driven models that represent the whole reservoir by considering information at several different levels such as grid-blocks, wells, and layers were successfully applied to a giant oil field (Mohaghegh et al. 2009), Marcellus shale (Amini and Mohaghegh 2019), and CO_2-EOR (Gholami et al. 2019, Sen et al. 2022). The fundamental requirement behind the development of all proxy models is the availability of a full-featured geology- and physics-based model (Schuetter et al. 2018). This must be considered when dealing with formations when there are concerns about the validity of the physical laws, as in the case with some types of unconventional reservoirs.

Although proxy modeling offers significant benefits in terms of computational speed-up, the real value of machine learning algorithms comes into the picture when (1) these algorithms are used to build models that are purely based on real observations or (2) modeling the phenomenon is not possible or practical using other approaches (i.e., numerically or analytically). In such kinds of situations, models that integrate different sources of data for forecasting purposes would be useful. An example is the integration of seismic data with well locations and well completion characteristics to train a neural network model that can predict gas recovery accurately (Thararoop et al. 2008). This application was validated with existing production history and forecasted the gas production at unexplored locations, resulting in a robust well placement optimization tool. In a similar study, seismic attributes, well logs, and production data were integrated to forecast the oil production profile, which included a characteristic plateau period followed by a decline period (Ertekin 2021). The integrated model included a seismic-to-log reservoir characterization model and a production-forecasting model that links reservoir characterization to production to generate a production heat map.

Artificial neural networks have been proven to be successful in the development of similar types of models and have been the algorithm of choice in many studies. Models that are very similar to numerical reservoir simulation models in terms of parametric representation were developed by using only real data, through a top-down modeling (TDM) approach (Mohaghegh 2011). This approach was successfully applied to a giant mature oil field in the Middle East by history-matching of the 26 years of production history, validating for 8 years and forecasting the production for the next 4 years (Mohaghegh et al. 2014).

The motivation for using real data to build data-driven models has gradually increased in the last decade due to the unconventional oil and gas boom and the difficulties faced in reservoir characterization and modeling. The TDM approach was applied to New Albany, Bakken, Huron, and Marcellus shale by incorporating reservoir properties as well as hydraulic-fracturing definitions as the controlling parameters for reservoir behavior (Kalantari-Dahaghi and Mohaghegh 2011, Mohaghegh et al. 2012, Esmaili and Mohaghegh 2016). These models were successfully history-matched by identifying flow regimes that appear in shale formations. Initial production rate, as well as cumulative production for the first 2 years in Marcellus shale were predicted from completion and fracture properties from 2,700 horizontal wells using partial least square regression (Al-Alwani et al. 2019). Extreme gradient boosting regressor (Chen and Guestrin 2016) with Bayesian optimization for hypertuning was successfully used to forecast the production in the liquid-rich Duvernay shale (Kong et al. 2021). The major strength of these models is the lack of assumptions related to the definition of the fluid-flow dynamics in the reservoirs. By using real observations that inherently include the physical laws, developed models also help to capture the physics and features that control the flow from shale reservoirs.

Decision trees, a rather simple but highly interpretable method, were successfully used to classify high- and low-performance Wolfcamp shale wells based on 1-year oil recovery from well and hydraulic fracture properties (Schuetter et al. 2018). The model demonstrated that depth, lateral length, amounts of proppant, and fracturing fluid are the important parameters that affect well productivity by also identifying ranges at which high and low categories change. Decision trees were also used to estimate monthly cumulative production for approximately 10,000 Middle Bakken and Three Forks wells from geological, completion, and spacing data, which included a total rock quality index (Cross et al. 2020). This helped to compare different operators with respect to how well their wells perform. Five thousand data points from 20 different fields in Western Siberia were used to develop a forecasting model based on gradient-boosted decision trees, namely CatBoost algorithm (Dorogush et al. 2017), for cumulative oil recovery (Morozov et al. 2020).

There have been some efforts to expedite or automate decline curve analysis through statistical learning approaches such as event detection and regression algorithms to either evaluate a large number of portfolios (Castiñeira et al. 2014) or to focus on a single asset to estimate well-by-well performance (Kianinejad et al. 2019). The same approach was applied to an unconventional reservoir to identify opportunities for value creation by integrating the forecast with an optimization routine (Darabi et al. 2020).

Recently, efforts have been focusing on the characterization of the time-series features of the production history and hybrid neural networks that combine multiple architectures to process different types of data, including images. These applications mainly used convolutional (CNN) and recurrent neural networks (RNN), which take conventional multilayer neural networks further by including additional forms of layers and units. These features make them ideally suited for image classification (in the case of CNN) and time-series problems (in the case of RNN) (James et al. 2021). Ensemble empirical mode decomposition based on long short-term memory neural network (a special version of RNN) was found to provide more accurate predictions than conventional neural networks and support vector machines (Liu et al. 2020), in which the hyperparameters were optimized using the genetic algorithm. A hybrid convolutional-recurrent neural network (c-RNN) will be used to train a model at the hydraulic-fracture-stage level to forecast cumulative oil recovery during the next 5 years in the Montney Formation in Alberta (Chaikine and Gates 2021). Using the leave-one-out method, the best combination of input parameters was determined consisting of rock mechanical properties, proppant amount, and spacing and completion of neighboring wells. A hybrid neural network that combines the multilayer perceptron (MLP) and CNN architectures to process multiple scales and types of data was developed to predict the oil rate at new wells (Yuan et al. 2021). The method showed improvement over the individual application of MLP and CNN. By considering production forecasting as a time-series problem, sliding window regression was applied to a waterflooded reservoir by considering integral (number of wells, injection rate), local (distance between wells), pressure, and autoregressive (time effects) features in the dataset (Davtyan et al. 2020).

10.3 METHODOLOGY

The generalized workflow to develop machine learning-based performance forecasting models for hydrocarbon reservoirs is summarized in Figure 10.1 in two parts: (1) pre-processing, which includes the definition of the problem and study objectives, followed by the preparation of the data that will be used for modeling,

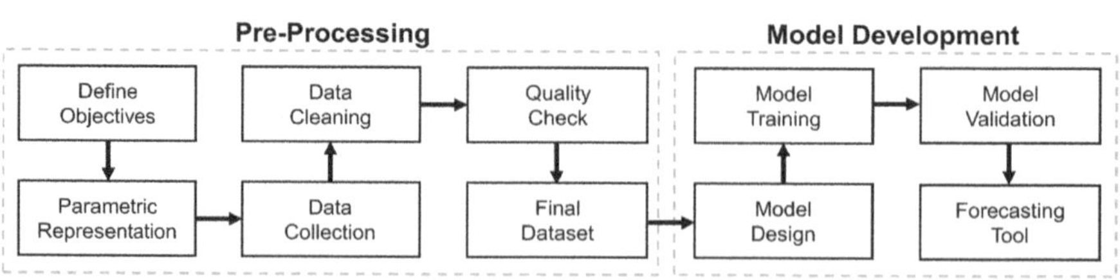

FIGURE 10.1 General workflow for the development of a machine learning-based forecasting model.

and (2) model development, which includes the actual modeling process, i.e., the way the model is designed, trained, validated, and deployed.

10.3.1 Pre-Processing

First, key objectives of the forecasting study should be clearly defined. According to the objectives, the parametric representation of the model can be defined. Supervised machine learning models are based on learning the relationship between inputs and outputs of a system by using a dataset (observations). Therefore, data availability and quality also play an important role in designing the parametric representation of the proposed modeling effort. In reservoir engineering problems, the parameters of interest can be broadly divided into three groups:

1. Reservoir characteristics (natural characteristics of the reservoir, which we typically do not have any control on such as porosity and permeability, etc.).
2. Operational (design) parameters (parameters that we can control in the field such as well completion interval, horizontal well length, number of hydraulic fractures, etc.)
3. Performance indicators (volumetric, economic, or environmental performance characteristics of a given well or reservoir or field that we are interested in forecasting)

Reservoir performance is a function of natural reservoir characteristics and operational parameters that we control. The exact parameters to be included in the model development depend on the problem and data availability. For instance, reservoir characteristics or operational parameters to include in a sandstone reservoir would be different than a hydraulically fractured shale reservoir. During primary oil recovery, we would be interested in forecasting oil rate or cumulative oil recovery. At later stages with saturated conditions or during secondary/enhanced recovery, multiple phases (i.e., water, gas) would come into the picture. As a result, phase ratios, breakthrough time, and incremental oil recovery would be the critical performance indicators. Therefore, the recovery mechanism and reservoir type play an important role in designing the parametric representation of the model. In some cases, for example, in studying field development problems, we may be interested in economic or environmental performance parameters.

After designing the parametric representation of the problem, appropriate data are collected. This stage may be completed synchronously with parametric representation because data availability may limit the number of parameters included. One of the common problems at this point is the missing data. Some features may not be available for some data points. However, removing those points may also lead to loss of valuable information in the dataset. In those cases, some approaches to deal with missing data, such as random forest imputation (Schuetter et al. 2018) and collaborative filtering (Morozov et al. 2020), can be utilized. Data must be cleaned, the quality of it must be checked, and any data point with questionable quality should be eliminated. The ultimate goal is to obtain an organized dataset that represents the problem, including both static and dynamic characteristics and the desired performance indicator. The dataset should include enough samples depending on the dimensionality of the problem (number of features involved) due to the *curse of dimensionality* (Bellman 1961). The number of features should be optimized through *feature selection* procedures (James et al. 2021). After this step, the development of the machine learning model starts.

10.3.2 Model Development

Model development starts with the selection of the machine learning algorithm and designing the application of that algorithm considering the parametric representation. Machine learning problems are divided into two as regression and classification problems. For each group, there are a number of algorithms that have been developed. Each algorithm has its own strengths and weaknesses, and each has its own theoretical foundations that require a certain level of knowledge of controlling parameters. Comparisons of machine learning algorithms are presented in other sources (Mishra and Datta-Gupta 2017, Schuetter et al. 2018). As the model complexity increases, the interpretability generally decreases. Therefore, this trade-off must be considered when selecting the algorithm.

Before model training, the dataset is resampled into two or three sets. A training set is used to introduce the patterns of data to the machine learning algorithm. During training, model parameters are optimized so that a desired level of performance is reached. This performance is defined by a loss function, such as the mean squared error (MSE) of the misfit between model predictions and actual data. Validation set and testing set are roughly the same sizes, and smaller than the training set. These two sets are called *hold-out* sets, since they cannot be used by the learning algorithm (James et al. 2021). Validation set is optional and application of the trained model to this set is used to further optimize some of the model parameters to improve the training performance. The testing set is completely hold-out and used for *blind-testing* purposes to see the model's capability to make predictions in different cases. Analysis of the model performance reveals some of the characteristics and robustness of the model. It is important to avoid *overfitting*, which happens when the model fits into the training patterns accurately, yet cannot generalize. To evaluate a model's robustness in terms of prediction and generalization capabilities, several metrics can be used. Accuracy of regression models can be evaluated via the average magnitude of the differences between real (y_i) and predicted responses ($\hat{y}_i$), i.e., the *average absolute error* or AAE (Mishra and Datta-Gupta 2017):

$$\text{AAE} = \frac{1}{n}\sum_{i=1}^{n}\left|y_i - \hat{y}_i\right| \tag{10.1}$$

MSE, is identical to *AAE*, except instead of the absolute value, it measures the average squared difference between observations and their associated predictions:

$$\mathrm{MSE} = \frac{1}{n}\sum_{i=1}^{n}\left(y_i - \hat{y}_i\right)^2 \tag{10.2}$$

The *root-mean-square error*, or *RMSE*, is a popular *MSE* variation that is just the square root of *MSE* (Mishra and Datta-Gupta 2017). The R^2 *statistic* is another way to assess the quality of the fit, representing the proportion of the variance explained in the model (James et al. 2021):

$$R^2 = 1 - \frac{\sum_{i=1}^{n}\left(y_i - \hat{y}_i\right)^2}{\sum_{i=1}^{n}\left(y_i - \overline{y}_i\right)^2} \tag{10.3}$$

Accuracy of classification models is evaluated via a tabular form called the *confusion matrix*, which shows the distribution of the cases based on their actual and predicted categories (Figure 10.2). Based on this distribution, three measures of interest are *accuracy*, *specificity*, and *precision* (James et al. 2021). Accuracy is the overall ratio of correct classification in the model. While accuracy is a good indicator of classification capability of the model for many types of problems, specificity and precision may also be included in the analysis. Precision is calculated as the number of correct positive predictions (TP) divided by the total number of positives. Specificity is calculated as the number of correct negative predictions (TN) divided by the total number of negatives.

During validation, it is also important to analyze variable importance in the model to see which parameters contribute to the model more. If the domain expertise allows us to comment about which variables would be more dominant in this problem, this analysis would allow us to further validate the model trained from a domain point of view. If not, this would help to understand the controlling parameters of the problem. After this, if the model is considered to be robust enough to make accurate predictions or classifications, then it can be deployed as

		ACTUAL	
		NO	YES
PREDICTED	NO	TN	FN
	YES	FP	TP

TN: True Negative
FN: False Negative
FP: False Positive
TP: True Positive

FIGURE 10.2 Confusion matrix definitions for evaluation of the binary-classification model.

a forecasting tool. Deterministic and probabilistic analyses can help to make a certain operational decision or optimize a certain operational process.

10.4 CASE STUDIES

10.4.1 Forecasting the Water-Cut in a Waterflooded, Fractured Carbonate Reservoir

In this study, the primary objective is to develop a model that can help to understand the water movement in a fractured-carbonate reservoir that has been waterflooded for 15 years. The reservoir parameters used are shown in Table 10.1. These properties are divided into two as (1) static variables (variables that do not change with time) and (2) dynamic variables (time-dependent variables that may change with time). Another classification is made based on the type of the well that the property belongs to, including the forecasted well location, the nearest production well, and the nearest injection well. *X*, *Y* are the latitude and longitude of the well that is being forecasted; $d_{well,p}$, $d_{well,i}$ are the distance between the well and the closest producer (*p*) or injector (*i*), respectively. *h* is the gross thickness; *NTG* is net-to-gross ratio; S_w is the water saturation; D_{top} is the formation top; WC_t, q_t are the initial water-cut and oil production rate at the time of forecasting, *t*; WC_i and q_i are the water-cut and oil flow rate at the initial time of the well production; q_w is well rate; and *EUDA* is the estimated ultimate drainage area and is estimated using Voronoi grids and image well theory (Voronoi 1908, Palagi and Aziz 1994, Mohaghegh and Bromhal, 2010). All these input parameters are used as inputs to classify a well as a high-water-cut (>50%) or low-water-cut (<50%) well at a given time, under given conditions. Only when there are at least two producers and one injector included was there a total of 6,500 data points. Eighty percent of these points were used for training and 20% were used for blind-testing to test the generalization capabilities. Three tree-based classification algorithms were considered: (1) C5.0 algorithm (a decision-tree based on the C4.5 algorithm, Quinlan 1993), (2) random forests (an ensemble tree approach, Breiman 2001), and (3) AdaBoost (Freund and Schapire 1997). Key model parameters for each algorithm were varied to see their impact on the results: number of trials in C5.0 and AdaBoost, and number of variables randomly sampled as candidates at each split for random forests. The accuracy results for training and testing sets are shown in Table 10.2. While

TABLE 10.1
Variables That Are Included in the Machine Learning Model for Water-Cut Classification

	Well	Closest Producer	Closest Injector
Static	X, Y, ϕ, h, NTG, S_w	$X, Y, \phi, h, NTG, S_w, D_{top}, d_{well}$	X, Y, d_{well}
Dynamic	$q_t, WC_t, EUDA$	$q_i, q_t, WC_i, WC_t, EUDA, \Delta t$	q_w

TABLE 10.2
Accuracy Results for Different Classification Algorithms with Different Model Parameters for the Water-Cut Classification Model

Algorithm		Training	Testing	Training	Testing	Training	Testing	Training	Testing
C5.0	R^2	0.91	0.89	0.93	0.89	0.95	0.90	0.97	0.91
	Parameter	1		5		10		50	
Random Forests	R^2	0.94	0.90	0.99	0.91	1.00	0.91	1.00	0.91
	Parameter	1		5		10		20	
AdaBoost	R^2	0.93	0.88	0.99	0.89	1.00	0.91	1.00	0.91
	Parameter	1		5		10		20	

TABLE 10.3
Confusion Matrices for the Water-Cut Classification Model Using AdaBoost Algorithm with 10 Trials

	Training					Testing			
		Real					Real		
		Low	High	TOTAL			Low	High	TOTAL
Predicted	Low	**0.84**	0	0.84	Predicted	Low	**0.80**	0.04	0.84
	High	0	**0.16**	0.16		High	0.05	**0.11**	0.16
TOTAL		0.84	0.16	**1.00**	TOTAL		0.85	0.15	**0.91**

satisfactory results were obtained with all algorithms, regardless of the model parameter, random forests with 10 and 20 sampled variables and AdaBoost with 10 and 20 trials resulted in the highest accuracy. Table 10.3 shows the confusion matrices and classification accuracies for training (1.00) and testing (0.91) sets, using the AdaBoost algorithm with 10 trials, which was the top-performer.

Figure 10.3 shows the ranking of variables in terms of mean decrease in the Gini Index (a variable's total decrease in node impurity, weighted by the proportion of samples reaching that node in each individual decision tree in the random forest). Water-cut and oil rate of the offset well, injection rate of and distance from the nearest injector, and estimated ultimate drainage areas are the most important variables to determine the water-cut classification of the well. Reservoir engineering domain knowledge confirms the validity of model findings. Also, the results show that embedding physics through derived parameters, such as estimated ultimate drainage area, was helpful in characterizing the variance in the output.

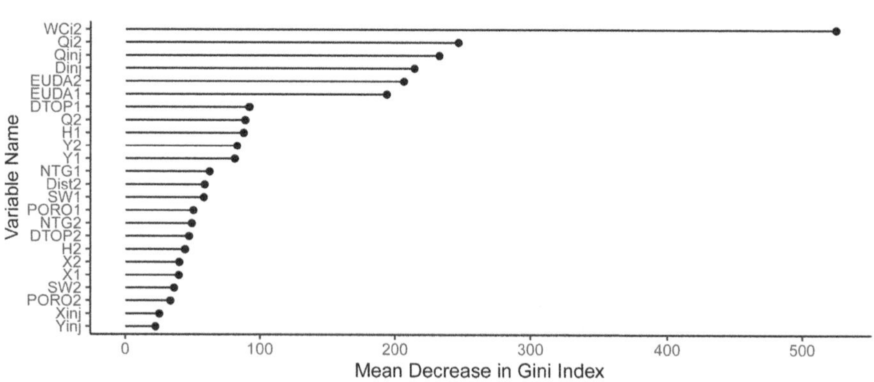

FIGURE 10.3 Ranking of variables in terms of mean decrease in Gini Index for water-cut classification.

10.4.2 Forecasting Incremental Oil Recovery and Economic Value of Single-Well Cyclic Nitrogen Injection

Cyclic injection of nitrogen is a promising EOR method for depleted and fractured reservoirs. The process includes cycles of nitrogen injection, soaking, and production period in a single well. Nitrogen diffuses from fractures into the matrix and displaces the oil in the matrix, which flows to the fracture network and then to the wellbore. Due to a number of design parameters involved and low-profit margins of depleted reservoirs, properly designing the process is a challenging optimization problem. In this study, a production lease area in the Appalachian Basin, which has gone through five cycles of injection, is considered. First, a representative reservoir model was built and history-matched. Then, 400 simulation scenarios were run by varying design parameters such as project life, production rate limit, injection rate, length of injection, and soaking periods. Uniformly distributed values of each parameter were generated within specified ranges. For each scenario, incremental oil recovery due to injection and net present value (NPV) were calculated using economic parameters such as oil price, nitrogen cost, operating cost, royalty, and tax. The machine learning model was designed to predict these two performance parameters from the design parameters. Eighty percent of the dataset was used for training and 20% was used for testing. Different neural network architectures were evaluated for training and testing performance. Table 10.4 shows the R^2 and

TABLE 10.4
Training and Testing Performances of Different Neural Network Architectures for the Cyclic Nitrogen Injection Problem

		Training				Testing			
		Net Present Value, (US$)		Incremental Recovery, bbl		Net Present Value, (US$)		Incremental Recovery, bbl	
No. of layers/neurons		R^2	RMSE	R^2	RMSE	R^2	RMSE	R^2	RMSE
1 layer	**10**	0.93	11,287	0.99	137.2	0.84	7,913	0.92	169.1
	20	0.95	9,364	0.99	114.2	0.89	6,530	0.95	137.7
	30	0.97	7,577	0.99	104.4	0.90	6,101	0.96	115.1
	40	0.97	7,613	1.00	82.5	0.85	7,579	0.94	149.8
2 layers	**10–5**	0.96	8,936	0.99	92.5	0.83	8,072	0.95	138.8
	20–10	0.90	5,087	0.99	92.4	0.68	11,071	0.99	69.1
	30–20	0.98	6,011	1.00	74.4	0.81	8,535	0.98	79.4
	40–25[a]	**0.98**	**6,192**	**1.00**	**74.5**	**0.90**	**6,301**	**0.99**	**66.5**
3 layers	**15–10-5**	0.97	7,260	0.99	100.8	0.87	7,013	0.96	117.8
	30–20–10	0.97	6,903	0.99	94.1	0.88	6,852	0.97	109.8
	40–30–20	0.98	5,899	1.00	72.9	0.83	8,081	0.98	93.3
	50–35–20	0.98	5,263	1.00	67.4	0.88	6,860	0.98	85.3

[a] Selected model.

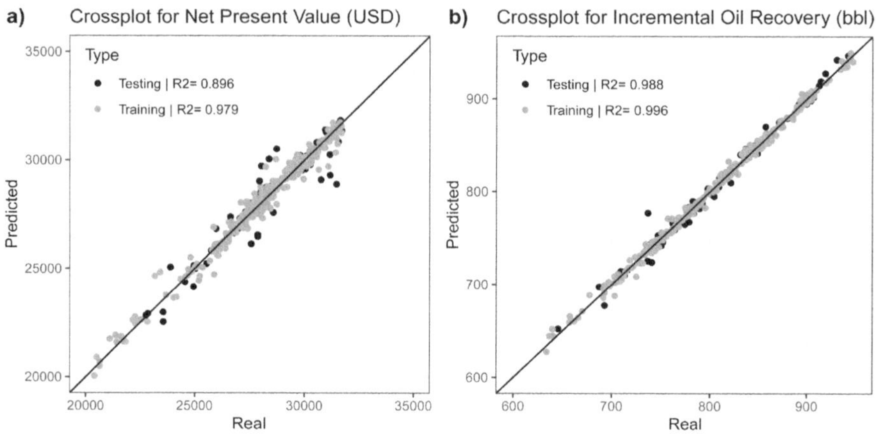

FIGURE 10.4 Real vs predicted cross-plots of (a) net present value (NPV) and (b) incremental oil recovery for the cyclic nitrogen injection problem.

RMSE statistics of these architectures. While all models had acceptable accuracies, the one with the highest accuracy for both variables, two-layer neural network with 40 and 25 neurons in each layer, was selected. Figure 10.4 shows real vs predicted cross-plots. The model's accuracy for NPV, as a derived property, was relatively lower. Incremental oil recovery was predicted with a high accuracy. Figure 10.5 shows the importance of variables as calculated by the *connection-weight approach* (Olden and Jackson 2002). For both parameters, injected volume per cycle and injection length are the most important variables. Length of the soaking period does not have a pronounced impact on the model. Production rate limit for each cycle is important for economics, and total project length affects the incremental recovery. These observations agree with the findings in the literature (Artun 2020).

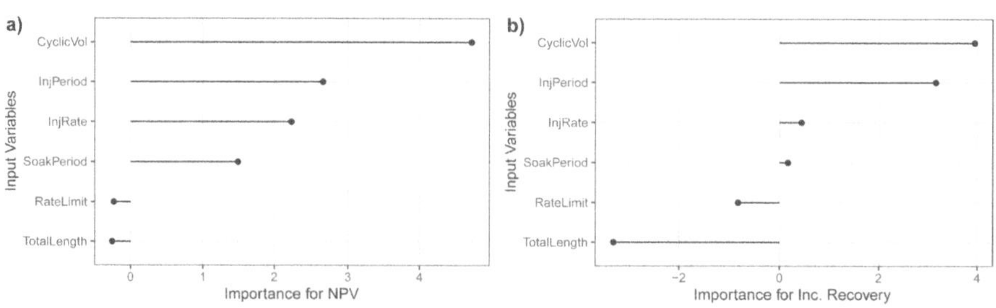

FIGURE 10.5 Variable importance based on neural network weights for the prediction of (a) net present value (NPV) and (b) incremental oil recovery for the cyclic nitrogen injection problem.

10.4.3 Forecasting Cumulative Gas Recovery from Shale Wells

As indicated earlier in this text, modeling shale reservoirs and forecasting its performance remains a challenging problem due to the complexities of these resources. Using a recently published dataset (SPE 2021), this study aimed to forecast cumulative gas recovery after 1 year of production in shale wells using multiple linear regression and random forests. The dataset included reservoir and well characteristics from 50 horizontal shale wells in Haynesville, Marcellus, and Eagle Ford formations including initial pressure, temperature, saturations, net pay thickness, porosity, API oil gravity, gas gravity, CO_2 and N_2 concentrations, total depth, fracture spacing, number of stages, clusters, lateral length, and perforation interval. The model included 25 input parameters including the type of the formation as a categorized variable. Table 10.5 shows the R^2 and *RMSE* results for both multiple linear regression and random forests with different numbers of variables randomly sampled at each split. The results indicate that random forests outperform multiple linear regression. With 25 sampled variables, best testing results were obtained with satisfactory training performance. The multiple linear regression model showed that lateral length and perforation interval were the most important variables that explained the variance in the cumulative gas recovery. According to variable importance (Figure 10.6), especially bottom perforation and lateral length are also critical parameters along with gas saturation, and hydraulic-fracture characteristics. These observations match with the findings in similar studies (Schuetter et al. 2018). Figure 10.7 shows the cross-plot of real vs predicted gas recovery for the random forest model with 25 variables sampled during training. These results indicate that even with a low number of data points, the relationships between input and output variables can be captured by machine learning algorithms as long as critical features exist in the dataset.

TABLE 10.5
Training and Testing Performances of Models Trained Using Random Forests with a Different Number of Variables Sampled, and Using Multiple Linear Regression

	R^2		*RMSE* (MMscf of gas)	
	Training	Testing	Training	Testing
RF with $n = 1$	0.85	0.74	8,177	4,203
RF with $n = 5$	0.94	0.88	5,287	2,885
RF with $n = 10$	0.95	0.90	4,969	2,584
RF with $n = 15$	0.95	0.89	4,979	2,772
RF with $n = 20$	0.95	0.90	4,884	2,567
RF with $n = 25$	0.94	0.91	5,086	2,468
Multiple Linear Reg.	0.94	0.68	5,195	4,653

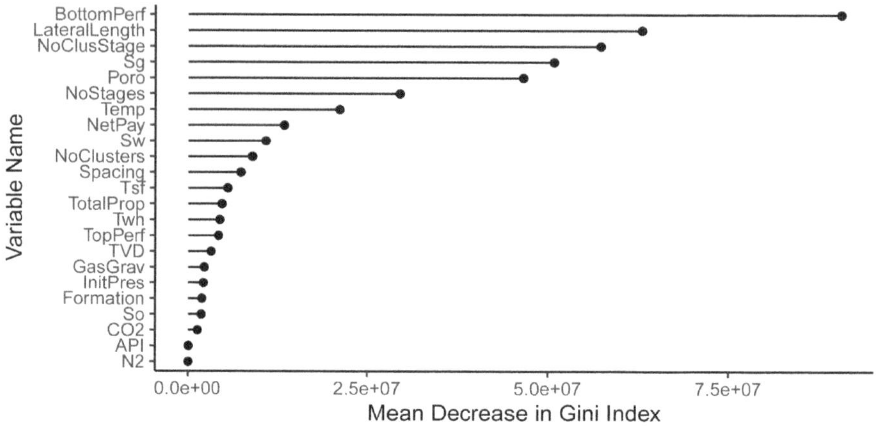

FIGURE 10.6 Ranking of variables in terms of mean decrease in Gini Index for the shale-gas recovery prediction problem.

10.5 SUMMARY, CONCLUSIONS, AND LESSONS LEARNED

Performance-forecasting problems in reservoir engineering have been solved in different ways with successful applications of different machine learning algorithms. Tree-based methods and neural networks have been the most widely used algorithms that proved to be successful in complex problems that include large datasets and a large number of parameters. In this chapter, after reviewing the applications in the literature, successful case studies for both classification and regression problems using both real and simulated data are presented. It is demonstrated that machine learning can be applied to a variety of forecasting problems related to unconventional reservoirs, fractured carbonates, and secondary

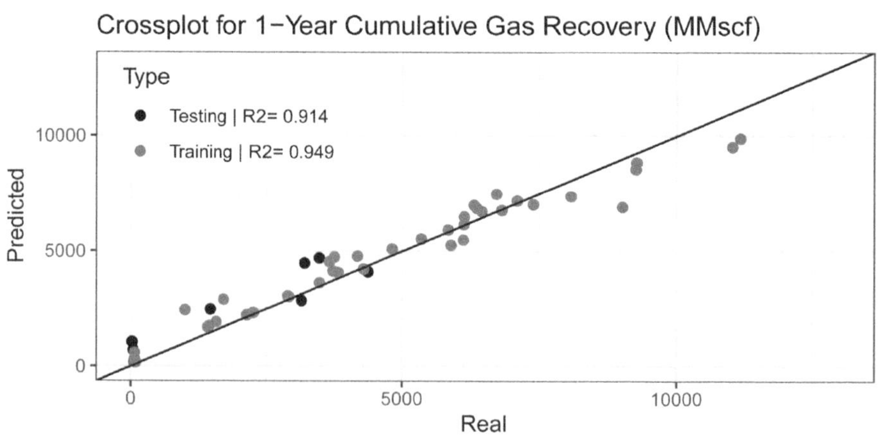

FIGURE 10.7 Real vs predicted cross-plot of 1-year cumulative gas recovery for the shale-gas recovery prediction problem.

and EOR. Some lessons learned in the presented case studies can be summarized as the following:

1. Testing different algorithms and model structures would help to optimize the final model. Performance of the model must be carefully analyzed before deploying it as a forecasting tool.
2. Variable importance analysis through model parameters helps to understand the key controlling factors in each problem.
3. Parametric representation of the problem should be carefully designed and performance indicators should represent the study objectives. Derived parameters that aid to improve the definition of the physical laws of the problem would increase the chances of successful model development.

10.6 FUTURE OUTLOOK

As the need for more efficient operations rises in every aspect of the petroleum industry, we should expect to see an increasing number of applications of machine learning for better forecasting of reservoir performance. While mostly volumetric and economic performance predictions have been the main focus of the industry, we may see an increasing number of studies that aim to focus on environmental performance in the future. Recently developed deep learning applications that integrate different sources and scales of data may be further improved for higher accuracy and practicality. Similarly, more advanced implementation of tree-based methods that include different forms of bagging and boosting applications can be applied. Physics-informed machine learning models have been an active area of development (Zhang et al. 2016, Raissi et al. 2017, Guo et al. 2018). These types of models are helpful to reduce the computational load, while also improving the accuracy. Another form of this hybridization takes place when data-driven models are integrated into numerical simulators to accelerate computationally demanding components of the fluid-flow formulation. For example, a machine learning-based model that can predict the pressure profile along the production was successfully developed using data generated using a wellbore hydraulics simulator (Ertekin 2021). Similar developments may increase with time to improve the capabilities in reservoir simulation.

The interest in the ability to explain machine learning models and their predictions has been gaining momentum in the machine learning community. Models with reasonable complexity can be explained through model parameters by analyzing variable importance. As models get more complex, it gets more difficult to interpret its results. It is expected to see more development in this area that can be applied to performance-forecasting models for oil and gas reservoirs.

One of the questions raised during the development of machine learning models for oil and gas fields is the universal applicability of a trained model (i.e., can we use a model trained with the dataset of a specific field for another field with similar features?). This question becomes important especially for fields with limited data. Some recent studies utilized the concept of *transfer learning* by developing a forecasting model using the data from a data-rich unconventional reservoir and applying the model to another reservoir with limited data (Cornello et al. 2021, Odi et al. 2021). Similar types of models can be further improved toward the development of more universal forecasting models.

Another potential development is the integration of machine learning models with other platforms such as databases and business intelligence software. While there are a large number of successful stand-alone model-development and visualization examples, there are not many applications that successfully link the raw database to a real-time visualization tool. Such an integration would bring a more user-centric application of the machine learning-based forecasting models to life. From the practical perspective, data sharing is very limited and should increase within the oil and gas community. With more publicly available data and comparative solution projects, more successful data-driven applications would be developed that serve the needs of the industry.

REFERENCES

Agarwal, R. G., R. Al-Hussainy, H. J. Ramey. 1970. "An investigation of wellbore storage and skin effect in unsteady liquid flow: i. analytical treatment," *SPE J.*, 10:279–290. doi:10.2118/2466-PA

Al-Alwani, M. A., L. Britt, S. Dunn-Norman, H. H. Alkinani, A. T. Al-Hameedi, A. Al-Attar. 2019. "Production performance estimation from stimulation and completion parameters using machine learning approach in the Marcellus Shale," In: Proceedings, 53rd U.S. Rock Mechanics/Geomechanics Symposium, 23–26 June, New York City, New York.

Alghazal, M., T. Ertekin. 2020. "Modeling of deep polymer gel conformance treatments using machine learning," In: Proceedings, Abu Dhabi International Petroleum Exhibition & Conference, 9–12 November, Abu Dhabi, UAE. doi:10.2118/203450-MS

Amini, S., S. Mohaghegh. 2019. "Application of machine learning and artificial intelligence in proxy modeling for fluid flow in porous media," *Fluids*, 4:126. doi:10.3390/fluids4030126

Arps, J. J. 1945. "Analysis of decline curves," *Trans. Soc. Pet. Eng.*, 160:228–247. doi:10.2118/945228-G

Artun, E. 2020. "Performance assessment and forecasting of cyclic gas injection into a hydraulically fractured well using data analytics and machine learning," *J. Pet. Sci. Eng.*, 195, 107768. doi:10.1016/j.petrol.2020.107768

Artun, E., T. Ertekin, R. Watson, B. Miller. 2010. "Development and testing of proxy models for screening cyclic pressure pulsing process in a depleted, naturally fractured reservoir," *J. Pet. Sci. Eng.*, 73(1):73–85. doi:10.1016/j.petrol.2010.05.009

Artun, E., M. Vanderhaeghen, P. Murray. 2016. "A pattern-based approach to waterflood performance prediction using knowledge management tools and classical reservoir engineering forecasting methods," *Intl. J. Oil, Gas and Coal Tech.*, 13(1):19–40. doi:10.1504/IJOGCT.2016.078046

Bellman R. E. 1961. "Adaptive Control Processes." Princeton University Press, Princeton, NJ, 276 pp.

Breiman, L., 2001, "Random forests," *Mach. Learn.* 45:5–32. doi:10.1023/A:1010933404324

Carter, R. 1985. "Type curves for finite radial and linear flow system," *SPE J.*, 25:719–728. doi:10.2118/12917-PA

Castiñeira, D., A. Mondal, S. Matringe. 2014. "A new approach for fast evaluations of large portfolios of oil and gas fields," In: Proceedings, SPE Annual Technical Conference and Exhibition, 27–29 October, Amsterdam, The Netherlands. doi:10.2118/170989-MS

Chaikine, I. A., I. D. Gates. 2021. "A machine learning model for predicting multi-stage horizontal well production," *J. Pet. Sci. Eng.*, 198, 108133. doi:10.1016/j.petrol.2020.108133

Chen, H., D. Sen. 2022. "A flow feature clustering-assisted uncertainty analysis workflow for optimal well rates in waterflood projects," *SPE J.*, 1–17. doi:10.2118/209608-PA

Chen, T., C. Guestrin. 2016 "XGBoost: a scalable tree boosting system," In: Proceedings, 22nd SIGKDD Conference on Knowledge Discovery and Data Mining, 13–17 August, San Francisco, California.

Cornello, J., S. M. Razak, A. Jahandideh, B. Jafarpour, Y. Cho. 2021. "Investigating transfer learning for characterization and performance prediction in unconventional reservoirs," In: Proceedings of SPE Middle East Oil & Gas Show and Conference, 28–1 December, Manama, Bahrain, doi:10.2118/204563-MS

Cross, T., K. Sathaye, K. Darnell, D. Niederhut, K. Crifasi. 2020. "Benchmarking operator performance in the Williston Basin using a predictive machine learning model," In: Proceedings, SPE/AAPG/SEG Unconventional Resources Technology Conference, Virtual, 20–22 July. doi:10.15530/urtec-2020-2750

Cullick, A., D. Johnson, G. Shi. 2006. "Improved and more rapid history matching with a nonlinear proxy and global optimization," In: Proceedings SPE Annual Technical Conference and Exhibition, 24–27 September. San Antonio, Texas. doi:10.2118/101933-MS

Darabi, H., X. Zhai, A. Kianinejad, Z. Ma, D. Castiñeira, V. S. Suicmez, R. Toronyi. 2020. "Augmented AI framework for well performance prediction and opportunity identification in unconventional reservoirs," In: Proceedings, International Petroleum Technology Conference, 13–15 January, Dhahran, Kingdom of Saudi Arabia. doi:10.2523/IPTC-20099-MS

Davtyan, A., A. Rodin, I. Muchnik, A. Romashkin. 2020. "Oil production forecast models based on sliding window regression," *J. Pet. Sci. Eng.*, 195, 107916. doi:10.1016/j.petrol.2020.107916

Dorogush, A. V., V. Ershov, A. Gulin. 2017. "CatBoost: gradient boosting with categorical features support," In: Proceedings, ML Systems Workshop at 31st Conference on Neural Information Processing Systems, 4–9 December, Long Beach, California.

Durzman, P. J., J. Leung, S. D. Zanon, E. Amirian. 2013. "Data-driven modeling approach for recovery performance prediction in SAGD operations," In: Proceedings, SPE Heavy Oil Conference, 11–13 June, Calgary, Alberta, Canada. doi:10.2118/165557-MS

Ebaga-Ololo, J., B. H. Chon. 2017. "Prediction of polymer flooding performance with an artificial neural network: a two-polymer-slug case," *Energies*, 10(7): 844. doi:10.3390/en10070844

Ersahin, A. T. Ertekin. 2020. "Artificial neural network modeling of cyclic steam injection process in naturally fractured reservoirs," *SPE Res. Eval. Eng.*, 23(3):979–991. doi:10.2118/195307-PA

Ertekin, T. 2021. "The efficacy and superiority of the expert systems in reservoir engineering decision making process," *Appl. Sci.*, 11, 6347. doi:10.3390/app11146347

Esmaili, S., S. Mohaghegh. 2016. "Full field reservoir modeling of shale assets using advanced data-driven analytics," *Geosci. Front.*, 7(1):11–20. doi:10.1016/j.gsf.2014.12.006

Fetkovich, M. J. 1980. "Decline curve analysis using type curves," *J. Pet. Technol.*, 32:1065–1077. doi:10.2118/4629-PA

Freund, Y, R. E. Schapire. 1997. "A decision-theoretic generalization of on-line learning and an application to boosting," *J. Comput. Syst. Sci.*, 55:119–139. doi:10.1006/jcss.1997.1504

Gholami, V., S. Mohaghegh, M. Mohammad. 2019. "Smart proxy modeling of SACROC CO_2-EOR," *Fluids*, 4:85; doi:10.3390/fluids4020085

Guo, Z., A. C. Reynolds, H. Zhao. 2018. "A physics-based data-driven model for history matching, prediction, and characterization of waterflooding performance," *SPE J.*, 23(2):367–395. doi:10.2118/182660-PA

Gupta, S., L. A. Saputelli, A. Verde, J. A. Vivas, G. A. Narahara. 2016. "Application of an advanced data analytics methodology to predict hydrocarbon recovery factor variance between early phases of appraisal and post-sanction in Gulf of Mexico deep offshore assets," In: Proceedings, Offshore Technology Conference, Houston, Texas, USA, 2–5 May. doi:10.4043/27127-MS.

James, G., D. Witten, T. Hastie, R. Tibshirani. 2021. "An Introduction to Statistical Learning: With Applications in R." 2nd Edition. Springer, Berlin, Germany, 622 pp.

Kalantari-Dahaghi, A., Mohaghegh, S., 2011, "A new practical approach in modelling and simulation of shale gas reservoirs: application to New Albany Shale," *Intl. J. Oil Gas Coal Tech.*, 4(2):103–133. doi:10.1504/IJOGCT.2011.038925

Kalantari-Dahaghi, A., S. Mohaghegh, S. Esmaili. 2015. "Data-driven proxy at hydraulic fracture cluster level: a technique for efficient CO_2-enhanced gas recovery and storage assessment in shale reservoir," *J. Nat. Gas Sci. Eng.*, 27:515–530. doi:10.1016/j.jngse.2015.06.039

Kianinejad, A., R. Kansao, A. Maqui, R. Kadlag, G. Hetz, F. Ibrahima, V. S. Suicmez, H. Darabi, D. Castiñeira. 2019. "Artificial-intelligence-based, automated decline curve analysis for reservoir performance management: a giant sandstone reservoir case study," In: Proceedings, Abu Dhabi International Petroleum Exhibition & Conference, 11–14 November, Abu Dhabi, UAE. doi:10.2118/197142-MS

Kong, B., Z. Chen, S. Chen, T. Qin. 2021. "Machine learning assisted production data analysis in liquid-rich Duvernay Formation," *J. Pet. Sci. Eng.*, 200:108377. doi:10.1016/j.petrol.2021.108377

Kulga, B., E. Artun, T. Ertekin. 2017. "Development of a data-driven forecasting tool for hydraulically fractured, horizontal wells in tight-gas sands," *Comput. Geosci.*, 103:99–110. doi:10.1016/j.cageo.2017.03.009

Liu, W., W. D. Liu, J. Gu. 2020. "Forecasting oil production using ensemble empirical model decomposition based long short-term memory neural network," *J. Pet. Sci. Eng.*, 189:107013. doi:10.1016/j.petrol.2020.107013

Mishra, S., A. Datta-Gupta. 2017. "Applied Statistical Modeling and Data Analytics: A Practical Guide for the Petroleum Geosciences." 1st Edition. Elsevier, Amsterdam, The Netherlands, 235 pp.

Mohaghegh, S. 2011. "Reservoir simulation and modeling based on artificial intelligence and data mining (AI&DM)," *J. Nat. Gas Sci. Eng.*, 3(6):697–705. doi:10.1016/j.jngse.2011.08.003

Mohaghegh, S., Y. Al-Mehairi, R. Gaskari, M. Maysami, Y. Khazaeni, M. Gashut, A. E. Al-Hammadi, S. Kumar. 2014. "Data-driven reservoir management of a giant mature oilfield in the Middle East," In: Proceedings, SPE Annual Technical Conference and Exhibition, 27–29 October, Amsterdam, The Netherlands. doi:10.2118/170660-PA

Mohaghegh, S., G. Bromhal. 2010. "Top-down modeling: practical, fast track, reservoir simulation & modeling for shale formations," In: Proceedings of AAPG/SPE/SEG/SPWLA Joint Hedberg Research Conference. December 2–10, Austin, Texas.

Mohaghegh, S., A. Modavi, H. Hafez, M. Haajizadeh, S. Guruswamy. 2009, "Development of Surrogate Reservoir Model (SRM) for fast track analysis of a complex reservoir," *Intl. J. Oil Gas Coal Tech.*, 2(1):2–23. doi:10.1504/IJOGCT.2009.023627

Mohaghegh, S., A. Modavi, H. Hafez, M. Haajizadeh, M. Kenawy, S. Guruswamy. 2006. "Development of surrogate reservoir models (SRM) for fast-track analysis of complex reservoirs," In: Proceedings, SPE Intelligent Energy Conference and Exhibition Proceedings. 11–13 April. Amsterdam, The Netherlands. doi:10.2118/99667-MS

Mohaghegh, S. D., O. Grujic, S. Zargari, A. Kalantari, G. Bromhal. 2012. "Top-down, intelligent reservoir modelling of oil and gas producing shale reservoirs; case studies," *Intl. J. Oil Gas Coal Tech.*, 5(1):3–28. doi:10.1504/IJOGCT.2012.044175

Morozov, A. D., D. O. Popkov, V. M. Duplyakov, R. F. Mutalova, A. A. Osiptsov, A. L. Vainshtein, E. V. Burnaev, E. V. Shel, G. V. Paderin. 2020. "Data-driven model for hydraulic fracturing design optimization: focus on building digital database and production forecast," *J. Pet. Sci. Eng.*, 194, 107504. doi:10.1016/j.petrol.2020.107504

Odi, U., K. Ayeni, N. Alsulaiman, K. Reddy, K. Ball, C. Temizel. 2021, "Applied transfer learning for production forecasting in shale reservoirs," In: Proceedings of SPE Middle East Oil & Gas Show and Conference, 28–1 December, Manama, Bahrain. doi:10.2118/204784-MS

Olden, J. D., D. A. Jackson. 2002. "Illuminating the 'black-box': a randomization approach for understanding variable contributions in artificial neural networks," *Ecol. Model.* 154:135–150. doi:10.1016/S0304-3800(02)00064-9

Palacio, C., T. Blasingame. 1993. "Decline-curve analysis using type-curves – analysis of gas well production data," In: Proceedings, Low Permeability Reservoirs Symposium, Denver, Colorado, 26–28 April. doi:10.2118/25909-MS

Palagi, C., K. Aziz. 1994. "Use of Voronoi grid in reservoir simulation," *SPE Adv. Tech. Series*, 2(2):69–77.

Parada, C. H., T., Ertekin. 2012. "A new screening tool for improved oil recovery methods using artificial neural networks," In: Proceedings, SPE Western Regional Meeting, 19–23 March, Bakersfield, California. doi:10.2118/153321-MS

Quinlan, J. R. 1993. "C4.5: Programs for Machine Learning." Morgan Kaufmann Publishers, Burlington, MA, 302 pp.

Raissi, M., P. Perdikaris, G. E. Karniadakis. 2017. "Physics informed deep learning (Part I): data-driven solutions of nonlinear partial differential equations." http://arxiv-export-lb.library.cornell.edu/abs/1711.10561 (accessed 03 September 2021).

Sankaran, S., S. Matringe, M. Sidahmed, L. Saputelli, X. Wen, A. Popa, S. Dursun. 2020. "Data Analytics in Reservoir Engineering." SPE Publications, Richardson, TX, 95 pp.

Schuetter, J., S. Mishra, M. Zhong, R. LaFolette. 2018. "A data-analytics tutorial: building predictive models for oil production in an unconventional shale reservoir," *SPE J.*, 23(4):1075–1089. doi:10.2118/189969-PA

Sen, D., H. Chen, A. Datta-Gupta, A. 2022. "Inter-well connectivity detection in CO_2 WAG projects using statistical recurrent unit models." *Fuel*, 311, 122600. doi:10.1016/j.fuel.2021.122600

Sen, D., H. Chen, A. Datta-Gupta, J. Kwon, S. Mishra. 2021. "Machine learning based rate optimization under geologic uncertainty," *J. Pet. Sci. Eng.*, 207, 109116. doi:10.1016/j.petrol.2021.109116

Silva, P., M. Clio, D. Schiozer. 2007. "Use of neuro-simulation techniques as proxies to reservoir simulator: application in production history matching," *J. Pet. Sci. Eng.*, 57(3–4):273–280. doi:10.1016/j.petrol.2006.10.012

Song, X., Y. Liu, L. Xue, J. Wang, J. Zhang, J. Wang, L. Jiang, Z. Cheng. 2020. "Time-series well performance prediction based on long short-term memory (LSTM) neural network model," *J. Pet. Sci. Eng.*, 186, 106682. doi:10.1016/j.petrol.2019.106682

SPE. 2021. SPE Data Repository: Data Set: 1. From URL." https://www.spe.org/datasets/ (accessed 10 July 2021).

Sun, Q., T. Ertekin. 2020. "Screening and optimization of polymer flooding projects using artificial-neural-network (ANN) based proxies," *J. Pet. Sci. Eng.*, 185:106617. doi:10.1016/j.petrol.2019.106617

Thararoop, P., Z. Karpyn, A. Gitman, T. Ertekin. 2008. "Integration of seismic attributes and production data for infill drilling strategies - A virtual intelligence approach," *J. Pet. Sci. Eng.*, 63(1–4):43–52. doi:10.1016/j.petrol.2008.08.002

Voronoi, G. 1908. "Nouvelles applications des paramètres continus à la théorie des formes quadratiques. Deuxième mémoire. Recherches sur les parallélloèdres primitifs." *J. für die Reine und Angew. Math.*, 134:198–287.

Xue, L., Y. Liu, Y. Xiong, Y. Liu, X. Cui, G. Lei. 2021. "A data-driven shale gas production forecasting method based on the multi-objective random forest regression," *J. Pet. Sci. Eng.*, 196, 107801. doi:10.1016/j.petrol.2020.107801

Yalgin, G., N. Zarepakzad, E. Artun, I. Durgut, M. V. Kok. 2018. "Design and development of data-driven screening tools for enhanced oil recovery processes," In: Proceedings, SPE Western Regional Meeting, 22–27 April, Garden Grove, California. doi:10.2118/190028-MS

Yeten, B., L. Durlofsky, K. Aziz. 2003. "Optimization of nonconventional well type, location, and trajectory," *SPE J.*, 8(3):200–210. doi:10.2118/86880-PA

Yuan, Z., H. Huang, Y. Jiang, J. Li. 2021. "Hybrid deep neural networks for reservoir production prediction," *J. Pet. Sci. Eng.*, 197, 108111. doi:10.1016/j.petrol.2020.108111

Zangl, G., M. Giovannoli, M. Stundner. 2006. "Application of artificial intelligence in gas storage management," In: Proceedings, SPE EUROPEC/EAGE Annual Conference and Exhibition. 12–15 June. Vienna, Austria. doi:10.2118/100133-MS

Zarepakzad, N., E. Artun, I. Durgut. 2021. "A comparative analysis and rapid performance prediction of polymer flooding process by coupling reservoir simulation with neural networks," *Intl. J. Oil Gas Coal Tech.*, 27(3): 227–246. doi:10.1504/IJOGCT.2020.10034580

Zhang, Y., N. Bansal, Y. Fujita, A. Datta-Gupta, M. J. King, S. Sankaran. 2016. "From streamlines to fast marching: rapid simulation and performance assessment of shale-gas reservoirs by use of diffusive time of flight as a spatial coordinate," *SPE J.*, 21(5):1883–1898. doi:10.2118/168997-PA

Zhao, H., Z. Kang, X. Zhang, H. Sun, L. Cao, A. Reynolds. 2015. "INSIM: a data-driven model for history matching and prediction for waterflooding monitoring and management with a field application," In: Proceedings, SPE Reservoir Simulation Symposium, 23–25 February, Houston, Texas. doi:10.2118/173213-MS

Section V

Reservoir Modeling Applications

11 An Efficient Deep Learning Based Workflow Incorporating a Reduced Physics Model for Drainage Volume Visualization in Unconventional Reservoirs

Tsubasa Onishi[1,2], *Hongquan Chen*[1], *Akhil Datta-Gupta*[1], *and Srikanta Mishra*[3]
[1]Texas A&M University,
College Station, Texas
[2]Chevron Technical Center,
Houston, Texas
[3]Battelle Memorial Institute,
Columbus, Ohio

CONTENTS

DOI: 10.1201/9781003207009-16

11.1 INTRODUCTION

Unconventional reservoirs with significant reserves in North America (EIA 2021) have led to extensive research on the development and application of fast modeling approaches for well completion optimization and performance predictions. To date, many different approaches incorporating diverse data types and varying levels of physics have been proposed. For time-series data, such as production data, decline curve analysis (DCA) (Arps 1945, Fetkovich 1980, Carter 1985, Fetkovich et al. 1987, Blasingame et al. 1991, Palacio and Blasingame 1993, Agarwal et al. 1999, Ilk et al. 2008, Duong 2010, Valkó and Lee 2010), and pressure-transient/rate-transient analysis (PTA/RTA) (Wattenbarger et al. 1998, Ilk et al. 2010, Song and Ehlig-Economides 2011, Nobakht and Clarkson 2012a, 2012b, Uzun et al. 2016, He et al. 2018) have been widely used for forecasting and reservoir characterization. While these methods provide significant benefits in terms of simplicity and computational efficiency, capturing the complex physics in unconventional reservoirs and heterogeneities in hydraulic fracture and reservoir properties, remains a challenge.

Numerical simulations are capable of capturing the complex nonlinear dynamics in unconventional reservoirs. Multiple continuum models (Warren and Root 1963, Kazemi et al. 1976), unstructured discrete fracture models (Cipolla et al. 2011, 2012, Sun and Schechter 2015, Marongiu-Porcu et al. 2016), multi-segment wells (Du et al. 2016), embedded discrete fracture models (EDFMs) (Li and Lee 2008, Moinfar et al. 2014, Ţene et al. 2017), and hybrid models (Jiang and Younis 2016, Yang et al. 2018) are some of the notable numerical models. Rather than the simple analytical models focusing on time-series data, the capability of the numerical models that allow us to "see" within the subsurface by visualizing the simulation results, such as pressure and stress fields, is particularly useful to better design field development plans including well spacing and treatment designs. However, numerical simulation for unconventional reservoirs involving complex multiphase flow and geomechanics in high-contrast systems with hydraulic and natural fractures can be computationally intensive involving history-matching, optimization, and uncertainty quantification.

To circumvent full physics simulations, the use of reduced physics models has been extensively investigated. Among many methods, a notable method for simulating unconventional reservoirs is the fast-marching method (FMM)-based rapid simulation (Zhang et al. 2016). This method aims to achieve faster computations via a coordinate transformation using the diffusive time of flight (DTOF). The DTOF represents the propagation time of the peak pressure disturbance caused by an impulse source and is given by a solution of the Eikonal equation, which governs the propagation of pressure fronts and generalizes the depth of investigation for heterogeneous reservoirs (Datta-Gupta et al. 2011). The Eikonal equation can be efficiently solved by the FMM, a class of front tracking algorithms (Sethian 1999). Once the DTOF is obtained, the original 2D or 3D problems can be transformed into equivalent 1D problems along the DTOF coordinate, leading to significant savings in computation time. The FMM-based simulation was originally developed for shale-gas reservoirs (Zhang et al. 2016), and has been extended to account for more complex problems including multiphase and multi-component flow (Fujita et al. 2016, Iino et al. 2017), non-isothermal flow (Cui et al. 2016), gravity (Onishi et al. 2020), as well as complex grid systems, such as unstructured grid (Yang et al. 2017) and EDFM (Xue et al. 2019). The robustness and efficiency of the method has been demonstrated by multiple field-scale applications (Iino et al. 2020a, Iino et al. 2020b).

The success of deep learning in a variety of fields including computer vision, natural language processing, and many others (Goodfellow et al. 2016) has led to the use of such techniques in subsurface modeling. In the context of reservoir simulation, a notable work was presented by Tang et al. (2020) who applied a residual U-Net (Ronneberger et al. 2015) and a convolutional long short-term memory recurrent neural network (Hochreiter and Schmidhuber 1997) to develop a surrogate model that predicts state maps and flow responses at wells given permeability maps in 2D channelized reservoir models. The developed surrogate model for oil-water two-phase flow was successfully applied to a probabilistic history-matching workflow. Recently, there has been intensive research efforts investigating the use of deep learning based methods to solve forward and inverse problems in science and engineering applications. Notable works that brought a new paradigm of data-driven modeling have been presented by Raissi et al. (2019) and Zhu et al. (2019). The method is often referred to as physics-informed neural network (PINN) or physics-informed machine learning (PIML). While standard deep learning models are typically trained solely using data, the loss function in the PIML framework consists of data misfit term(s) as well as regularization term(s) based on some physical laws, which are typically residuals of governing partial differential equations to ensure that the trained model is consistent with the underlying physical laws. In the field of reservoir modeling, applications of the PIML framework to the Buckley Leverett problem (Buckley and Leverett 1942) have been investigated by several authors (Fraces et al. 2020, Fuks 2020, Xu et al. 2021). Despite the widespread use of deep learning based methods for subsurface modeling in recent years, the applications are mostly limited to simple cases like two-phase flows in 2D or simple 3D grid systems. The challenge here is that the workflow becomes computationally expensive as the complexity of the problem increases in terms of size in space and time and underlying nonlinearities. There are two components contributing to the expensive run time: generating datasets for training by running a large number of full physics simulations and the training (solving the minimization problem) to find optimal weights and biases of the neurons. The latter tends to be more evident in the PIML framework due to the physics term(s) in the loss function that typically results in slower convergence, as in other minimization problems with regularizations.

In this chapter, we put forth a novel deep learning based workflow incorporating a reduced physics model that can efficiently visualize the well drainage volume and pressure propagation in unconventional reservoirs. The workflow generates a subsurface image (e.g., pressure propagation) given routinely measured data from the field (e.g., production data, distributed data along horizontal wells, etc.). The expensive run time of full physics simulations is addressed by replacing the forward model by the FMM-based rapid simulation. Furthermore, we leverage an autoencoder for dimensionality reduction to further enhance the workflow. The architecture of the deep learning model is designed to be simple so that the choices of the forward model and measured data are flexible. The proposed approach is simple, efficient, and applicable to 3D field-scale unconventional reservoir models that involve a large number of hydraulic fractures and multiphase flow.

11.2 METHODOLOGY

To illustrate our proposed approach, we consider an example where measured data, including multiphase production rates, bottom hole pressure, and distributed flow rate along a multi-fractured horizontal well in an unconventional reservoir are available. Our aim is to develop an efficient workflow that allows us to image the evolution of the flow field given the measured data (Figure 11.1).

The most straightforward machine learning approach to achieve our objective will be a deep learning based regression model using a combination of feed-forward neural network and convolutional neural networks (CNNs). Specifically, the inputs are first passed to the feedforward neural network containing several hidden layers. The last hidden layer is connected to a stack of CNNs that eventually outputs a 3D subsurface image. This simple architecture, however, may be difficult to train because of the high-output dimension that corresponds to the large number time steps and corresponding spatial images of the flow field (for example, pressure and saturation maps). In practical applications, it can easily

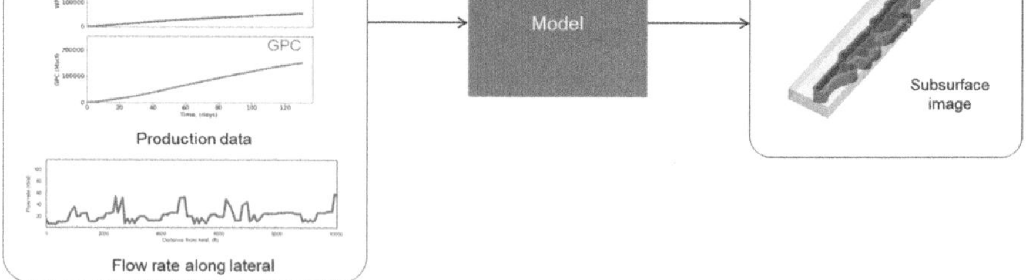

FIGURE 11.1 Overview of the proposed approach.

exceed the order of millions. A viable remedy would be the use of dimensionality reduction techniques, such as principal component analysis (PCA). PCA-based techniques have found various applications in geological parameterization (Oliver 1996, Reynolds et al. 1996, Sarma et al. 2006). A well-known limitation of the PCA-based parameterization techniques is for more complex parameter distributions, such as those found in permeability distributions in channelized reservoirs (non-Gaussian). Several extensions including the kernel PCA (KPCA) (Sarma et al. 2008) and the optimization-based PCA (O-PCA) (Vo and Durlofsky 2014) have been proposed. Recently, applications of deep learning based generative models, such as (variational) autoencoders (Canchumuni et al. 2017, Laloy et al. 2017), generative adversarial network (GAN) (Chan and Elsheikh 2017, Laloy et al. 2018), and CNN-PCA (Liu et al. 2019), to geological parameterization are being actively explored. These techniques are essentially nonlinear generalizations of PCA (Baldi and Hornik 1989) and have been shown to outperform the existing PCA-based techniques in capturing complex features. In our application, the problem settings involve high-contrast systems due to hydraulic and natural fractures and hence, and the output images are likely to exhibit sharp contrasts (similar to channelized reservoirs). Therefore, we adopt an autoencoder whose capability in capturing complex features is well suited for our problems of interest. Among the deep learning based generative methods, an autoencoder is one of the simplest methods.

A step-by-step workflow of the proposed approach is summarized in Figure 11.2, which is based on the well-known Design of Experiment (DoE)-based techniques (Bhark and Dehghani 2014, Olalotiti-Lawal et al. 2018, Li et al. 2019). More detailed descriptions of the workflow are as follows:

1. An objective function representing the data misfit is defined and parameters that might have influence on the objective function are identified, typically based on prior knowledge and engineering judgments. In this work, the objective function is defined as root-mean-square error over n_d data points,

$$O_j(\mathbf{m}) = \sqrt{\sum_{i=1}^{n_d} \left(d_i^{obs} - d_i^{sim}(\mathbf{m})\right)^2 / n_d} \tag{11.1}$$

where $O_j(\mathbf{m})$ is the objective function for metric j (oil production rate [OPR], water production rate [WPR], gas production rate [GPR], cumulative oil production [OPC], etc., at the well) given parameter $\mathbf{m}$, d_i^{obs} is observed data, and $d_i^{sim}(\mathbf{m})$ is the simulation response given parameter $\mathbf{m}$ at step i. In this study, the one-variable-at-a-time design is applied to generate tornado plots, from which the influential parameters are selected for the subsequent steps. It should be noted that the simple tornado plots have limitations as interactions between parameters may not be accounted for (Li et al. 2019). Nevertheless, they provide a good starting point for parameter screening.

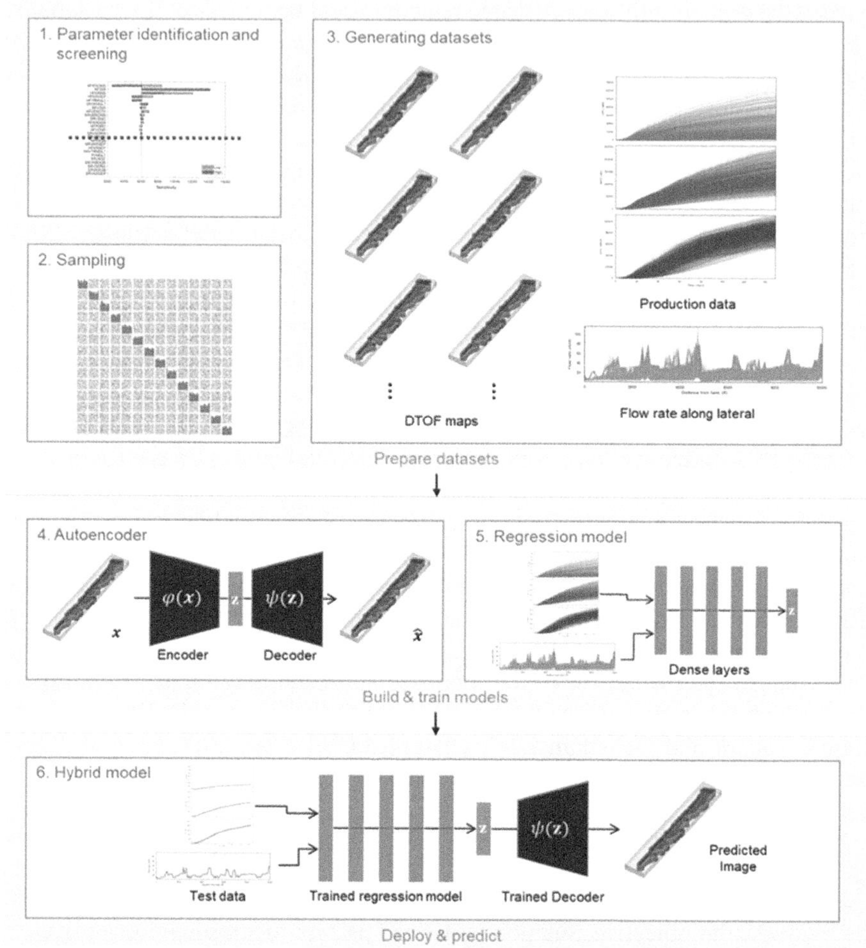

FIGURE 11.2 The proposed deep learning based workflow incorporating the FMM-based rapid simulation.

2. The selected influential parameters from step 1 will be used in sampling via a Latin hypercube design (McKay et al. 2000) to generate cases for the FMM-based simulation. The number of samples is typically in the order of hundreds or thousands, depending on the complexity of the problem.
3. The FMM-based rapid simulation is applied to the samples from step 2 to generate datasets for training and validation. The rapid simulation capability allows users to explore a variety of parameters, even for field-scale applications within realistic simulation times that would be difficult to achieve with full physics simulations.
4. The DTOF maps are used to train an autoencoder, which learns a non-linear mapping of an input image ($\mathbf{x}_i$) to a low-dimensional latent space

via $\mathbf{z}_i = \varphi(\mathbf{x}_i)$ (encoder), then back via $\hat{\mathbf{x}}_i = \psi(\mathbf{z}_i)$ (decoder) by solving the minimization problem,

$$\underset{\theta_{AE}}{\arg\min} \frac{1}{n_s} \sum_{i=1}^{n_s} \left\| \mathbf{x}_i - \hat{\mathbf{x}}_i \right\|_2^2 \tag{11.2}$$

where θ_{AE} are the parameters to be tuned (weights and biases) in the autoencoder during the training, and n_s is the number of samples. The loss function is defined as a mean square error between an original DTOF image ($\mathbf{x}_i$) and a reconstructed DTOF image ($\hat{\mathbf{x}}_i$). Here, the DTOF τ_i is normalized to $\mathbf{x}_i$ [0 1] using minimum and maximum values of τamong the training and validation datasets,

$$\mathbf{x}_i = \frac{\tau_i - \tau_{\min}}{\tau_{\max} - \tau_{\min}} \tag{11.3}$$

The encoder and decoder include CNNs and dense layers, and more details of the architecture will be discussed in the following sections. In our application, instead of multiple pressure and saturation maps, a single DTOF map is used as a representative subsurface image over the simulation period. This means that each sample has only one image of DTOF, rather than a set of pressure and saturation maps over the time steps. The intention behind this is as follows:

- The use of a single DTOF image per realization is a reasonable simplification, as reported by Iino et al. (2020b), and it allows us to work with much simpler autoencoder architecture compared with the use of a set of the state maps that may require more complex architectures, making the training more difficult and time-consuming.
- Once we predict the DTOF, it can be readily used as a 1D spatial coordinate in the rapid FMM-based simulation, from which we can obtain pressure and saturation maps. The run time of the FMM-based simulation is typically less than a few minutes. Alternatively, we could use a more complex autoencoder for a set of pressure and saturation maps to directly predict them. However, the computational cost for the training is expected to be more expensive.

5. In this step, we train a deep learning based regression model consisting of a feedforward neural network that predicts a vector of the latent variable given the measured data (e.g., production data and flow rate along a lateral) by solving the following minimization problem,

$$\underset{\theta_{reg}}{\arg\min} \frac{1}{n_s} \sum_{i=1}^{n_s} \left\| \mathbf{z}_i - \hat{\mathbf{z}}_i \right\|_2^2 \tag{11.4}$$

where θ_{reg} are the weights and biases on the neurons in the regression model, $\mathbf{z}$ is the known latent variable (normalized to [0 1] in a similar manner to Eq. (11.3)) from the trained autoencoder, and $\hat{\mathbf{z}}$ is the predicted latent variables. While the multiphase production data and flow rate along the lateral are collected and saved for this step in Figure 11.2, the workflow has flexibility to incorporate different data types, and it is applicable to various asset types and associated measured data.

6. We have already trained and saved the autoencoder (step 4) and the regression model (step 5) from the previous steps. In this step, the two models are combined to construct a hybrid model that consists of the regression model and the decoder (part of the autoencoder). The hybrid model predicts a DTOF map given the measured data. An extension of the workflow to account for prediction uncertainties will be discussed in the next section.

It should be noted that there are recent works in the literature that present deep learning based methods combining a regression model with an autoencoder (Champion et al. 2019, Razak and Jafarpour 2020) and an approach combining a reduced physics model and a response surface design (Tanaka et al. 2020), and part of our workflow is inspired by them. The novelty of our approach is the use of the FMM-based simulation and the concept of the DTOF that substantially accelerates the workflow and simplifies the machine learning architecture, making it suitable for field-scale applications. Additionally, the flexible workflow is applicable to various types of measured data.

11.3 FIELD APPLICATION

In this section, we present an application of our approach to an unconventional reservoir to demonstrate practicability of the proposed approach.

11.3.1 Case Description

The case is based on a field-scale hydraulic fracturing research program in the West Texas Midland Basin, the HFTS-I (Ciezobka and Reeves 2021). The primary targets of the study are 11 horizontal wells with 400+ stages drilled in the upper and middle Wolfcamp formation. A variety of field data, such as core data and microseismic surveys, has been made publicly available by the operators. In our application, we focus on one of the wells, and build a sector geologic model (Figure 11.3a). The size of the model is 1886.5 × 11,237 × 618 ft^3 with 23 × 137 × 47 grid discretization for the matrix domain. More than 100 hydraulic fractures are added on top of the matrix domain via the EDFM approach, and their lateral extent and orientations are set based on the available microseismic data (Maity 2018, Stegent and Candler 2018). The fluid model is the same as the one described in Onishi et al. (2021). For the matrix domain, homogeneous permeability and porosity ($k_m = 0.05$ nD and $\phi_m = 0.05$) are assumed, and uniform values of 3,000 psi, 0.3, 0, and 0.987 Mscf/stb are assigned for initial pressure, water saturation, gas

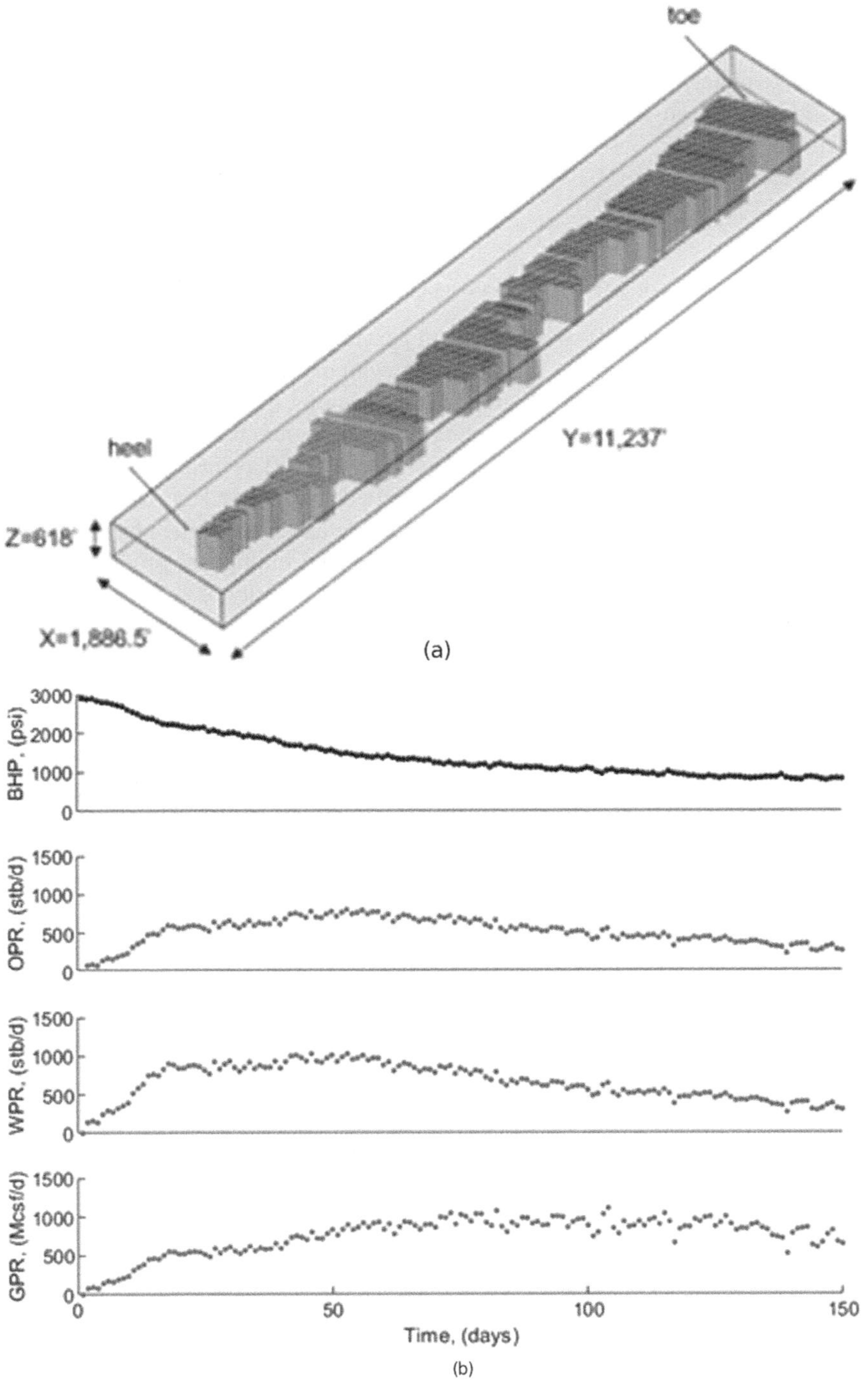

FIGURE 11.3 (a) Field-scale reservoir with 100+ hydraulic fractures described by EDFMs and (b) the observed data.

saturation, and solution gas-oil ratio. These values are based on the available data and a previous simulation study Forrest et al. (2018). The bubble point pressure is approximately 2,800 (psi); therefore, the reservoir is initially undersaturated. In addition, because limited observed data during production phase is publicly available, synthetic observed data, including multiphase production and bottom hole pressure data for 150 days, is generated with a 3D reference simulation model (Figure 11.3b).

11.3.2 Parameter Identification and Sampling

The first step is to identify an initial set of parameters that are thought to be influential on the observed data. Based on the available measurements and reports, uniform multipliers for the hydraulic fracture lengths, conductivity, and porosity; relative permeability-related parameters; and parameters for rock compaction for both hydraulic fractures and SRV regions were selected as uncertain parameters. Next, a one-variable-at-a-time design is applied to identify key parameters that will be used for sampling to generate training datasets. Here, the objective functions were defined as differences between computed multiphase production rates and the observed data (Eq. (11.1)). As a result, the majority of parameters for the hydraulic fractures and several parameters for the stimulated reservoir volume (SRV) region were found to be sensitive. Because the number of identified key parameters was still relatively large, we performed a global history-matching, where large-scale parameters, such as uniform multipliers for the hydraulic fracture length and the reservoir properties, were tuned using the multi-objective genetic algorithm (MOGA), to further reduce the number of parameters and to obtain a reasonable starting point for the next step. In the history-matching, we validated the FMM-based rapid simulation by comparing simulation results with 3D simulation and used it as a fast forward model during the history-matching. The run time of the FMM-based simulations is 2–3 minutes on average, which is roughly a fivefold reduction compared with the 3D simulation. The speedup factor becomes more significant as size and complexity of the problem increase (Iino et al. 2020b). Although the history-matching was performed in this application, it is not a requirement for our proposed workflow. An application of our workflow without history-matching is presented in our recent work (Onishi et al. 2021). After the history-matching, most of the parameters were well constrained except for hydraulic fracture conductivity and exponents for oil-water relative permeability curves in the hydraulic fracture region, which remained relatively uncertain. Following Figure 11.2, the next step is to conduct parameter sampling. The parameters of interest include the three relatively uncertain parameters from the history-matching and more local parameters for the hydraulic fracture lengths. According to the available data, there are 37 stages along the lateral length, and 11 of them are representative stages, at which measurement errors are minimal, and the data exhibits a high degree of confidence (Warpinski 2009, Stegent and Candler 2018). While lengths of the hydraulic fractures were uniformly tuned during the history-matching, in the next step, the hydraulic fractures will be adjusted using stage-level length multipliers [0.75 1.25] for each of the representative stages (×11 multipliers). Subsequently, 2,000 samples were generated through the application of a Latin hypercube design.

11.3.3 Generating Datasets

The FMM-based simulation is applied to all the 2,000 samples to generate datasets. The simulations are constrained by the observed bottom hole pressure. The DTOF maps, multiphase production data, and distributed flow rate data along the lateral at a given time, which can be obtained from production logging tools in practice, are saved for the trainings as shown in Figure 11.4. In regard to multiphase production responses, we observe wider ranges in oil and water production than gas production. This is likely because the exponents for oil-water relative permeability curves are the parameters of interest, whereas fixed oil-gas relative permeability curves are used for all the samples. The other observation here is that the flow rate profiles show higher values at stages with longer length, because of greater surface area of the hydraulic fractures. The samples are divided into 1,400, 300, and 300, for training, validation, and test, respectively.

11.3.4 Training an Autoencoder

Next, we trained an autoencoder. Architecture of the autoencoder includes CNNs and dense layers as shown in Figure 11.5, and more detailed settings are provided in Appendix A. The 3D CNNs that are typically used with 3D image data, such as images from computerized tomography (CT) scans (Zheng et al. 2020), are employed for the 3D DTOF maps.

For the training, batch size of 256, the maximum number of epochs of 10,000, and patience of 50, are used. The other settings and the procedures are the same as the example case. The run time for the training process was approximately 110 minutes, which had stopped at around 4,800 epochs. The trained model is then applied to the test dataset (300 samples). As a result, an overall relative error of 4.92e-3 is obtained. The low relative error value validates accuracy of the trained autoencoder.

11.3.5 Training a Regression Model

Next, a regression model is trained to relate the measured data and the latent variables. Architecture of the model is displayed in Figure 11.6. In this application, the dimensions of the input data are different, specifically, the dimensions of the production data and the distributed flow rate data correspond to the number of time steps and the number of data points along the lateral. To accommodate the two different data types, two separate input layers are used for each of them. The input layers are connected with the intermediate layers that will be

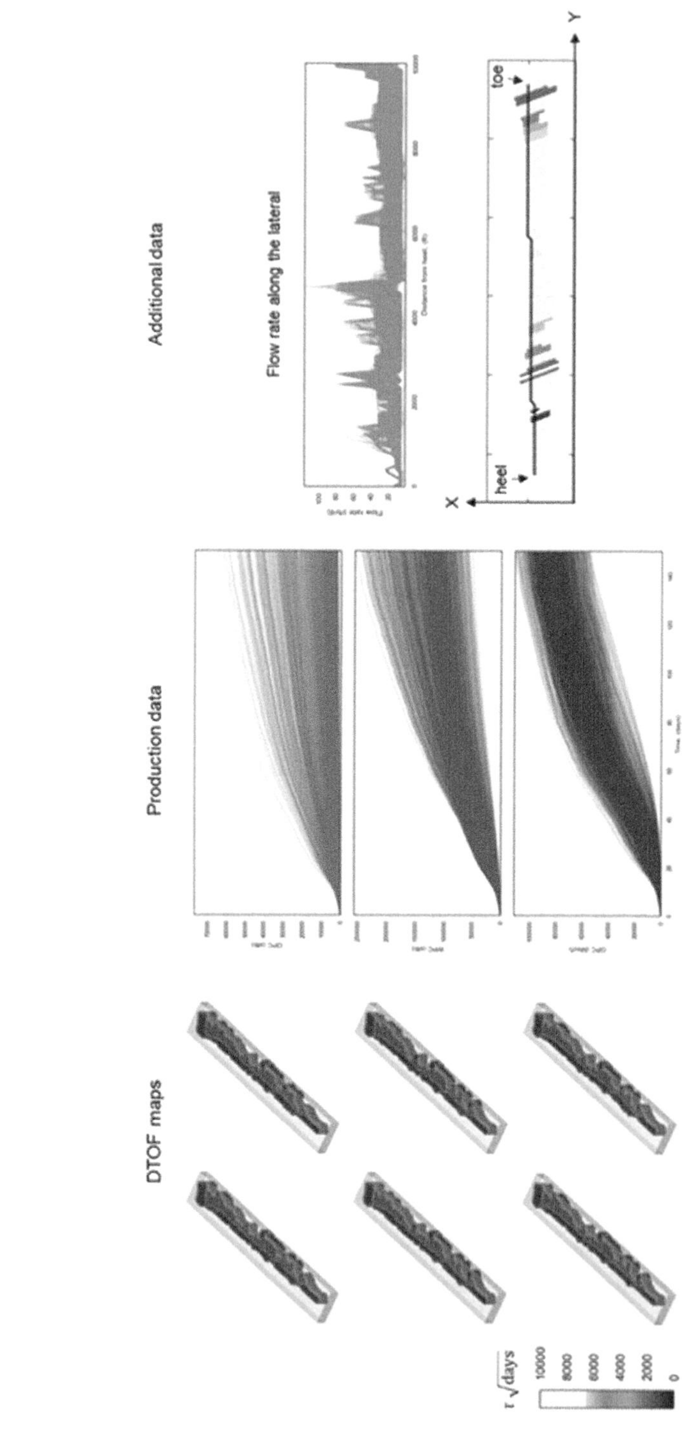

FIGURE 11.4 Datasets generated using the FMM-based rapid simulation for the field-scale application.

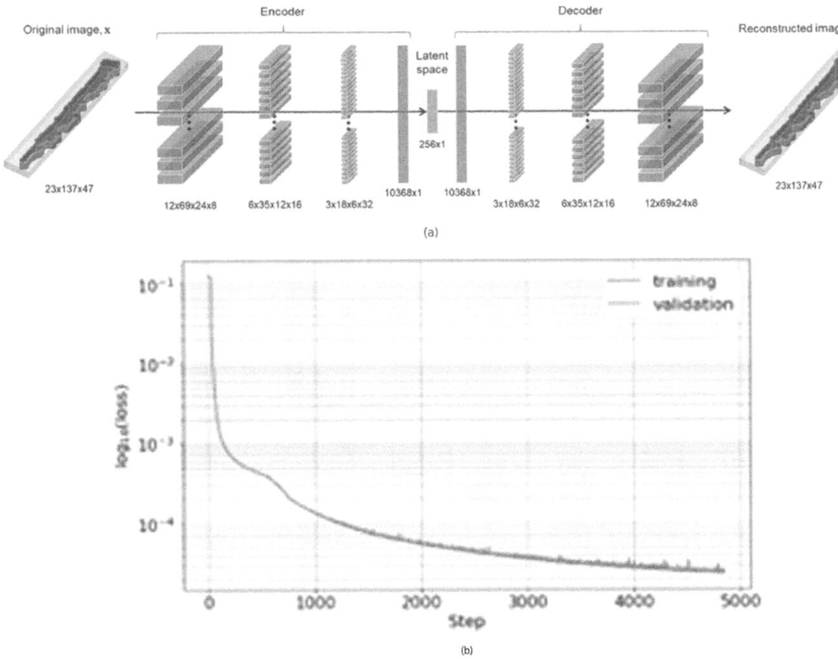

FIGURE 11.5 (a) Schematic of the autoencoder for the field-scale application and (b) model loss on training and validation datasets.

concatenated to form a single hidden layer connected with a set of other hidden layers. More details of the architecture are provided in Appendix A. The training procedure is the same as the example case and is finished at around 150 epochs, as shown in Figure 11.6. The run time was approximately 38 seconds.

11.3.6 Prediction of DTOF Maps

Finally, the trained regression model and the decoder are combined to build a hybrid model. The hybrid model is applied to the measured data from the 3D reference simulation model. Figure 11.7a shows architecture of the deep learning model and a comparison between the predicted DTOF map and the true DTOF map. The predicted DTOF map visually shows good agreement with the true DTOF. The relative error between the predicted DTOF and true DTOF maps was 1.91e-3.

At this point, the prediction is deterministic, and we would like to obtain multiple plausible realizations of DTOF maps to account for subsurface uncertainty. To this end, we aim to find samples among the training and validation datasets that are similar to the predicted DTOF map using multidimensional scaling (MDS). The MDS is a technique to project a dissimilarity matrix consisting of pairwise dissimilarities between individuals associated with high-dimensional data into a configuration of points in n-dimensional Euclidian space (Borg and Groenen 2005). The visually intuitive representation of the individual realizations in a

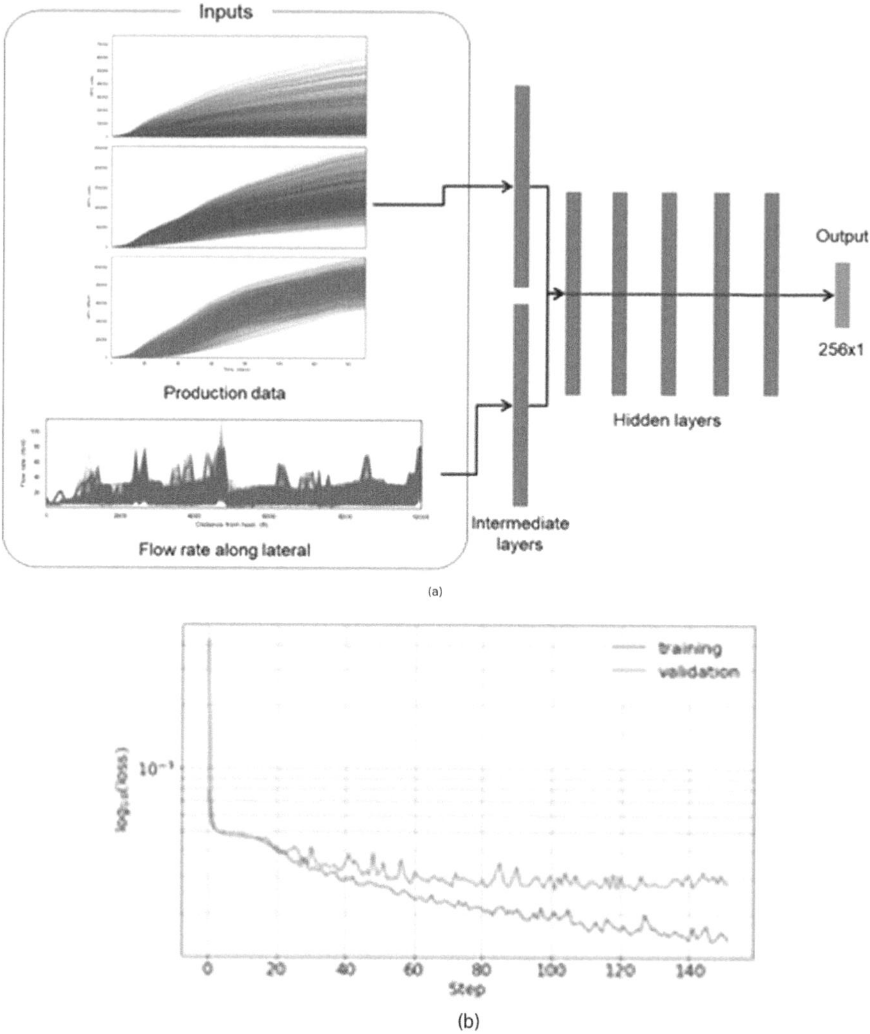

FIGURE 11.6 (a) Architecture of the regression model and (b) model convergence.

low-dimensional map is useful for understanding dissimilarities between them. In the context of subsurface modeling, MDS has found many applications, for example, uncertainty quantification in reservoir performance (Scheidt and Caers 2009). In our application, we utilized MDS to map the training and validation samples together with the test case into a low-dimensional (2D) space, whereby the measure of dissimilarity to compute a dissimilarity matrix is values of the latent variables (**z**) associated with each of the samples. The resultant 2D map is shown in Figure 11.7b, in which each point represents a sample, and distances between points indicate dissimilarities between the samples in terms of the values of latent variable (**z**). The data points in close proximity to the X in the figure

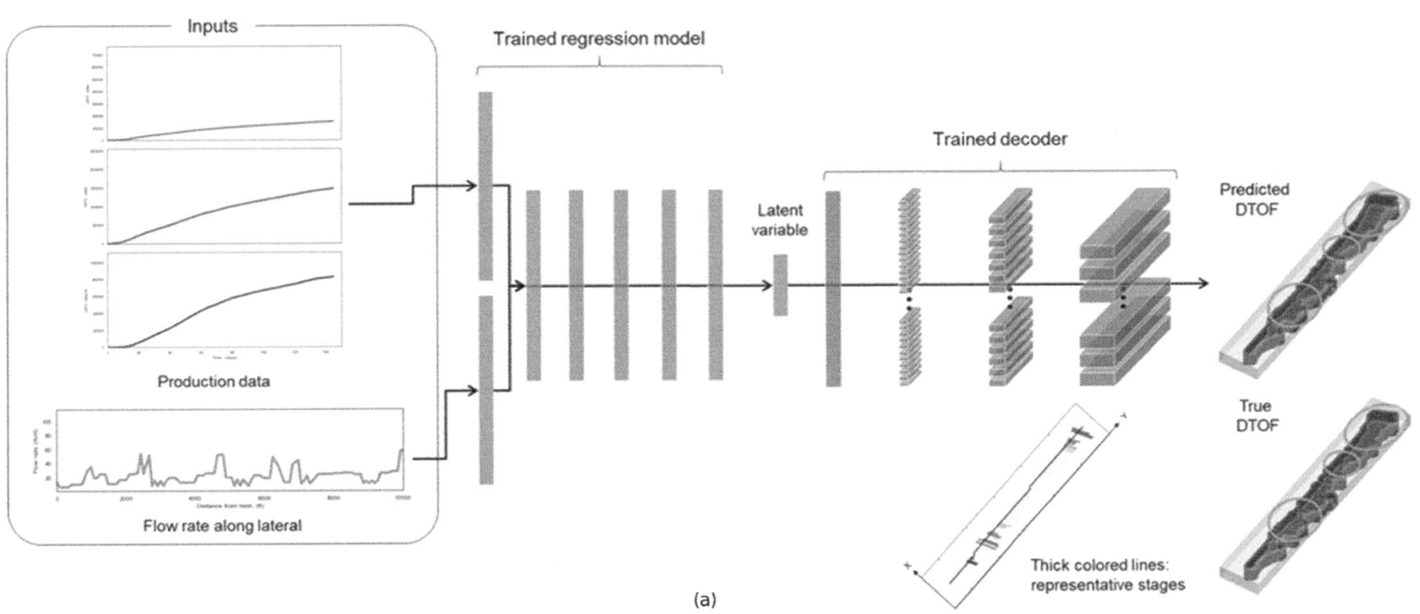

FIGURE 11.7 (a) Architecture of the hybrid model for the field-scale application and (b) prediction of DTOF maps accounting for subsurface uncertainty: 2D MDS space where each point represents a sample. *(Continued)*

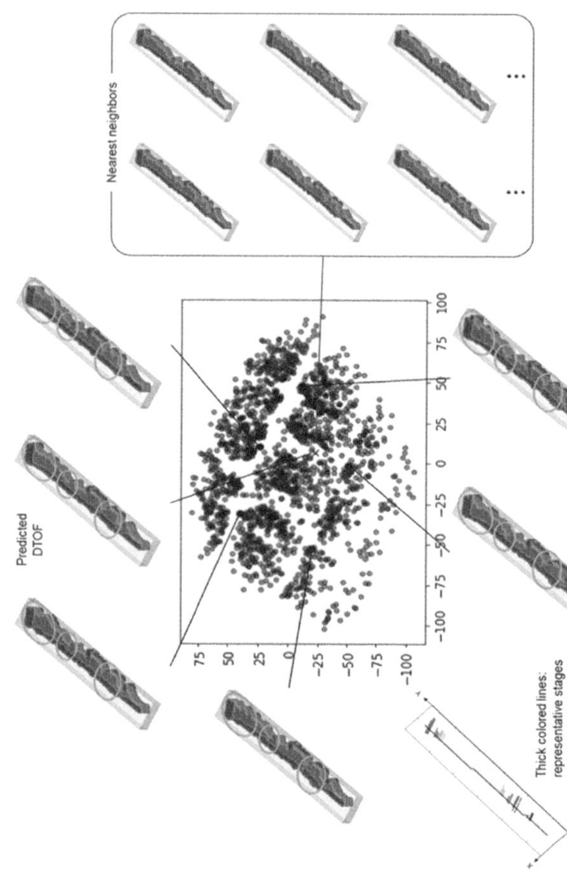

FIGURE 11.7 *(Continued)*

light gray points) indicate similarity of the latent variables (**z**) for the DTOF maps. In contrast, data points that are located further apart have very little similarity, especially within the circled region of the DTOF maps. The nearest neighbors can be viewed as multiple plausible realizations of subsurface images in this example and can be utilized for a variety of applications, such as visualization and rapid simulations for forecasting, and uncertainty quantification. For example, an application the workflow to quickly evaluating data acquisition programs has been demonstrated by Onishi et al. (2021).

11.4 CONCLUSIONS AND LESSONS LEARNED

We presented a deep learning based workflow incorporating the FMM-based rapid simulation. The workflow has been successfully applied to a 3D field example. The following conclusions and lessons can be drawn from this study.

- A novel deep learning based workflow that can efficiently visualize subsurface flow has been developed. The approach incorporates FMM-based rapid simulation and the concept of DTOF to accelerate the workflow while capturing key physics. Specifically, training datasets can be efficiently generated using the FMM-based rapid simulation, and the use of DTOF as a representative image allows us to use simple and robust architectures of the deep learning based models.
- The dimensionality reduction via an autoencoder and the use of latent variables for regression modeling further enhances the proposed workflow and facilitates large scale field application.
- A field-scale application of the approach to subsurface imaging accounting for subsurface uncertainties was presented. The successful application demonstrated the robustness and efficacy of the approach.

11.5 FUTURE OUTLOOK

- While the multiphase production data and distributed flow rate data along the lateral length were considered in the application, with the flexibility of deep learning based model architectures, the workflow can be readily extended to incorporate other data types. An interesting data type to investigate would be higher dimensional data, such as distributed temperature sensing data.
- Finally, to further improve our proposed workflow, the parameter screening procedure could use more sophisticated statistical analysis, such as *t*-tests, instead of the one-variable-at-a-time design. For deep learning models, we adopted an autoencoder and a relatively simple feedforward neural network. Replacing part of the architecture with more advanced methods could potentially enhance the workflow. For example, the use of a probabilistic model (e.g., a variational autoencoder) alone can account for subsurface uncertainty, and the postprocessing step using MDS would be no longer necessary.

ACKNOWLEDGMENT

This work is partly supported by the U.S. Department of Energy initiative on Science Informed Machine Learning for Accelerating Real Time Decisions for Subsurface Applications (SMART) and Texas A&M Joint Industry Project "Model Calibration and Efficient Reservoir Imaging (MCERI)." The views and opinions of authors expressed herein do not necessarily state or reflect those of the United States Government or any agency thereof.

REFERENCES

Agarwal, Ram G, Gardner, David C, Kleinsteiber, Stanley W et al. 1999. Analyzing well production data using combined-type-curve and decline-curve analysis concepts. *SPE Reservoir Evaluation & Engineering* **2** (5): 478–486.

Arps, Jan J. 1945. Analysis of decline curves. *Transactions of the AIME* **160** (1): 228–247.

Baldi, Pierre and Hornik, Kurt. 1989. Neural networks and principal component analysis: Learning from examples without local minima. *Neural Networks* **2** (1): 53–58.

Bhark, Eric and Dehghani, Kaveh. 2014. Assisted history matching benchmarking: Design of experiments-based techniques. *Proceedings, SPE Annual Technical Conference and Exhibition*. https://doi.org/10.2118/170690-MS.

Blasingame, TA, McCray, TL, and Lee, WJ. 1991. Decline curve analysis for variable pressure drop/variable flowrate systems. *Proceedings, SPE Gas Technology Symposium*.

Borg, Ingwer and Groenen, Patrick JF. 2005. *Modern multidimensional scaling: Theory and applications*. Springer Science & Business Media.

Buckley, SE and Leverett, MC. 1942. Mechanism of fluid displacement in sands. *Transactions of the AIME* **146** (1): 107–116.

Canchumuni, Smith Arauco, Emerick, Alexandre A, and Pacheco, Marco Aurelio. 2017. Integration of ensemble data assimilation and deep learning for history matching facies models. *OTC Brasil*.

Carter, Robert D. 1985. Type curves for finite radial and linear gas-flow systems: constant-terminal-pressure case. *Society of Petroleum Engineers Journal* **25** (5): 719–728.

Champion, Kathleen, Lusch, Bethany, Kutz, J Nathan et al. 2019. Data-driven discovery of coordinates and governing equations. *Proceedings of the National Academy of Sciences* **116** (45): 22445–22451.

Chan, Shing and Elsheikh, Ahmed H. 2017. Parametrization and generation of geological models with generative adversarial networks. *arXiv preprint arXiv:170801810*.

Ciezobka, Jordan and Reeves, Scott. 2021. Overview of hydraulic fracturing test sites (HFTS) in the Permian basin and summary of selected results (HFTS-I in Midland and HFTS-II in Delaware). *Proceedings, Latin America Unconventional Resources Technology Conference*, 16–18 November 2020, 93–102.

Cipolla, Craig L, Fitzpatrick, Tony, Williams, Michael John et al. 2011. Seismic-to-simulation for unconventional reservoir development. *Proceedings, SPE Reservoir Characterisation and Simulation Conference and Exhibition*.

Cipolla, Craig L, Maxwell, Shawn C, and Mack, Mark Gavin. 2012. Engineering guide to the application of microseismic interpretations. *Proceedings, SPE Hydraulic Fracturing Technology Conference*.

Cui, Jingyuan, Yang, Changdong, Zhu, Ding et al. 2016. Fracture diagnosis in multiple-stage-stimulated horizontal well by temperature measurements with fast marching method. *SPE Journal* **21** (6): 2289–2300.

Datta-Gupta, Akhil, Xie, Jiang, Gupta, Neha et al. 2011. Radius of investigation and its generalization to unconventional reservoirs. *Journal of Petroleum Technology* **63** (07): 52–55. https://doi.org/10.2118/0711-0052-JPT.

Du, Song, Yoshida, Nozomu, Liang, Baosheng et al. 2016. Application of multi-segment well approach: Dynamic modeling of hydraulic fractures. *Journal of Natural Gas Science and Engineering* **34**: 886–897.

Duong, Anh N. 2010. An unconventional rate decline approach for tight and fracture-dominated gas wells. *Proceedings, Canadian Unconventional Resources and International Petroleum Conference.*

EIA. 2021. Proved Reserves of Crude Oil and Natural Gas in the United States, Year-End 2019.

Fetkovich, MJ. 1980. Decline curve analysis using type curves. *Journal of Petroleum Technology* **32** (6): 1065–1077. https://doi.org/10.2118/4629-PA.

Fetkovich, MJ, Vienot, ME, Bradley, MD et al. 1987. Decline curve analysis using type curves: Case histories. *SPE Formation Evaluation* **2** (4): 637–656. https://doi.org/10.2118/13169-PA.

Forrest, James, Zhu, Tiejun, Xiong, Hongjie et al. 2018. The effect of initial conditions and fluid PVT properties on unconventional oil and gas recoveries in the Wolfcamp formation in the Midland basin. *Proceedings, Unconventional Resources Technology Conference*, Houston, TX, 23–25 July 2018, 895–908.

Fraces, Cedric G, Papaioannou, Adrien, and Tchelepi, Hamdi. 2020. Physics informed deep learning for transport in porous media. Buckley Leverett problem. *arXiv preprint arXiv:200105172.*

Fujita, Yusuke, Datta-Gupta, Akhil, and King, Michael J. 2016. A comprehensive reservoir simulator for unconventional reservoirs that is based on the Fast Marching method and diffusive time of flight. *SPE Journal* **21** (6): 2276–2288.

Fuks, Olga. 2020. *Physics informed machine learning and uncertainty propagation for multiphase transport in porous media*. Stanford University.

Goodfellow, Ian, Bengio, Yoshua, Courville, Aaron et al. 2016. *Deep learning*, Vol. 1. MIT Press Cambridge.

He, Youwei, Cheng, Shiqing, Rui, Zhenhua et al. 2018. An improved rate-transient analysis model of multi-fractured horizontal wells with non-uniform hydraulic fracture properties. *Energies* **11** (2): 393.

Hochreiter, Sepp and Schmidhuber, Jürgen. 1997. Long short-term memory. *Neural Computation* **9** (8): 1735–1780. https://doi.org/10.1162/neco.1997.9.8.1735.

Iino, Atsushi, Jung, Hye Young, Onishi, Tsubasa et al. 2020a. Rapid simulation accounting for well interference in unconventional reservoirs using fast marching method. *Proceedings, Unconventional Resources Technology Conference*, 20–22 July 2020, 4725–4743.

Iino, Atsushi, Onishi, Tsubasa, and Datta-Gupta, Akhil. 2020b. Optimizing CO_2-and field-gas-injection EOR in unconventional reservoirs using the fast-marching method. *SPE Reservoir Evaluation & Engineering* **23** (1): 261–281.

Iino, Atsushi, Vyas, Aditya, Huang, Jixiang et al. 2017. Rapid compositional simulation and history matching of shale oil reservoirs using the fast marching method. *Proceedings, Unconventional Resources Technology Conference*, Austin, TX, 24–26 July 2017, 2834–2854.

Ilk, D, Anderson, DM, Stotts, GWJ et al. 2010. Production-data analysis—challenges, pitfalls, diagnostics. *SPE Reservoir Evaluation & Engineering* **13** (3): 538–552. https://doi.org/10.2118/102048-PA.

Ilk, Dilhan, Rushing, Jay Alan, Perego, Albert Duane et al. 2008. Exponential vs. hyperbolic decline in tight gas sands: Understanding the origin and implications for reserve estimates using Arps' decline curves. *Proceedings, SPE Annual Technical Conference and Exhibition.*

Jiang, Jiamin and Younis, Rami M. 2016. Hybrid coupled discrete-fracture/matrix and multicontinuum models for unconventional-reservoir simulation. *SPE Journal* **21** (3): 1009–1027.

Kazemi, H., Merrill, L.S., Jr., Porterfield, K.L. et al. 1976. Numerical simulation of water-oil flow in naturally fractured reservoirs. *Society of Petroleum Engineers Journal* **16** (6): 317–326. https://doi.org/10.2118/5719-PA.

Laloy, Eric, Hérault, Romain, Jacques, Diederik et al. 2018. Training-image based geostatistical inversion using a spatial generative adversarial neural network. *Water Resources Research* **54** (1): 381–406.

Laloy, Eric, Hérault, Romain, Lee, John et al. 2017. Inversion using a new low-dimensional representation of complex binary geological media based on a deep neural network. *Advances in Water Resources* **110**: 387–405.

Li, Boxiao, Bhark, Eric W, Billiter, Travis C et al. 2019. Best practices of assisted history matching using design of experiments. *SPE Journal* **24** (4): 1435–1451.

Li, Liyong and Lee, Seong H. 2008. Efficient field-scale simulation of black oil in a naturally fractured reservoir through discrete fracture networks and homogenized media. *SPE Reservoir Evaluation & Engineering* **11** (4): 750–758. https://doi.org/10.2118/103901-PA.

Liu, Yimin, Sun, Wenyue, and Durlofsky, Louis J. 2019. A deep-learning-based geological parameterization for history matching complex models. *Mathematical Geosciences* **51** (6): 725–766.

Maity, Debotyam. 2018. Microseismicity analysis for HFTS pad and correlation with completion parameters. *Proceedings, Unconventional Resources Technology Conference*, Houston, TX, 23–25 July 20183, 810–3821.

Marongiu-Porcu, Matteo, Lee, Donald, Shan, Dan et al. 2016. Advanced modeling of interwell-fracturing interference: an Eagle Ford shale-oil study. *SPE Journal* **21** (5): 1567–1582.

McKay, Michael D, Beckman, Richard J, and Conover, William J. 2000. A comparison of three methods for selecting values of input variables in the analysis of output from a computer code. *Technometrics* **42** (1): 55–61.

Moinfar, Ali, Varavei, Abdoljalil, Sepehrnoori, Kamy et al. 2014. Development of an efficient embedded discrete fracture model for 3D compositional reservoir simulation in fractured reservoirs. *SPE Journal* **19** (2): 289–303.

Nobakht, Morteza and Clarkson, Christopher R. 2012a. A new analytical method for analyzing linear flow in tight/shale gas reservoirs: Constant-flowing-pressure boundary condition. *SPE Reservoir Evaluation & Engineering* **15** (3): 370–384.

Nobakht, Morteza and Clarkson, Christopher R. 2012b. A new analytical method for analyzing linear flow in tight/shale gas reservoirs: constant-rate boundary condition. *SPE Reservoir Evaluation & Engineering* **15** (1): 51–59.

Olalotiti-Lawal, Feyi, Onishi, Tsubasa, Datta-Gupta, Akhil et al. 2018. Post-combustion CO_2 WAG pilot in a mature field: Model calibration and optimization. *Proceedings, SPE Annual Technical Conference and Exhibition.*

Oliver, Dean S. 1996. Multiple realizations of the permeability field from well test data. *SPE Journal* **1** (2): 145–154.

Onishi, Tsubasa, Chen, Hongquan, Datta-Gupta, Akhil et al. 2021. An efficient deep learning-based workflow incorporating a reduced physics model for subsurface imaging in unconventional reservoirs. *Proceedings, SPE Annual Technical Conference and Exhibition.*

Onishi, Tsubasa, Iino, Atsushi, Jung, Hye Young et al. 2020. Fast marching method based rapid simulation accounting for gravity. *Proceedings, Asia Pacific Unconventional Resources Technology Conference*, Brisbane, Australia, 18–19 November 2019, 1093–1111.

Palacio, JC and Blasingame, TA. 1993. Decline curve analysis using type curves: Analysis of gas well production data. *Paper SPE* **25909**: 12–14.

Raissi, Maziar, Perdikaris, Paris, and Karniadakis, George E. 2019. Physics-informed neural networks: A deep learning framework for solving forward and inverse problems involving nonlinear partial differential equations. *Journal of Computational Physics* **378**: 686–707.

Razak, Syamil Mohd and Jafarpour, Behnam. 2020. Convolutional neural networks (CNN) for feature-based model calibration under uncertain geologic scenarios. *Computational Geosciences* **24** (4): 1625–1649.

Reynolds, Albert C, He, Nanqun, Chu, Lifu et al. 1996. Reparameterization techniques for generating reservoir descriptions conditioned to variograms and well-test pressure data. *SPE Journal* **1** (4): 413–426.

Ronneberger, Olaf, Fischer, Philipp, and Brox, Thomas. 2015. U-net: Convolutional networks for biomedical image segmentation. *Proceedings, International Conference on Medical image computing and computer-assisted intervention*. 234–241.

Sarma, Pallav, Durlofsky, Louis J, and Aziz, Khalid. 2008. Kernel principal component analysis for efficient, differentiable parameterization of multipoint geostatistics. *Mathematical Geosciences* **40** (1): 3–32.

Sarma, Pallav, Durlofsky, Louis J, Aziz, Khalid et al. 2006. Efficient real-time reservoir management using adjoint-based optimal control and model updating. *Computational Geosciences* **10** (1): 3–36.

Scheidt, Céline and Caers, Jef. 2009. Uncertainty quantification in reservoir performance using distances and kernel methods–application to a west Africa deepwater turbidite reservoir. *SPE Journal* **14** (4): 680–692.

Sethian, James A. 1999. Fast marching methods. *SIAM Review* **41** (2): 199–235.

Song, Bo and Ehlig-Economides, Christine. 2011. Rate-normalized pressure analysis for determination of shale gas well performance. *Proceedings, North American Unconventional Gas Conference and Exhibition*. https://doi.org/10.2118/144031-MS.

Stegent, Neil and Candler, Cody. 2018. Downhole microseismic mapping of more than 400 fracturing stages on a multiwell pad at the hydraulic fracturing test site (HFTS): Discussion of operational challenges and analytic results. *Proceedings, Unconventional Resources Technology Conference*, Houston, Texas, 23–25 July 2018, 3754–3781.

Sun, Jianlei and Schechter, David. 2015. Optimization-based unstructured meshing algorithms for simulation of hydraulically and naturally fractured reservoirs with variable distribution of fracture aperture, spacing, length, and strike. *SPE Reservoir Evaluation & Engineering* **18** (4): 463–480.

Tanaka, Shusei, Onishi, Tsubasa, Kam, Dongjae et al. 2020. Application of combined streamline based reduced-physics surrogate and response surface method for field development optimization. *Proceedings, International Petroleum Technology Conference*. https://doi.org/10.2523/IPTC-19958-MS.

Tang, Meng, Liu, Yimin, and Durlofsky, Louis J. 2020. A deep-learning-based surrogate model for data assimilation in dynamic subsurface flow problems. *Journal of Computational Physics* **413**: 109456.

Ţene, Matei, Bosma, Sebastian BM, Al Kobaisi, Mohammed Saad et al. 2017. Projection-based embedded discrete fracture model (pEDFM). *Advances in Water Resources* **105**: 205–216.

Uzun, Ilkay, Kurtoglu, Basak, and Kazemi, Hossein. 2016. Multiphase rate-transient analysis in unconventional reservoirs: theory and application. *SPE Reservoir Evaluation & Engineering* **19** (4): 553–566.

Valkó, Peter P and Lee, W John. 2010. A better way to forecast production from unconventional gas wells. *Proceedings, SPE Annual Technical Conference and Exhibition.*

Vo, Hai X and Durlofsky, Louis J. 2014. A new differentiable parameterization based on principal component analysis for the low-dimensional representation of complex geological models. *Mathematical Geosciences* **46** (7): 775–813.

Warpinski, Norm. 2009. Microseismic monitoring: Inside and out. *Journal of Petroleum Technology* **61** (11): 80–85.

Warren, JE and Root, PJ. 1963. The behavior of naturally fractured reservoirs. *Society of Petroleum Engineers Journal* **3** (3): 245–255. https://doi.org/10.2118/426-PA.

Wattenbarger, Robert A, El-Banbi, Ahmed H, Villegas, Mauricio E et al. 1998. Production analysis of linear flow into fractured tight gas wells. *Proceedings, SPE Rocky Mountain Regional/Low-Permeability Reservoirs Symposium.*

Xu, Rui, Zhang, Dongxiao, Rong, Miao et al. 2021. Weak form theory-guided neural network (TgNN-wf) for deep learning of subsurface single-and two-phase flow. *Journal of Computational Physics* **436**: 110318.

Xue, Xu, Yang, Changdong, Onishi, Tsubasa et al. 2019. Modeling hydraulically fractured shale wells using the fast-marching method with local grid refinements and an embedded discrete fracture model. *SPE Journal* **24** (6): 2590–2608.

Yang, Changdong, Xue, Xu, King, Michael J. et al. 2017. Flow simulation of complex fracture systems with unstructured grids using the fast marching method. *Proceedings, SPE/AAPG/SEG Unconventional Resources Technology Conference.* https://doi.org/10.15530/URTEC-2017-2691393.

Yang, D, Xue, X, and Chen, J. 2018. High resolution hydraulic fracture network modeling using flexible dual porosity dual permeability framework. *Proceedings, SPE Western Regional Meeting.*

Zhang, Yanbin, Bansal, Neha, Fujita, Yusuke et al. 2016. From streamlines to fast marching: Rapid simulation and performance assessment of shale-gas reservoirs by use of diffusive time of flight as a spatial coordinate. *SPE Journal* **21** (5): 1883–1898.

Zheng, Chuansheng, Deng, Xianbo, Fu, Qing et al. 2020. *Deep learning-based detection for COVID-19 from chest CT using weak label. IEEE Transactions on Medical Imaging* https://doi.org/ 10.1109/TMI.2020.29959650000106.

Zhu, Yinhao, Zabaras, Nicholas, Koutsourelakis, Phaedon-Stelios et al. 2019. Physics-constrained deep learning for high-dimensional surrogate modeling and uncertainty quantification without labeled data. *Journal of Computational Physics* **394**: 56–81.

APPENDIX A: MODEL SETTING DETAILS

Detailed model settings in terms of operations at each of the layers and output shapes are summarized in Tables A1–A4 below. "None" in the "Output Shape" columns corresponds to the batch size during the trainings. The "Relu" activation function is used for all the layers, except for the output layers, where "Sigmoid" is used to scale the values to [0, 1].

TABLE A1
Architecture of the Autoencoder for the 2D Synthetic Case

Model	Layer	Output Shape
Encoder	Input	(None, 100, 100, 1)
	Conv2D layer #1 (8 filters, kernel size 3 × 3, stride 2)	(None, 50, 50, 8)
	Conv2D layer #2 (16 filters, kernel size 3 × 3, stride 2)	(None, 25, 25, 16)
	Conv2D layer #3 (32 filters, kernel size 3 × 3, stride 2)	(None, 13, 13, 32)
	Flatten	(None, 5,048)
	Dense layer (latent space)	(None, 32)
Decoder	Decoder dense layer	(None, 5,408)
	Reshape	(None, 13, 13, 32)
	Conv2D transpose layer (16 filters, kernel size 3 × 3, stride 2)	(None, 26, 26, 16)
	Conv2D transpose layer (8 filters, kernel size 3 × 3, stride 2)	(None, 52, 52, 8)
	Conv2D transpose layer (1 filters, kernel size 3 × 3, stride 2)	(None, 104, 104, 1)
	Cropping2D	(None, 100, 100, 1)

TABLE A2
Architecture of the Regression Model for the 2D Synthetic Case

Model	Layer	Output Shape
Feedforward	Input	(None, 8, 201)
	Dense #1	(None, 8, 40)
	Dense #2	(None, 8, 40)
	Dense #3	(None, 8, 40)
	Dense #4	(None, 8, 40)
	Dense #5	(None, 8, 40)
	Flatten	(None, 320)
	Dense #6 (output)	(None, 32)

TABLE A3
Architecture of the Autoencoder for the 3D Field-Scale Application

Model	Layer	Output Shape
Encoder	Input	(None, 23, 137, 47, 1)
	Conv3D layer #1 (8 filters, kernel size 3 × 3 × 3, stride 2)	(None, 12, 69, 24, 8)
	Conv3D layer #2 (16 filters, kernel size 3 × 3 × 3, stride 2)	(None, 6, 35, 12, 16)
	Conv3D layer #3 (32 filters, kernel size 3 × 3 × 3, stride 2)	(None, 3, 18, 6, 32)
	Flatten	(None, 10,368)
	Dense layer (latent space)	(None, 256)
Decoder	Decoder dense layer	(None, 10,368)
	Reshape	(None, 3, 18, 6, 32)
	Conv3D transpose layer (16 filters, kernel size 3 × 3 × 3, stride 2)	(None, 6, 36, 12, 16)
	Conv3D transpose layer (8 filters, kernel size 3 × 3 × 3, stride 2)	(None, 12, 72, 24, 8)
	Conv3D transpose layer (1 filters, kernel size 3 × 3 × 3, stride 2)	(None, 24, 144, 48, 1)
	Cropping3D	(None, 23, 137, 47, 1)

TABLE A4
Architecture of the Regression Model for the 3D Field-Scale Application

Model	Layer	Output Shape
Feedforward	Input #1	(None, 3, 151)
	Input #2	(None, 1, 112)
	Intermediate layer #1	(None, 3, 100)
	Intermediate layer #2	(None, 1, 100)
	Concatenate	(None, 4, 100)
	Dense #1	(None, 4, 100)
	Dense #2	(None, 4, 100)
	Dense #3	(None, 4, 100)
	Flatten	(None, 400)
	Dense #4 (output)	(None, 256)

12 Reservoir Modeling Using Fast Predictive Machine Learning Algorithms for Geological Carbon Storage

Seyyed Hosseini[1], *Richard Larson*[1], *Parisa Shokouhi*[2], *Vikas Kumar*[2], *Sumedha Prathipati*[2], *Dan Kifer*[2], *Jonathan Garcez*[2], *Luis Ayala*[2], *Michael Riedl*[3], *Brandon Hill*[3], *Sanjay Tamrakar*[3], *Jared Schuetter*[3], *and Srikanta Mishra*[3]
[1]University of Texas at Austin, Austin, Texas
[2]Pennsylvania State University, State College, Pennsylvania
[3]Battelle Memorial Institute, Columbus, Ohio

CONTENTS

DOI: 10.1201/9781003207009-17

12.1 INTRODUCTION

Geological sequestration of CO_2 is a process that mitigates anthropogenic CO_2 from large point sources to reduce their contributions to the greenhouse gas effect by injecting it underground. Knowledge and understanding of a site's capacity for, and response to, CO_2 injection are critical to a successful project typically defined by its stability and containment. Numerical modeling is necessary for predicting such information. Modeling multiphase flow through porous media, such as this, involves spatial, temporal, multi-component, nonlinear calculations involving various degrees of nonuniformity and uncertainty. Due to the complex nature of these systems, numerical models are most widely used (Orr, 2005; Wen and Benson, 2019). The current standard is to numerically simulate reservoir behavior to predict the storage capacity, pressure response, and recovery.

Perhaps the most well-established analogue to modeling CO_2 sequestration is the modeling performed in petroleum exploration and production. In the petroleum industry, a long history of activity and data acquisition has led to vast experience in computational methods and an accumulation of data.

In contrast to the petroleum industry, the experience related to operating and monitoring large CO_2 storage sites is very limited. Uncertainties caused by this lack of existing data may increase the computational intensity. Large-scale, high-fidelity, dynamic simulations considering such uncertainties are necessary to achieve the appropriate accuracy; however, they are cost prohibitive and slow, limiting their utility for real-time commercial-scale CO_2 storage management.

Recent advances in data collection, cheap data storage, and high-performance computing capabilities have created the perfect conditions for the application of machine learning (ML) and artificial intelligence (AI) to derive fast-predictive surrogate models from numerically derived data (Schmidhuber, 2015). These data-driven simulators can create relationships between the inputs and outputs that aren't based on the governing equations of the system and therefore have less cumbersome calculations and in reduced dimensions (Hinton and Salakhutdinov, 2006; Ben Taieb, 2012). This can be applied to CO_2 injection and storage, giving operators and stakeholder easy-to-use tools for practical purposes that could be used on common hardware. Power of Machin learning has not been limited only to the storage part of carbon capture and storage (CCS) projects but widely applied in many elements of carbon capture, utilization, and storage (CCUS) projects including capture and utilization (Yan et al., 2021).

In reservoir simulations, ML has been explored as a means to create these data-driven simulators. In oil and gas reservoirs, this has been done in a number of different applications in reservoir characterization (Pang et al., 2022) and reservoir fluid characterization (Sambo and Feng, 2021; Mahdiani et al., 2020). Studies have also applied these techniques to CO_2 injection for enhanced oil recovery (EOR) (Ampomah et al., 2017; Gholami et al., 2019; Artun, 2020), CO_2 pure storage plays (Zhong et al., 2019), and tracking fluid migration in subsurface (Amini and Mohaghegh, 2019; Zhong et al., 2019; Wang et al., 2020; Yan et al., 2020).

The method selection is application based considering accuracy, training time, and simulation time. Previously, fundamental artificial neural networks (ANNs) have been tested in simple situations to address a number of reservoir simulation problems (Shahkarami, 2015; Shahkarami and Mohaghegh, 2020). On the other hand, more complex models are being developed as well. Physics-informed neural networks (PINN) have physics-based constraints imposed upon them in addition to the input data (Shokouhi et al., 2021). Furthermore, so called neural operators such as Fourier neural operators (FNOs and U-FNOs) have also been studied regarding their performance and practicality (Wen et al., 2021), and deep reinforcement learning is demonstrated to be able to find optimal carbon sequestration reservoir management solutions (Sun, 2020). High-level networks can create very accurate simulation proxies, but the extent of their benefits may be lost due to higher training times and/or higher run times. Ultimately, the goals of these types of methods are the same, to deliver suitably accurate results in a fraction of the time.

In this chapter, a few different methods and implementations of deep learning networks are applied to modeling CO_2 storage for a specific Gulf Coast geological case, and the results are reviewed. Numerical simulations are run under various conditions based on the same static geological reservoir model to generate datasets. Each ML algorithm is then trained from this generated data. The inputs are coordinate data, permeability, porosity, injection rate, pressure, and time. The model's performance is then tested through the accuracy of target outputs such as phase saturation, pressure distribution, and production of brine with testing data.

12.2 DATA – RESERVOIR MODEL

The High Island 24L (HI 24L) field is in offshore Texas state waters, as represented by the red rectangle shown in Figure 12.1a. The geology of the sandstone interval studied here has been characterized, and assessment of its CO_2 storage capacity was conducted previously (Ruiz, 2019). The studies also assessed the CO_2 storage capacity of an upper sandstone interval, named the storage interval of interest (SIOI). Figure 12.1b depicts the stratigraphic chart for the HI 24L field, explicitly identifying the key stratigraphic packages and regionally interpreted surfaces. The SIOI is a thick unit of aggradational sands with an approximate thickness of 520 m (1,700 ft).

12.2.1 Numerical Model Conceptualization

The original static geological model constructed at the Bureau of Economic Geology has approximately 30 million grid cells. The large number of grid blocks would have required significant computational time for the required number of numerical simulations. This led to a model upscaling step to reduce the number of grid cells from 30,528,00 to 596,440 (a 52 × 62 × 185 grid). After upscaling, the grid dimensions increase to 750 ft and 10 ft in aerial and vertical dimensions, respectively. Further trimming in the reservoir model by removing shallower layers and grids beyond sealing faults reduced the number of active cells to 237,584.

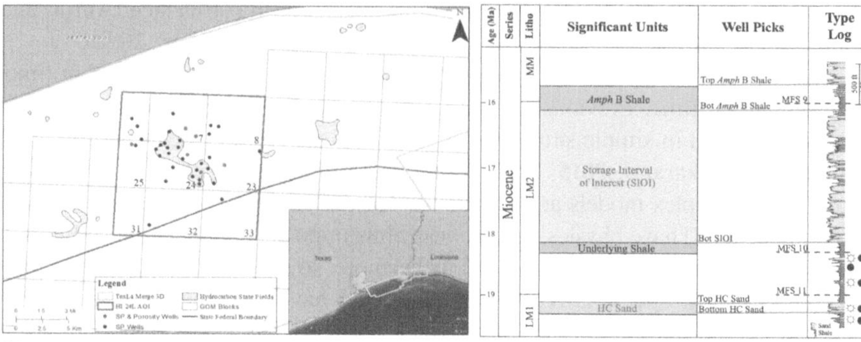

(a) (b)

FIGURE 12.1 (a) The location of blocks and wells used for model characterizations. (b) Stratigraphic chart for the interval of interest.

The available geological data for the GoM-HL24 field consisted of 50 effective, equally probable porosity realizations. These 50 realizations were ranked to identify P10, P50, and P90 porosity realizations using the concept of diffusive time of flight (DTOF) — a proxy that provides good relative estimate changes in connected volumes with heterogeneity levels based on quantification of pressure wave travel times through each realization (Hosseini et al., 2010). The *WellOpt* Petrel plugin was used to execute these rankings based on DTOF.

Numerical simulations to predict CO_2 storage behavior were performed using the compositional reservoir simulator GEM by Computer Modeling Group (CMG). The predicted pressure and CO_2 saturation distributions in the reservoir during the injection and post-injection periods were provided as input to the ML models. The model was initialized as a fully saturated brine system and set to hydrostatic equilibrium, with a reference pressure of 2,900 psia at 7,500 ft depth. Reservoir temperature was assumed to be constant (isothermal process) and was set at 149°F. Relative permeability curves were obtained using the Brooks-Corey correlation with empirical parameters extracted from analogous formation/brine properties reported in the literature (Bachu, 2013). The Carter-Tracy analytical aquifer function was used and tuned to ensure constant-pressure constraint at the lateral boundaries of the model.

A total of four vertical injection wells with constant injection rate specification were used to attain the targets for cumulative stored CO_2 within a 30-year injection period. Over pressurization of the reservoir was prevented by using three water production wells that were operated assuming a constant bottom hole pressure (BHP) constraint, set at the initial reservoir pressure. Figure 12.2 exhibits the numerical model and spatial locations of wells (injectors and producers). All the wells are only perforated at the bottom 20 layers. Given the large lateral extension of the grid cells in the x and y directions (750 × 750 ft), we employed local grid refinement (LGR; each large grid refined to a 7 × 7 × 1 grid) for cells with injection wells and neighboring cells.

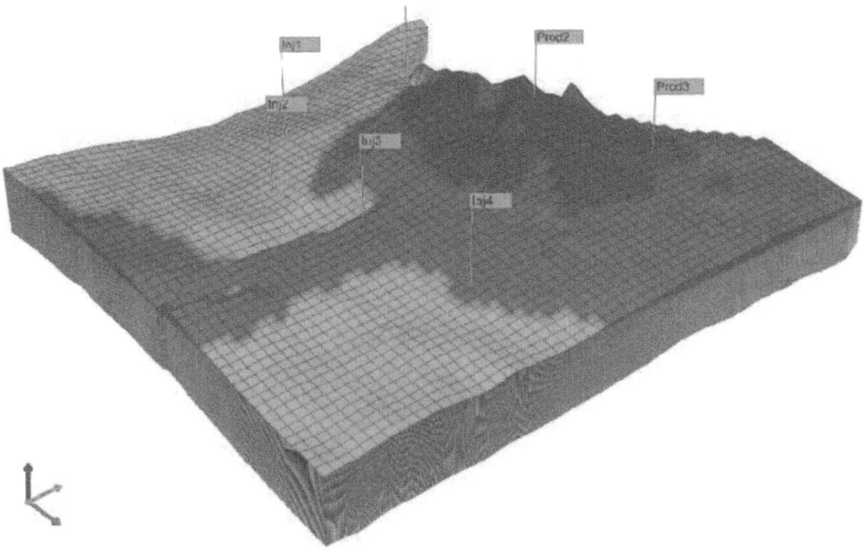

FIGURE 12.2 GoM discretized model and well locations.

12.2.2 Simulation Case Matrix

Simulation cases were designed with the purpose of covering a broad set of combinations to ensure effective training of ML-based algorithms. The cumulative amount of injected CO_2 was varied between 60, 50, 40, and 30 million tons along with three different porosity (and permeability) realizations (P10, P50, and P90).

The target injection total was allocated by permuting the injection wells in different pairs, as depicted in Figure 12.3. Once the well pairs were defined, i.e., "Pair 1 and Pair 2," the total amount of CO_2 allocated to each pair changed according to the following proportions: *Pair 1 = 0.4/Pair 2 = 0.6*, *Pair 1 = 0.5/Pair 2 = 0.5*, and *Pair 1 = 0.6/Pair 2 = 0.4*. A total of 108 cases were generated for ML training purposes. An additional set of 12 cases (cases 109–120, about 10% of the training data set size) were randomly generated by the Latin hypercube sampling (LHS) method for developing the testing data to validate the trained ML models. Simulation time includes 720 monthly time steps where the first half of the time steps (360 months) are the injection period and the second half are for the post-injection period where all the wells are shut-in.

12.3 METHODS

12.3.1 CNN/MLP Implementation

In our first method, we used a combination of a convolutional neural network (CNN) and multilayer perceptron (MLP) to train the model. In this application, to save training time, we have only used 50% of training data (randomly selected)

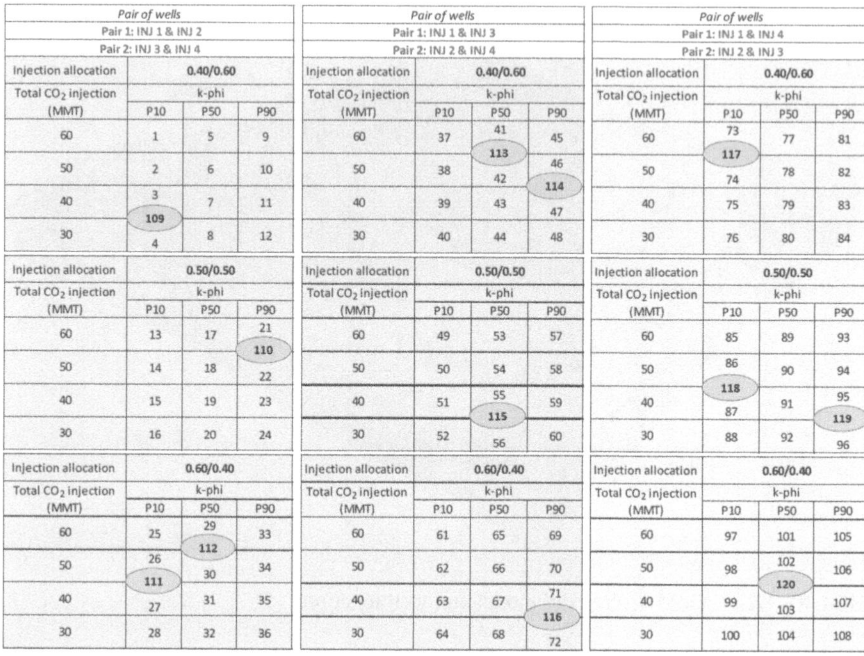

FIGURE 12.3 Simulation matrix with testing and training cases. The sequential numbers within the boxes denote assigned training case numbers. Numbers within ovals represent the relative location of the additional testing case scenarios with respect to the 108 training sets.

to train the models. We have used a combination of CNN, to forecast pressure and saturation, and MLP, to forecast production rate. Our CNN model has around 207 million parameters, whereas the MLP model only has approximately 8,000 parameters in their neural networks.

The structure that we used to feed the data into the CNN and MLP models is very similar to the previous work on developing deep learning structures (Gundersen et al., 2020). We reorganized the input data for injection rates into five columns where the first column represents the time. The next four columns are cumulative injection rates that are monotonically increasing for the first 360 time steps and then become flat when injection stops. This implementation ensures that at least one variable is changing, therefore, ML algorithms can use that to learn the system behavior.

Permeability and porosity data are fed into the model with their original shape. First, we took $\log_{10}$ of the permeability values and then normalized (between 0,1) it across all the realizations. We only have three different permeability and porosity realizations for P10, P50, and P90.

After converting all the datasets into the shape of (number of samples, number of time steps, x elements, y elements, z elements), these data were fed into neural networks for training. We used the CNN model for predicting pressure

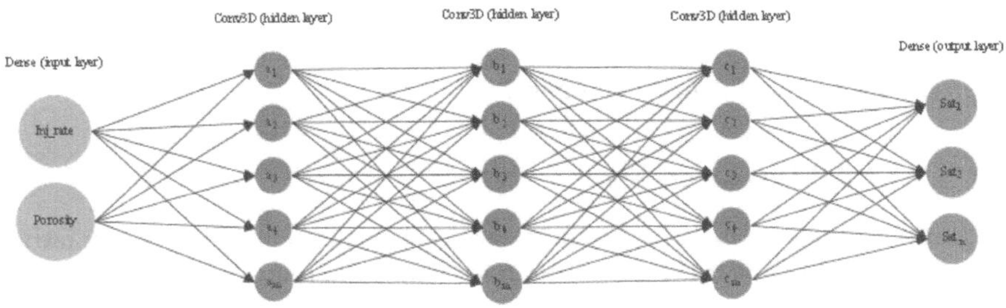

FIGURE 12.4 Architecture used in CNN models to predict pressure and saturation values.

and saturation where model architecture simply includes three consecutive 3D convolutional layers (Figure 12.4). We trained two separate models for pressure and saturation predictions where model architecture and parameters are exactly the same. Prediction of the production rate seemed much simpler so we decided to use a simple MLP model for prediction that provided us with a good result. The MLP model has only two hidden layers.

12.3.2 GNN Implementation

In addition to the CNN/MLP method, a graph neural network (GNN) for predicting pressure and saturation for the Gulf of Mexico model was developed. The GNN takes, as input, the properties of a cell, properties of a cell's neighbors, and properties of the neighborhood at the previous time step to predict the pressure and saturation of the cell at the current time point. The GNN is trained to predict a single time point into the future and is applied to test data by recursively feeding previous predictions back into the GNN to predict the next time step. One advantage of the GNN is that it has the potential to generalize to other well models with little to no retraining since it makes its predictions based on a local neighborhood of cells. Another advantage is the possibility of incorporating physical constraints into the choice of the local neighborhood used for prediction. The size and shape of the neighborhood can dynamically change over the space of the well based on the local permeability and porosity. One disadvantage of the GNN model is the recursive nature of the predictions; prediction errors can compound over time with the possibility of the largest errors occurring at the end of the well model simulation.

Current GNN models were adopted from a model developed for a toy problem, and we do not believe the results here are a complete characterization of the GNN's performance. There are still known issues with the implementation (i.e., incorrect neighbor distances) that once corrected, could significantly improve performance. The GNN operates on a local neighborhood of data to make predictions. The model takes as input a 2D data cube where the first dimension is the spatial dimension of the well-flattened to linear indices, and the second dimension is the property dimension. The model also takes another 2D input that consists of all the possible distances from one cell to another. The final two inputs to the model are index maps that describe how to select the neighborhoods of cells and their distances from the first two inputs.

12.3.3 MLP Implementation

In this implementation, due to the large size of the dataset and increased training time, only a portion of the dataset is used for training to reduce training time. The training data is chosen sparsely, with the spatial coordinates covering the grid boundaries across the three axes, the production well locations, and a range of randomly chosen grid points for every 10 rows across the three axes. This is done for each time step in a realization per epoch. The results are converted back into the original scale once the model is evaluated.

The network has an input layer, five hidden layers, and an output layer containing 128 nodes, 64 nodes, 32 nodes, 16 nodes, 8 nodes, and 3 nodes (output layer), respectively. The tunable hyperparameters of the models are learning rate, batch size, training epochs, number of hidden layers, nodes in each layer, optimizer, dropout, activation function, and batch normalization. The input is normalized and given to the model batch-wise using a data generator. Model architecture is shown in Figure 12.5.

12.3.4 LSTM Implementation

in our next implementation with the long short-term memory (LSTM) algorithm, (Hochreiter and Schmidhuber, 1997; Hochreiter, 1998), the network has an input layer, six LSTM layers, four fully connected layers, and an output layer using the

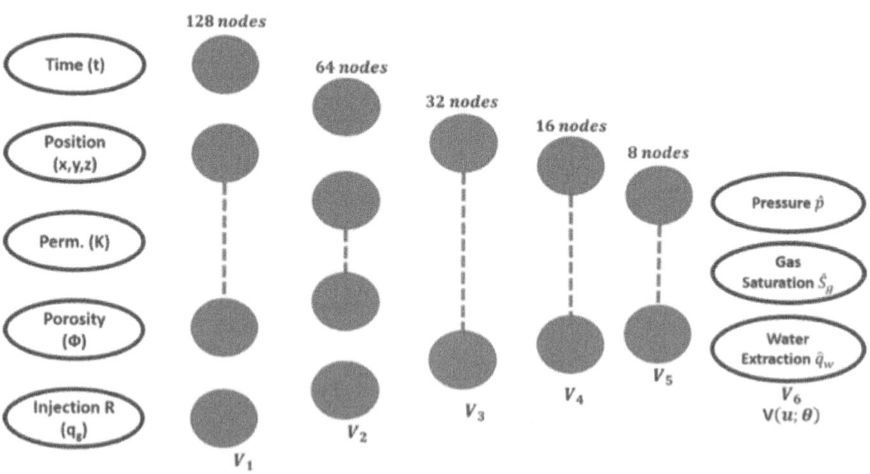

FIGURE 12.5 MLP architecture for joint prediction of pressure, gas saturation, and water production rates at the three wells.

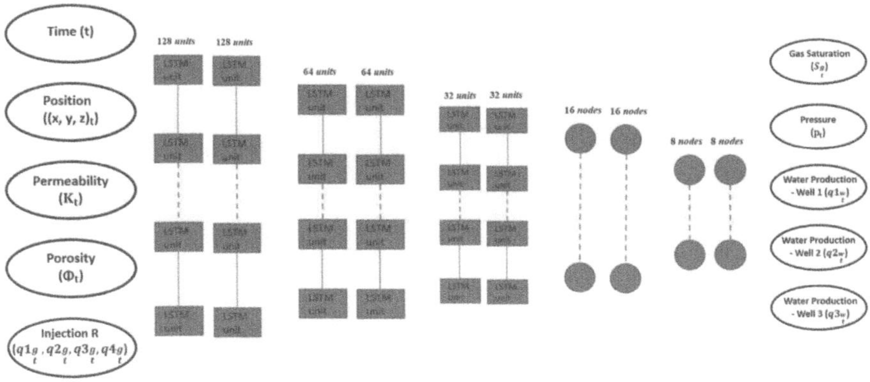

FIGURE 12.6 LSTM architecture for joint prediction of pressure, gas saturation, and water production rates at the three wells.

ReLU activation function. The gas saturation, pressure and the water production rates at each of the three production wells predicted jointly. Therefore, the shape of the input is (10,1) and the output is a vector of five variables for a time step t. The input is normalized and given to the model batch-wise using a data generator. Model architecture is shown in Figure 12.6.

12.4 RESULTS

In the reservoir simulation models, injection wells are perforated in layers 71–91, which are at the bottom of the geological formation. We have selected layer 81 (middle perforation) to visualize the pressure and saturation maps. Figure 12.7 shows the porosity and permeability distributions in layer 81 in reference to the following 2D maps for pressure and saturation.

The plots in Figure 12.8 show the model performance in predicting the saturation in layer 81. Expectedly, some saturation prediction errors can be found around the injection wells while most of the errors are toward the edge of the plume. Model performance is consistent across various layers and selected time steps. Post-injection performance does not show any specific deterioration compared with injection time results.

To further investigate the performance of the model as a whole, we have made a plot of root-mean-squared error (RMSE) (Figure 12.9) of the results both per any given time step or given geological layers. Results show that RMSE for saturation prediction is increasing over time until injection stops and becomes pretty much steady afterward. Results are color coded for the 12 test realizations.

Our next series of results represents the same plots for pressure data prediction (Figure 12.10). Two-dimensional pressure maps also show a very good match between true and predicted value where error in most cases is less than 30 psi. Unlike saturation error, pressure error is distributed throughout most of the grid blocks and does not correspond directly to the plume extension.

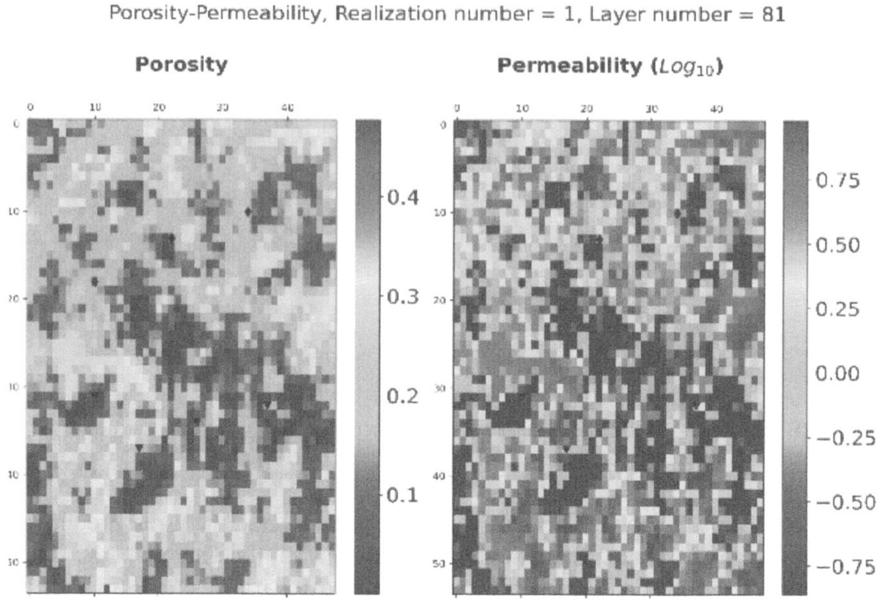

FIGURE 12.7 Porosity and permeability distribution in layer 81 in Gulf of Mexico model.

Evolution of the RMSE values for pressure prediction with time is plotted in Figure 12.11.

In the next set of results, we will focus on performance of the models in forecasting production rates. Our production wells are far from injection wells so they only produce brine, and CO_2 will not breakthrough to these wells. Depending on the realizations, production rates vary from very small to large numbers. For example, well 3's ultimate production rates vary from 100 bbl/day to over 4,000 bbl/day. Our forecast model is capable of capturing this behavior across all the ranges. RSME variation with time for all three wells aggregated across all 12 test realizations as shown in Figure 12.12.

12.5 SUMMARY AND CONCLUSIONS – GoM

Results of the reported ML models demonstrate that initial models are reasonably proficient at achieving the goals of the project while there is still room to improve current models both for accuracy and speedup. The overall computational performance of these models is summarized in Table 12.1. Compositional simulations on average took about 138 minutes to run. Training times range from 330–2520 minutes, whereas forecast time for one single realization ranged from 5–204 seconds. Based on provided data we can achieve a speedup of 50–2,000×.

In addition to the computational performance of the models, we have summarized the accuracy of the models in predicting pressure, saturation, and production

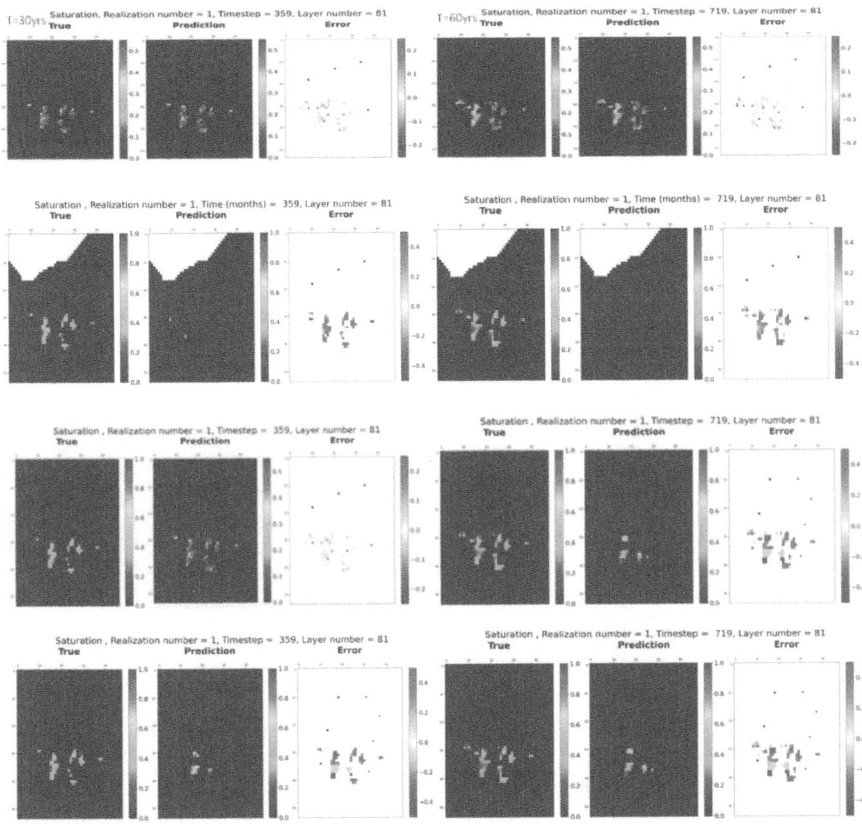

FIGURE 12.8 Saturation distribution plots for true, prediction, and error in layer 81. Plots are provided for two time steps covering end of injection (left) and end of post-injection (right). Rows from top to bottom correspond to CNN/MLP, GNN, MLP, and LSTM methods.

rate in Table 12.2. We are reporting a single RMSE value for all models that is calculated across all the test datasets for a given geological model.

We have models that are reasonably capable of producing low RMSE values with low forecast times. The GNN model is an outlier in the models. In general, RMSE values for pressure are less than a few psi and RMSE values for saturation are below 0.04.

Our application of a simple CNN/MLP architecture and MLPs shows very promising results. Performance of the CCN/MLP models is adequate for brine production prediction because there are not too many output parameters.

In GNN implementation, the prediction errors compound quickly, resulting in predictions that are not well aligned with the truth. This appears to stem from high error cells occurring along the top boundary of the reservoir. We believe there are two potential sources of this compounding error. First, the GNN is

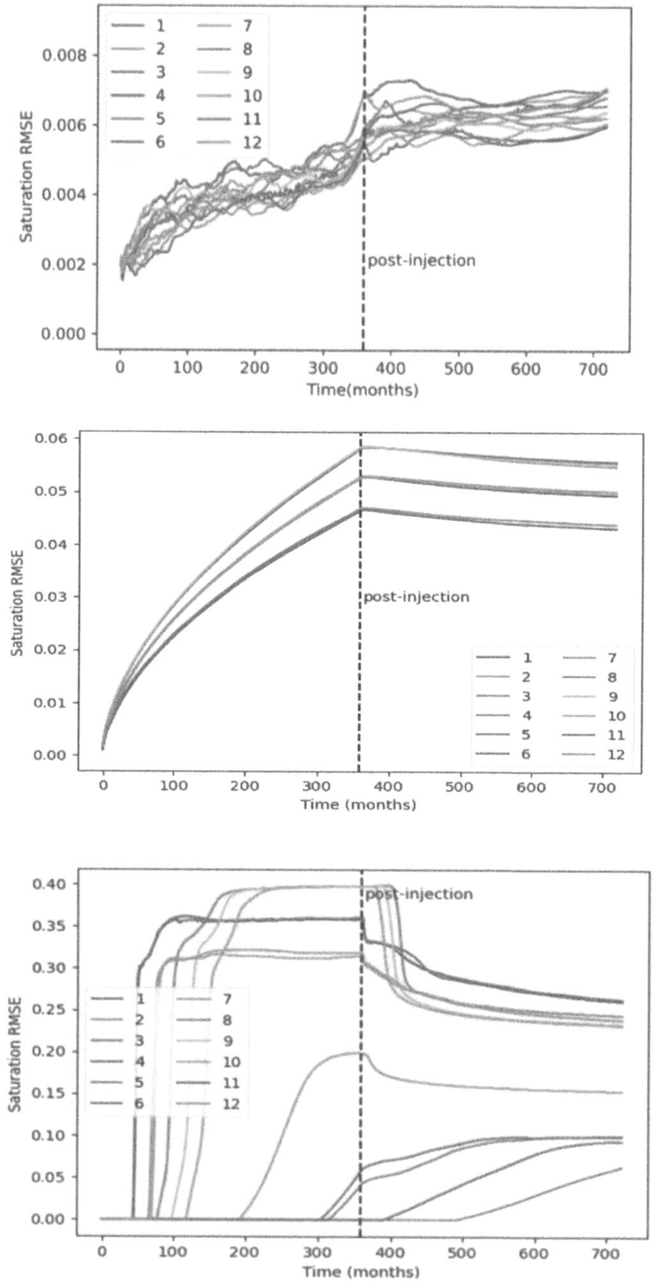

FIGURE 12.9 RMSE for saturation prediction for 720 time steps plotted for 12 testing realizations. Performance of the models are consistent across all realizations and time steps. From top to bottom, CNN/MLP, GNN and LSTM results. The MLP model has not reported this output.

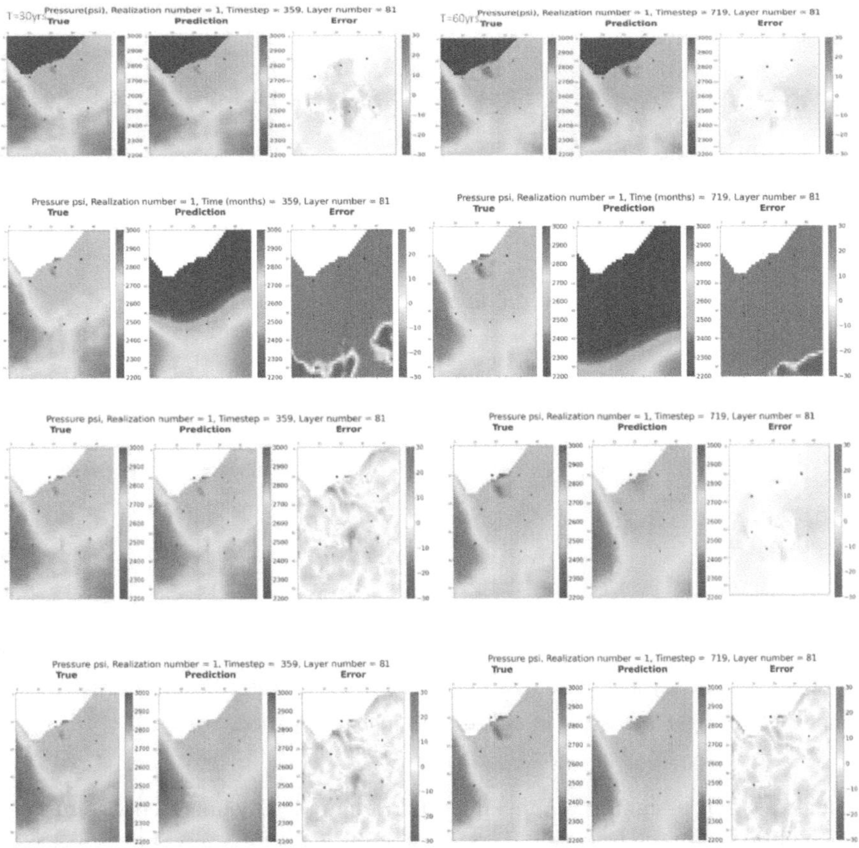

FIGURE 12.10 Pressure distribution plots for true, prediction, and error in layer 81. Plots are provided for two time steps covering end of injection (left) and end of post-injection (right). Rows from top to bottom correspond to CNN/MLP, GNN, MLP, and LSTM methods.

trained on only a subset of the complete training data, which may create a bias toward predicting zero-valued saturation. A potential solution to this training bias is to weigh the non-zero saturation voxels higher in the MSE training loss function. The second source of error may be due to the distance calculation for the neighboring cells. The distance is currently calculated in the same way as the 3D toy problem. However, in the GoM model the data is in a cube, but the locations of those cells in space are not uniform. Our GNN does not currently take this into account. We believe that addressing these two known sources of error could improve the multi-step performance significantly.

It appears that the LSTM model is not effectively predicting the gas saturation at the right places. It predicts the zeroes at the boundaries but fails to predict the saturation toward the dynamic parts of the reservoir. For LSTM models this can

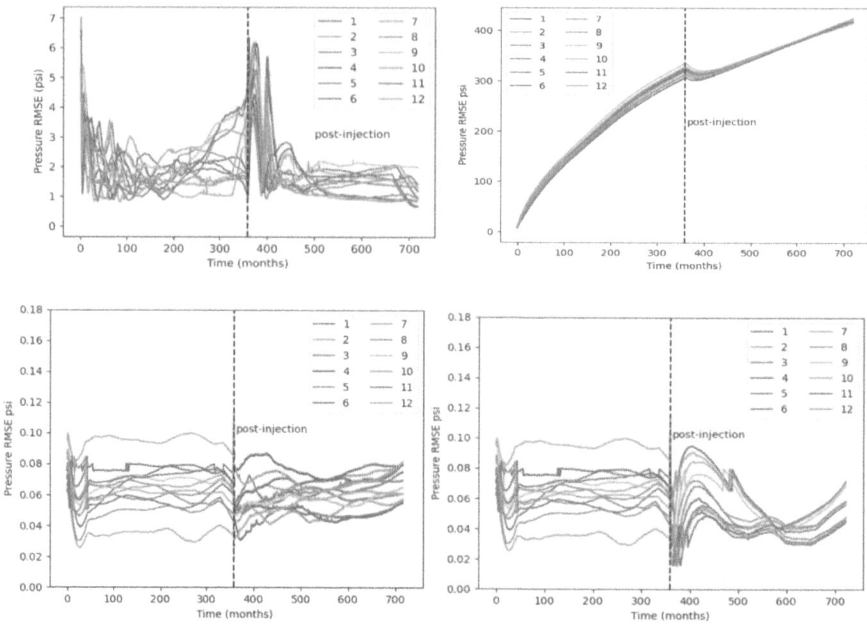

FIGURE 12.11 RMSE for pressure prediction for 720 time steps plotted for 12 testing realizations. Performance of the model is consistent across all realizations and there is a jump in RMSE right at the start of the post-injection time period.

be addressed by assigning weights in proportion to the amount of change in saturation observed in the reservoir.

MLP and LSTM models, although very different, produce incredibly similar results. However, the LSTM model is much slower to train. The performance of the CNN/MLP model is similar to that of LSTM. The LSTM models seem to be more accurate in predicting pressure, whereas the CCN/MLP model seems to be faster to train and test, although a direct comparison of training and testing times is not warranted due to the different computational resources employed. All models except the GNN model are performing well. CNN/MLP is providing reasonable results, whereas the LSTM model shows some improvement in predicting the production rates.

In summary, the model performances are satisfactory. Because of the large dataset size, the training has been limited to a portion of the available data. Future efforts will seek to improve the model performances and training/prediction times. The former may be achieved by including more training data or applying weights to data points, where pressure/saturation values are high (for example, near wells). Another direction is to include fundamental physical laws in the model architecture as previously done for the smaller problems (Shokouhi et al., 2021). To reduce training/prediction time, one may use dimensionality reduction strategies such as autoencoders. Simple post-processing methods such as interpolation may also

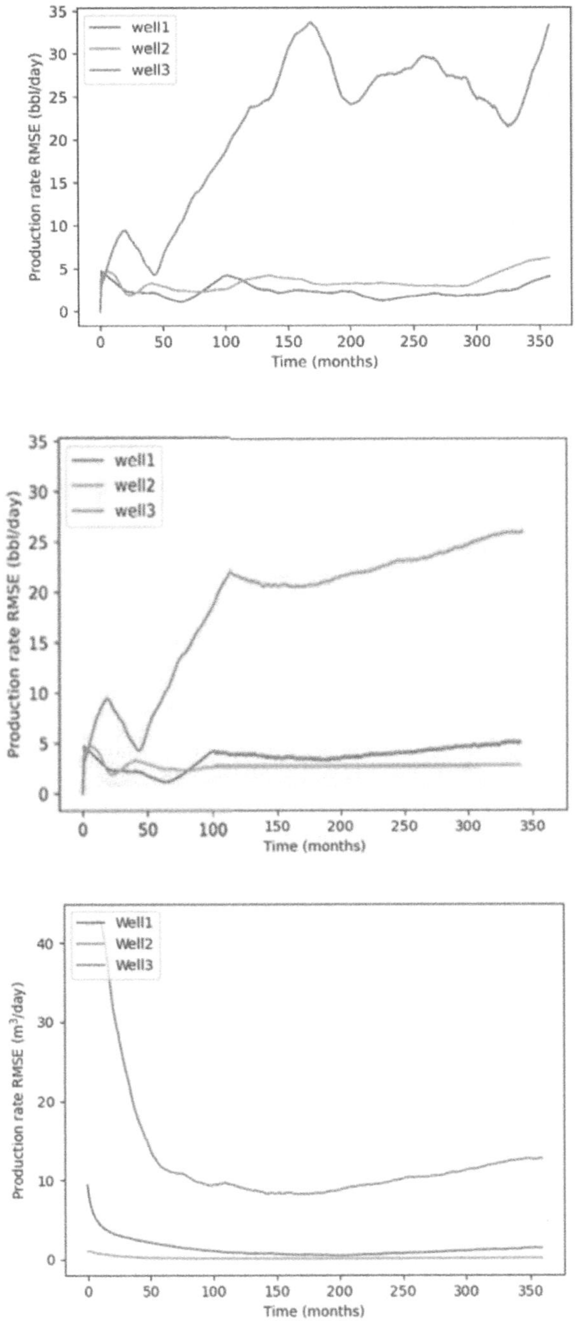

FIGURE 12.12 RMSE in production rate prediction for three producers. Well 3 shows the highest RMSE because it has higher absolute production rates. In terms of percentage of RMSE all three wells performing in the same range. From top to bottom: CCN/MLP, MLP, and LSTM model results. The GNN model has not reported this output.

TABLE 12.1
Summary of Computational Data for all Models

Model	~Training Time (mins)	Forecast Time (secs)	~Simulation Time (mins)	~Speedup	System Info
CNN/MLP	330	5	160	2000×	Intel Xeon CPU E5-268W0
GNN	420.	204.	160	50×	2× 32GB V100 GPUs
MLP	1,260	165.	160	60×	NVIDIA K80
LSTM	2,520	190.	160	50×	Intel Xeon E3-1200

be used to speed up the process. In the long run, we may use transfer learning by initializing the model using weights trained on a different set to improve both accuracy and speed of modeling. It seems that models with poor performance are still doing a good job in capturing the plume shape but not capturing the accurate saturation values. There is a huge potential in using data reduction techniques to reduce training and forecast time but further research is needed. Data reduction techniques in range of spatiotemporal techniques (using lower resolution data followed by interpolation of the outputs), and pure data reduction techniques such as principal component analysis (PCA), autoencoders, etc., has been tested. Linear data reduction methods can have some drawback given the nonlinear behavior of the system. In addition, data reduction techniques require an additional step to transform the data back to its original domain/size, which limits application of some of the methods (some methods are not reversible) and adds to the computational cost. Another frontier is developing fast proxy models that can fully replace the full-physics simulations but train the models on the higher resolution datasets created by full physics simulations (training on interaction among grid blocks so models can simulate any other discretized gridded static model) and make the models capable of running on a static model from the start and skip the requirement to have access to multiple full-physics simulations. This may be achievable for specific uses such as models that are specifically designed for CCS projects (Wen et al., 2021).

TABLE 12.2
Summary of RMSE for All Models

Model	Pressure (psi)	Saturation (–)	Production Rate (bbl/day)	Forecast Time (secs)
CNN/MLP	2.06	0.0053	13.86	5
GNN (multi)	296.62	0.0068	N/A	204
MLP	0.16	0.0043	6.50	165.
LSTM	0.12	0.0429	9.09	190.

ACKNOWLEDGMENT

The authors are grateful to anonymous and internal reviewers. Our gratitude is extended to the Computer Modeling Group, Calgary, Canada (CMG) for providing CMG-GEM and associated software. This work was completed as part of the Science-informed Machine learning to Accelerate Real Time decision making for Carbon Storage (SMART-CS) Initiative (edx.netl.doe.gov/SMART). Support for this initiative was provided by the U.S. Department of Energy's (DoE) Office of Fossil Energy's Carbon Storage Research program through the National Energy Technology Laboratory (NETL).

REFERENCES

Amini, S., & Mohaghegh, S., 2019. Application of machine learning and artificial intelligence in proxy modeling for fluid flow in porous media. Fluids, 4(3). https://doi.org/10.3390/fluids4030126

Ampomah, W., Balch, R. S., Cather, M., Will, R., Gunda, D., Dai, Z., & Soltanian, M. R., 2017. Optimum design of CO_2 storage and oil recovery under geological uncertainty. Applied Energy, 195, 80–92. https://doi.org/10.1016/j.apenergy.2017.03.017

Artun, E., 2020. Performance assessment and forecasting of cyclic gas injection into a hydraulically fractured well using data analytics and machine learning. Journal of Petroleum Science and Engineering, 195, 107768. https://doi.org/10.1016/j.petrol.2020.107768

Bachu, S., 2013. Drainage and imbibition CO_2/brine relative permeability. Energy Procedia, 37, 4428–4436.

Ben Taieb, S. H. R., 2012. Recursive and Direct Multi-Step Forecasting: The Best of Both Worlds. Brussels, Belgium: Monash University. (ISSN 1440-771X ed.).

Gholami, V., Mohaghegh, S., & Maysami, M., 2019. Smart proxy modeling of SACROC CO2-EOR. Fluids, 4(2), 85. https://www.mdpi.com/2311-5521/4/2/85

Gundersen, K., Hosseini, S., Oleynik, A., & Alendal, A., 2020. A variational auto-encoder for reservoir monitoring. arXiv:2009.11693.

Hinton, G. E., & Salakhutdinov, R. R., 2006. Reducing the dimensionality of data with neural networks. Science, 313(5786), 504–507.

Hochreiter, S., 1998. The vanishing gradient problem during learning recurrent neural nets and problem solutions, fuzziness and knowledge-based systems. International Journal of Uncertainty, 6, 107–116.

Hochreiter, S., & Schmidhuber, J., 1997. Long short-term memory. Neural Computation, 9(8), 1735–1780.

Hosseini, S.A., Kang, S., & Datta-Gupta, A., 2010. Qualitative well placement and drainage volume calculations based on diffusive time of flight. Journal of Petroleum Science and Engineering, 75(1–2), 178–188.

Mahdiani, M. R., Khamehchi, E., Hajirezaie, S., & Hemmati-Sarapardeh, A., 2020. Modeling viscosity of crude oil using k-nearest neighbor algorithm. Advances in Geo-Energy Research, 4(4), 435–447. https://doi.org/10.46690/ager.2020.04.08

Orr, F. M., 2005. Theory of Gas Injection Processes. Vol. 5, Copenhagen: Tie-Line Publications.

Pang, H., Meng, H., Wang, H., Fan, Y., Nie, Z., & Jin, Y., 2022. Lost circulation prediction based on machine learning. Journal of Petroleum Science and Engineering, 208. https://doi.org/10.1016/j.petrol.2021.109364

Ruiz, I., 2019. Characterization of the High Island 24L Field for Modeling and Estimating CO_2 Storage Capacity in the Offshore Texas State Waters, Gulf of Mexico. Master Thesis. The University of Texas, Austin.

Sambo, C., & Feng, Y., 2021. Physics Inspired Machine Learning for Solving Fluid Flow in Porous Media: A Novel Computational Algorithm for Reservoir Simulation. Paper presented at the SPE Reservoir Simulation Conference, On-Demand, October 2021, SPE-203917-MS.

Schmidhuber, J., 2015. Deep learning in neural networks: An overview. Neural Networks, 6, 85–117.

Shahkarami, A., 2015. Paper SPE-173405-MS Assisted History Matching Using Pattern Recognition Technology.

Shahkarami, A., & Mohaghegh, S., 2020. Applications of smart proxies for subsurface modeling. Petroleum Exploration and Development, 47(2), 400–412. https://doi.org/10.1016/S1876-3804(20)60057-X

Shokouhi, P., Kumar, V., Prathipati, S., Hosseini, S. A., Giles, C. L., & Kifer, D., 2021. Physics-informed deep learning for prediction of CO_2 storage site response. Journal of Contaminant Hydrology, 241. https://doi.org/10.1016/j.jconhyd.2021.103835

Sun, A., 2020. Optimal carbon storage reservoir management through deep reinforcement learning. Applied Energy, 278. https://doi.org/10.1016/j.apenergy.2020.115660

Wang, Z., Dilmore, R. M., & Harbert, W., 2020. Inferring CO_2 saturation from synthetic surface seismic and downhole monitoring data using machine learning for leakage detection at CO_2 sequestration sites. International Journal of Greenhouse Gas Control, 100, https://doi.org/10.1016/j.ijggc.2020.103115.

Wen, G., & Benson, S. M., 2019. CO_2 plume migration and dissolution in layered reservoirs. International Journal of Greenhouse Gas Control, 87, 66–79. https://doi.org/10.1016/j.ijggc.2019.05.012

Wen, G., Hay, K., & Benson, S.M., 2021. CCSNet: a deep learning modeling suite for CO_2 storage, arXiv:2104.01795v1.

Wen, G., Li, Z., Azizzadenesheli, K., Anandkumar, A., & Benson, S. M., 2021. U-FNO-an enhanced Fourier neural operator based-deep learning model for multiphase flow. arXiv:2109.03697 [physics.geo-ph]

Yan, Yongliang, Borhani, Tohid N., Subraveti, Sai Gokul, Pai, Kasturi Nagesh, Prasad, Vinay, Rajendran, Arvind, Nkulikiyinka, Paula, Asibor, Jude Odianosen, Zhang, Zhien, Shao, Ding, Wang, Lijuan, Zhang, Wenbiao, Yan, Yong, Ampomah, William, You, Junyu, Wang, Meihong, Anthony, Edward J., Manovic, Vasilije, & Clough, Peter T., 2021. Harnessing the power of machine learning for carbon capture, utilization, and storage (CCUS) "a state-of-the-art review." The Royal Society of Chemistry, 14(12), 6122–6157.

Yan, B., Harp, R., Chen, B., & Pawar, R., 2020. Improving deep learning performance for predicting large-scale porous-media flow through feature coarsening. arXiv preprint.

Zhong, Z., Sun, A. Y., & Jeong, H., 2019. Predicting CO_2 plume migration in heterogeneous formations using conditional deep convolutional generative adversarial network. Water Resources Research, 55, 5830–5851.

13 Physics-Embedded Machine Learning for Modeling and Optimization of Mature Fields

Pallav Sarma
Tachyus,
Berkeley, California

CONTENTS

13.1 INTRODUCTION

The beginnings of the "digital transformation" effort of the oil and gas industry can be traced back at least two decades, when many of the oil majors started their initial efforts on "smart fields." Different oil majors had different names for this concept: Shell called it Smart Fields, Chevron called it Intelligent Oil Fields, BP called it Field of the Future and so forth, but the basic premise was the same: data combined with software and technology will enable autonomous, continuous optimization of oil fields. The underlying technology enabling smart

DOI: 10.1201/9781003207009-18

fields is known as closed-loop reservoir management [Jansen et al., 2009], and is described by the schematic below:

In Figure 13.1, the "Real Reservoir" box represents the real oil field over which one or more objectives need to be optimized. In a typical application, there might be net present value (NPV) or cumulative oil produced. The optimization is carried out by finding the optimal values of a set of control parameters including, for example, well rates for injection optimization and well locations for infill drilling optimization. The "Models" box represents the approximate models that are mathematical representations of the real reservoir. These could be complex simulation models, analytical models, or machine learning models that relate the control variables to the objective functions. Because our knowledge of the reservoir is generally uncertain and data is noisy, the models and their output are also uncertain. The closed-loop process starts with an optimization performed over the current models to maximize or minimize the objectives. The optimization provides optimal settings of the controls for the next control step. These controls are then applied to the real reservoir over the control step, which impacts the outputs from the reservoir (such as water-cuts, bottom hole pressures [BHPs], etc.) which are measured. These measurements provide new information about the reservoir, and therefore enable the models to be updated (and model uncertainty to be reduced) by application of data assimilation algorithms. The optimization can then be again performed on the updated model over the next control step, and the process is repeated over the life of the field.

For these crucial decisions, operators traditionally rely on either semi-quantitative approaches such as decline curve analysis and simple analytical models, or on sophisticated simulation models. Analytical approaches are capable of using only a small portion of available data and function as rules of thumb, resulting in qualitatively optimized reservoir management decisions at best. On the other

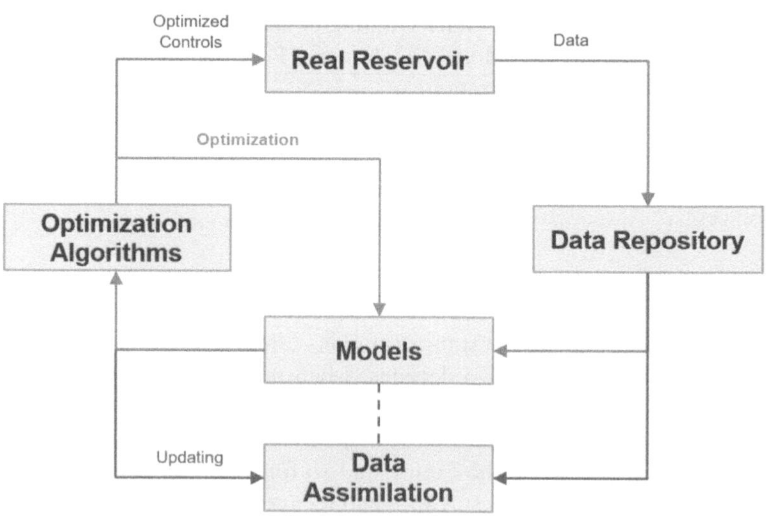

FIGURE 13.1 Closed-loop reservoir management.

extreme, reservoir simulation is the most advanced technique available today in that it is capable of integrating disparate data sources and predicting over long time horizons [Aziz and Settari, 1979]. While reservoir simulation is an excellent tool for field studies and long-term planning, certain limitations prevent operators from leveraging simulation for day-to-day decision-making:

1. Reliance on geo-modeling workflows requiring months to years of manual setup.
2. Simulation requires hours or days to run a single scenario due to computational complexity.
3. It is impractical to obtain optimal solutions given a limited set of scenarios.
4. Lack of uncertainty quantification because of an insufficient number of models.
5. Sequential data integration results in inconsistencies and failure to honor all the data.
6. Traditional models cannot process the large number of parameters required to increase resolution.
7. It is difficult to integrate data for which physical models are unknown or unclear.

In essence, reservoir simulation enables operators to precisely model the physics of a small number of scenarios; however, it is not designed to produce thousands or millions of scenarios and to automatically update those scenarios based on real-time data. Even for operators with very sophisticated simulators, there is a need for faster models to leverage real-time data, achieve prescriptive analytics, and optimize day-to-day operations.

Purely data-driven models lie on the opposite end of the spectrum compared with physics-based reservoir simulation. While simulation models require months to set up and days to run, machine learning models can be built in days and run for a full field in real time to rapidly explore thousands of scenarios and identify optimal solutions. Although such models offer a significant speed advantage and enable predictive analytics in domains such as unconventionals where the reservoir physics are relatively more difficult to model, the absence of underlying physics prevents machine learning's use for quantitative optimization. The main drawbacks of data-driven approaches include the following:

1. Models can only accurately predict already measured response in the reservoir, i.e., difficulty in extrapolating beyond historical data.
2. Susceptibility to significant prediction errors due to data quality issue, which can also lead to overfitting.
3. Predictions can often be non-physical and unreliable, given lack of physics in these models.
4. These models are usually unable to differentiate between surface and subsurface causes of production issues.
5. Poor predictivity over longer time horizons and changing reservoir conditions.

Physics-based reservoir simulation and data-driven machine learning offer complementary strengths. An ideal predictive model would combine the speed and flexibility of machine learning with the predictive accuracy of reservoir simulation so that operators could integrate data in real time to quantitatively optimize reservoir management decisions continuously. Such a model can be used to quantitatively optimize any future performance indicator such as NPV within a closed-loop reservoir management/optimization framework. Absent fast-running physically accurate models with quantitative production estimates and statistical confidence, such a closed-loop process has never been practically possible. With a closed-loop system where the model is constantly re-run and re-tuned, the reservoir model remains true to actual field conditions at all times and does not fall "out of date."

Data Physics models aim to bridge this gap between simulation and machine learning by including physical equations into machine learning models. The next section describes different approaches to do so and the benefits of such an approach. This is followed by a discussion of data assimilation approaches used in this work for assimilating data into these *Data Physics* models. The trained *Data Physics* models can then be used to optimize multiple objectives by application of multi-objective optimization (MOO) algorithms, which is discussed in the next section. The final section describes the application of the proposed approach to a real-world complex waterflood in Argentina, where this technology has been implemented since 2019.

13.2 DATA PHYSICS

The primary motivation of *Data Physics* is to create a framework in which petroleum engineers can quantitatively optimize oil production from a reservoir. The goal of quantitative optimization is to predict which actions–for example, on a daily, weekly, or monthly basis–will produce the maximum quantified (predicted and risk-adjusted) return in oil production. Investigating tens of thousands of scenarios that quantitatively simulate specific sets of activities (e.g., increased injection, changes in drilling programs, etc.) allows engineers to select operational plans that meet specific optimization criteria, such as preserving current production rates or optimizing long-term reserves growth. Often, it is necessary or desired to optimize multiple criteria simultaneously.

To achieve this goal, one must have both a quantitatively accurate and statistically proven predictive model of reservoir mechanisms. One must accurately be able to predict not only well-by-well behavior, but also correctly analyze well and pattern interdependencies. In each case, the physics must drive the predicted outcome, and statistical confidence must be calculated so that relative risk can be properly accounted for. These are the goals of the *Data Physics* approach.

To understand this approach, it will be useful to understand the key difference between traditional modeling and machine learning, as described in Figure 13.2.

In traditional modeling (or simulation), input data x (e.g., injection rates) and a functional mapping f (equations or instruction to apply on x) are given (e.g., Darcy's law), and the result of application of f on x is the output response $y = f(x)$

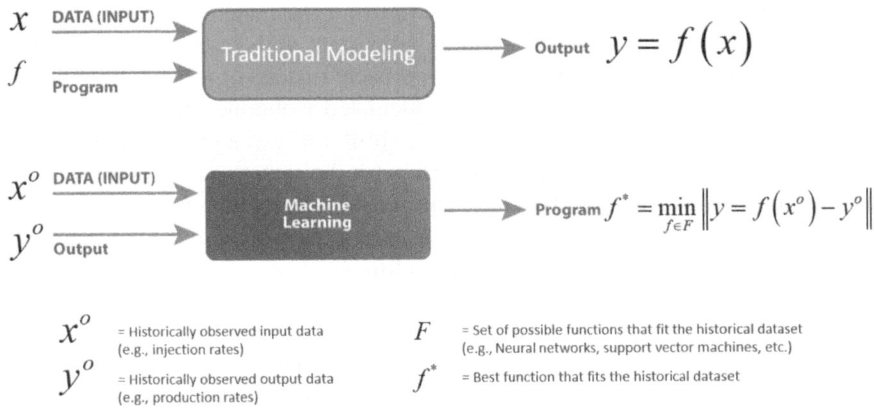

FIGURE 13.2 Traditional modeling vs machine learning.

(e.g., oil production rates). In machine learning, an opposite approach is taken, wherein, historical or observed inputs x^o and outputs y^o are given, and the objective is to find the best functional mapping f^* between them. Then f^* is found from a possible universe of general functions like neural networks [Haykin, 1999], support vector machines [Smola and Schölkopf, 1998], etc., by minimizing the error between the observed y^o and the calculated y obtained by applying f^* on x^o. As explained earlier, both approaches have different benefits and limitations, and the objective of *Data Physics* is to create a unified model that has the benefits of both approaches.

Figure 13.3 explains the basic idea behind the *Data Physics* approach. At their core, *Data Physics* models are machine learning models constrained by physics: the underlying physical equations (partial differential equations [PDEs] for mass

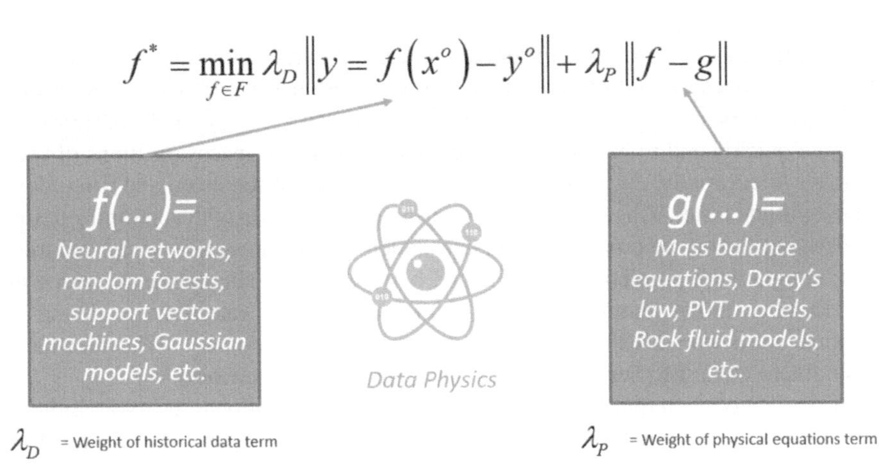

FIGURE 13.3 Combining machine learning and physics.

balance, heat balance, Darcy's law, etc.) constrain the machine learning models and thus limit solutions to those that are consistent with reservoir physics. The approach extends the machine learning approach: instead of just minimizing the difference between observed y^o and the calculated y obtained by applying f^* on x^o, the objective is to also find an f^* that is similar to given physical equations g (e.g., PDEs of fluid flow) that are appropriate for the system being modeled. This is done by adding another term to the minimization objective as seen in the figure; λ_D is the weight of the usual data term, while λ_P is the weight of the physics term. If $\lambda_P = 0$, this reverts back to traditional machine learning, if $\lambda_D = 0$, then this will be equivalent to traditional simulation.

The main benefits of the *Data Physics* approach are that, like machine learning models, these require only days to set up and can be run in real time as model evaluation is very fast. Additionally, because they include all the same physics as a reservoir simulation model, they offer excellent long-term predictive capacity and physically realistic responses, even when historical data is sparse, missing, or noisy.

While the previous approach describes one way to combine physics and machine learning [Sarma et al., 2017; Raissi et al., 2019], there are many related approaches [Kansa, 1990; Lagaris, 1998]. While the approach implemented here is similar to physics-informed neural networks (PINNs) [Raissi et al., 2019], a key difference is in the treatment of the physical equations as constraints rather than adding them to the objective function. While superficially the *Data Physics* approach might look similar to machine learning, the actual implementation is quite different and involved. Purely data-driven approaches such as traditional implementations of neural networks only utilize input-output behavior present in training data and do not involve any actual physical equations. Such methods ignore the physics and (incorrectly) assume the neural network will "learn" the equations. They consequently fail either by generating "non-physical" output solutions (i.e., output that is physically implausible or impossible) or when presented with data not anticipated in the training dataset (which causes the models to predict incorrectly).

Additionally, a significant difference when compared with traditional simulation is that, because *Data Physics* models seek to predict the production impact of certain activities—not to characterize the reservoir per se—they can be designed to operate quickly and to be used to explore wide possibility spaces (millions of scenarios). Accordingly, *Data Physics* models are optimized for speed and run orders of magnitude faster. This speed permits repeated comparison with and tuning to the underlying history of production and consequently permits a statistically quantifiable comparative prediction performance between many alternative scenarios in service of quantitative optimization. The PDEs solved for waterflood modeling as applied in this work are the black oil equations [Aziz and Settari, 1979], which have been found to be sufficient to model waterfloods all over the world.

There is no intent that *Data Physics* would replace traditional reservoir simulators for those tasks that require a geomodel with a high degree of accuracy. For example, for the waterflood optimization problem, *Data Physics* models define "accuracy" to mean "does the model correctly predict oil production in a certain scenario." *Data Physics* models are highly accurate in this regard and are easily re-run with real-time field and production data. The assumption remains that those tasks that require a full geo-model will still rely on existing discretized simulations. For example, the process of understanding geological scenarios, drive mechanisms, enhanced oil recovery (EOR) method screening, and green-field development decisions generally would still require simulations produced by existing, discretized simulation methods.

Figure 13.4 describes the *Data Physics* workflow. The process is divided into three steps. The first step is fitting or training the model. During this step, a portion of the available data is used to train the model. A variant of the ensemble Kalman filter (EnKF) described in the next section is used to assimilate the historical data into the model. The remaining data not used in training is used next to validate the model by comparing the actual data with the predicted data from the model as in a usual back test. Statistical measures like correlations can be used to understand the quality of the model. Additionally, other measures like posterior histograms of reservoir properties, connectivity, etc., can also be used to validate the model. This process usually takes a few weeks. After model validation, the model is deemed predictive, and large-scale MOO is performed to obtain the pareto front of optimal scenarios. This step is also discussed in detail in a later section.

13.3 DATA ASSIMILATION

The data assimilation approach used in this work is a combination of a modified EnKF with quadratic programming that enables assimilation of data from thousands of wells and different data sources with a relatively small ensemble. Additionally, because the EnKF is known to underestimate uncertainty, statistical techniques are used to correct the uncertainty estimates to conform with empirical estimates.

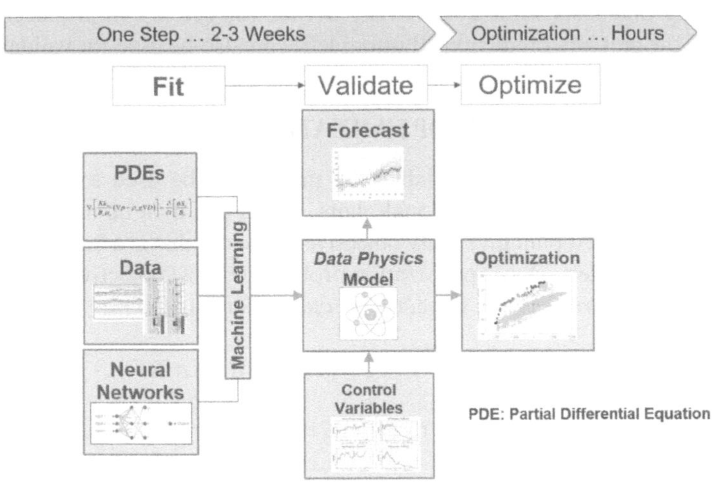

FIGURE 13.4 *Data Physics* workflow.

The predictive capacity of the calibrated models is demonstrated through a statistical back-test process, wherein, the model is fitted to part of the historical data (training data), and the predictions from the model are compared with the rest of the historical data (test data) at field and well level statistically. Details of the EnKF and its variants can be found in Evensen [2003], Naevdal et al. [2003], Wen and Chen [2005], Sarma and Chen [2009], and Lorentzen et al. [2012], among many others.

In our experience fitting fields with thousands of wells with EnKF, localization and inflation were not sufficient to address the previously mentioned issues to a sufficient degree of accuracy. As such, we applied two additional steps: (1) we solved an additional quadratic programming problem after the Kalman update to update local variables to fit the observed data within the specified error bounds and (2) we inflated the empirical ensemble variance of the predicted response (e.g., cumulative oil production) during the back-test phase such that the P90 to P10 range of the inflated ensemble contains the true cumulative production 80% of the time across all wells. This ensures that the uncertainty estimate of future performance is consistent with the historical data.

Once the *Data Physics* model is calibrated to the observed data as previously seen, the next step is to understand its predictive capacity. To do so, we divide the historical data into a training and a test set, fit the model only using the training data, and test the predictive capacity using the test data. As such, we can then compare the predicted response with the actual response, and thus statistically verify the predictive capacity of the model across many wells.

Figure 13.5 shows other means of validating the model. One of them is property histograms. The top left figure shows parameter convergence before and after training for permeability. Before training, the unconverged prior probabilities (gray) are broadly distributed. After training, the converged probabilities (black) are tightly clustered among the ensembles as would be expected. Additionally, other reservoir engineering diagnostics like connectivity between injectors and producers (right) and reservoir pressure predictions (bottom left) can also be used for validation.

13.4 MULTI-OBJECTIVE OPTIMIZATION

Once a predictive model is available, the model can be used to predict future performance for different operational plans, which can then be used to formally optimize for the best plan given any objective. As such, this is a "multi-objective" optimization problem. Any optimization problem with M objectives (cast in vector $\vec{F}$) and K constraints (cast in vector $\vec{C}$) can be formulated as:

$$min \ \vec{F}(\vec{x}) = (f_1(\vec{x}), f_2(\vec{x}), \ldots, f_M(\vec{x})) \in \Re^M$$

$$\text{subject to } c_k(\vec{x}) \leq d_k, \quad k = 1, K$$

where $\vec{x} \in X \leq \Re^N$ is the design vector and X the design space. If equality constraints $c^*(x) = d^*$ are to be imposed, these can be transformed into inequality

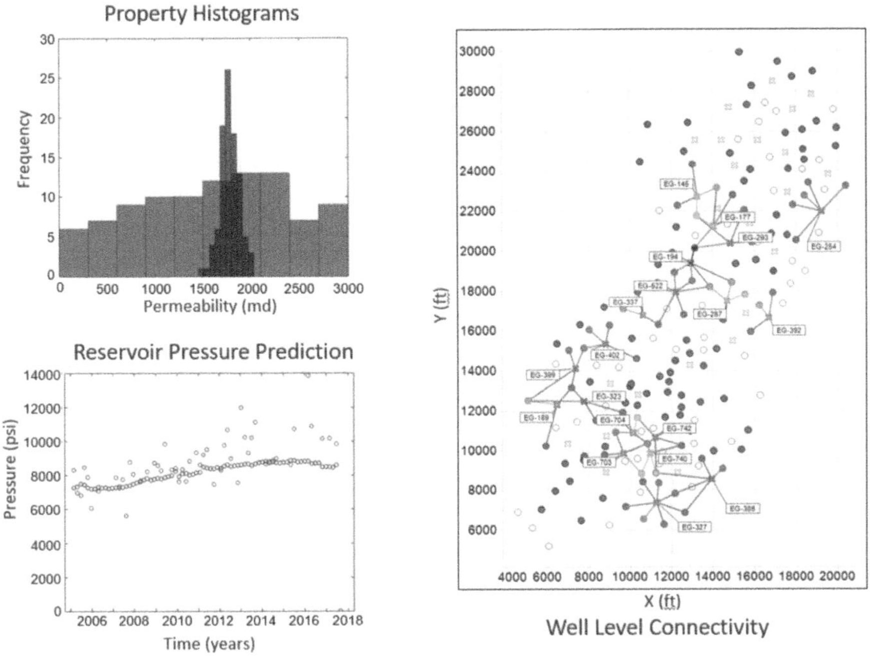

FIGURE 13.5 Different ways to validate model predictive ability.

ones $c(x) = \left\| c^*(x) - d^* \right\| \le d$, where $d \in \Re$ is an infinitesimally small number. If $M > 1$, then the above equation represents a MOO problem; in such a case, the notion of Pareto dominance [Zitzler et al., 2000] is used to associate a scalar cost value to each individual, based on which an algorithm solving single-objective optimization (SOO) problems can be employed. In MOO problems, a single run of the optimization algorithm should be capable of delivering a Pareto front of non-dominated individuals, rather than a single "optimal" solution, standing for an offset among the various objective functions.

Nowadays, evolutionary algorithms (EAs) have become very attractive across a wide range of scientific fields/disciplines, including engineering optimization, and will be the tool used herein. EAs can efficiently handle complex, constrained, single and multi–objective optimization problems, such as steam flood optimization, by accommodating any ready-to-use model as the aforementioned *Data Physics* models. Details on the EAs used in this work can be found in Kampolis and Giannakoglou [2008], Kyriacou [2013], and extensions using meta models can be found in Ong et al. [2003], Buche et al. [2005], Karakasis and Giannakoglou [2006], and Emmerich et al. [2002]. Additionally, further speedup is obtained by utilization of multi-processor systems. giving rise to the so-called parallel evolutionary algorithm (PEA) [Herrera et al., 1999; Alba and Troya, 2001; Karakasis et al., 2003; Ong and Keane, 2004; Asouti et al., 2009; Kampolis and Giannakoglou 2009]. Validation and results from this approach are presented in Kyriacou et al. [2017].

13.5 WATERFLOOD OPTIMIZATION CASE STUDY

Pan American Energy (PAE) operates several fields in Argentina, including the Cerro Dragon field, the largest in the country, with production of approximately 85,000 barrels of oil per day. Located in the Gulf of San Jorge basin (SJGB), the field has over 3,000 producers and more than 800 injectors. A pilot area with close to 140 wells (including 27 active injectors) was selected to test the *Data Physics* methodology with two key objectives: (1) validate its predictive capacity for complex geological environments and (2) generate opportunities for optimization and production increase. The pilot zone has been under water injection since the 1990s, and producing wells include different lifting mechanisms, including electrical submersible pump (ESP), progressive cavity pump (PCP), and mechanical pumps. At the beginning of the project the pilot zone produced about 500 m^3/day of oil, and injection was 10,000 m^3/day. Figure 13.6 shows the historical production and injection up to October 2018.

San Jorge Gulf is one of the most important hydrocarbon producing basins in Argentina. SJGB is characterized by multiple stacked reservoirs created by fluvial and lacustrine depositional environments. The most productive formations within this basin are Comodoro Rivadavia and Mina El Carmen located in the middle and upper Cretaceous. These formations are created in alluvial and

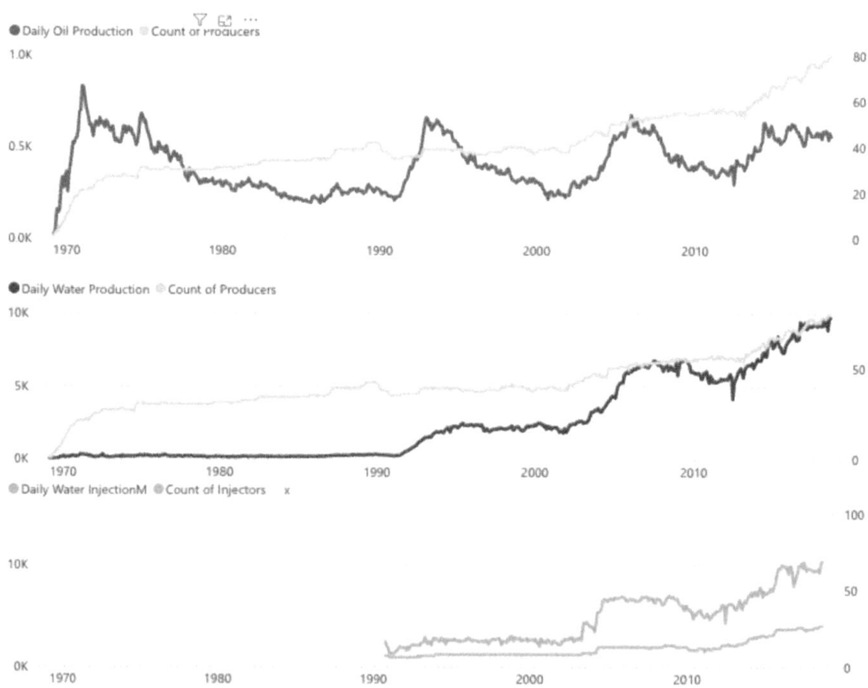

FIGURE 13.6 Oil production (top), water production (center), and water injection (bottom). Vertical axis is m^3/day.

river depositional systems and consist of sandstone bodies with varying thicknesses ranging from 2–10 m. Their effective porosity and permeablity are also variable. The porosity values are between 17% and 27%. The average permeability value is around 100 mD (with high degree of deviation); however, there are areas with higher permeability values of 500–1000 mD, which is believed to be the main reason for the different flow characteristics in the different geological units.

The traps are mainly sedimentary and structural (faults, intrusions, and folds). The main structures are an extensional type with descending blocks along regional lystric faults, generally of high angle and with associated minor faults. The depositional environment and the trap system gave rise to a system of differentiated units with variations in pressure, fluid properties, and rock types. These reservoirs are distributed along the sedimentary column and, depending on the location, depths can vary from 400–3,500 m. These hydraulic units are relatively small with an approximate width and length of 200–700 m and varying thicknesses. Due to their small sizes, drilled wells can pass through several reservoirs (up to 30) with non-productive layers in between. These non-productive layers are mainly clays that act as a flow barrier. On the other hand, these bodies are not vertically aligned; a well can go through different segments in different parts, which is why the well configurations are irregular to ensure more efficient contact with productive bodies. The aforementioned anisotropy in rock and fluid properties plus well layouts have a significant impact on flow behavior and consequently waterflood performance.

13.6 IMPLEMENTATION OF DATA PHYSICS IN THE ZORRO FIELD

Having analyzed the applicability of the technology, PAE selected the Zorro field, a portion of Cerro Dragon, for a pilot [Gomez et al., 2021]. To ensure that the benefits could be accurately assessed, other projects like polymer injection, in-fill drilling, etc., were suspended, thus ensuring that the impact on production would only be due to the injection changes prescribed. The Zorro field has over 80 production layers, typical for the SJGB. After discussing with the geologists and ensuring there would be enough data to train a multilayered model (i.e., enough number of wells drilled through each layer), these were grouped into eight flow units and the model trained in a multilayered manner with eight layers. This means that the optimization prescriptions would be provided for each of the eight individual layers.

The project was laid down with two project engineers, one from Tachyus and one from PAE, and weekly meetings were arranged to monitor progress. A Steering Committee was also named including management from both companies to align objectives, monitor progress, and ensure resources would be available when needed. The Steering Committee was to meet quarterly.

13.6.1 Step 1: Data Loading and Preparation

Initial data was provided around November 2018. Monthly production and injection data (flowrates and pressures when available) were exported from the existing databases and uploaded into the platform, together with well locations, formation tops and bottoms, information of known fault blocks, and deviation data for non-vertical wells.

Data from existing PVT analysis and ranges of the petro-physical parameters were also determined. The data from the 80+ layers was then consolidated in eight flow units and the production and injection time series adjusted accordingly. Several quality control (QC) verifications were performed ensuring data integrity over the whole field and time interval.

13.6.2 Step 2: Initial Model Training or Back Test

Once the data preparation was completed, a back test was performed. Although nearly 50 years of data were available, the last 14 years were selected for this step, as it is generally sufficient to use 10–15 years of data, and older data can have significant noise and is sometimes unreliable. The objective is to train and validate the model, in a single step. The ensemble size of the EnKF was selected as 96, meaning 96 independent models will be trained simultaneously. The training period from Jan 2004 to Dec 2013 and the prediction period from Jan 2014 to Oct 2018 (latest data available at the time) were selected. A single back test takes around 2 hours for this field on 32 core 48GB VMs on Google cloud.

Figure 13.7 shows the result of the back test for both oil (top) and gross (bottom) production. The black dots represent historical data (field production rate during each time step), the light gray dots represent the field production calculated for each ensemble member during the training period, and the dark gray lines represent the field production calculated during the prediction period. The black lines are the mean prediction of the ensemble. The light gray line in the back is the total number of active wells.

As can be seen, both oil and gross tendencies are well predicted by the model, indicating good predictivity even when the number of wells is increasing during the prediction period, an indication that the tool is able to predict the behavior of new wells that were not present during the training period. Predictivity is finally measured by means of correlation coefficients of cumulative production at the well level, and the results obtained were satisfactory with both Pearson and rank correlation coefficients above 0.7 for both oil and water. One further verification is to ensure that the petro-physical parameters inferred represent the knowledge of the corresponding reservoirs, which they did in this case.

13.6.3 Step 3: Final Model Creation and Full Fit

The next step was to extend the training of the model, to include the whole time interval, keeping the global petro-physical static parameters found in the previous

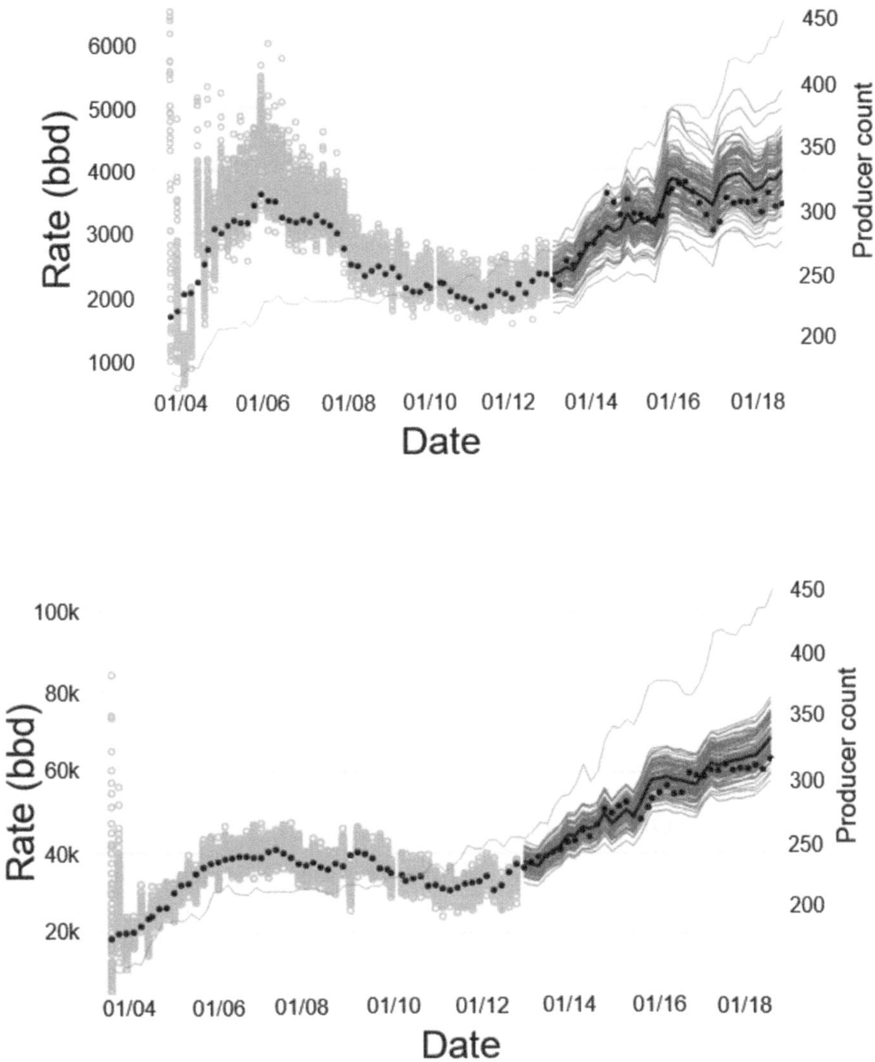

FIGURE 13.7 Ensemble training (dots) and prediction (lines) for oil (top) and gross (bottom).

step, but refining the local and dynamic parameters to further improve the overall predictive capacity of the model. Figure 13.8 shows the final result with correlation coefficients.

The plot shows predicted (*Y*-axis) vs real (*X*-axis) cumulative production per well during the last two-time steps to verify predictivity. Overall, the correlation coefficients obtained are above 0.7 for both oil and gross, showing good predictivity. This is the model that will be used for optimization. A single full fit takes about 2.5 hours for this field on 32 core 48GB VMs on Google cloud.

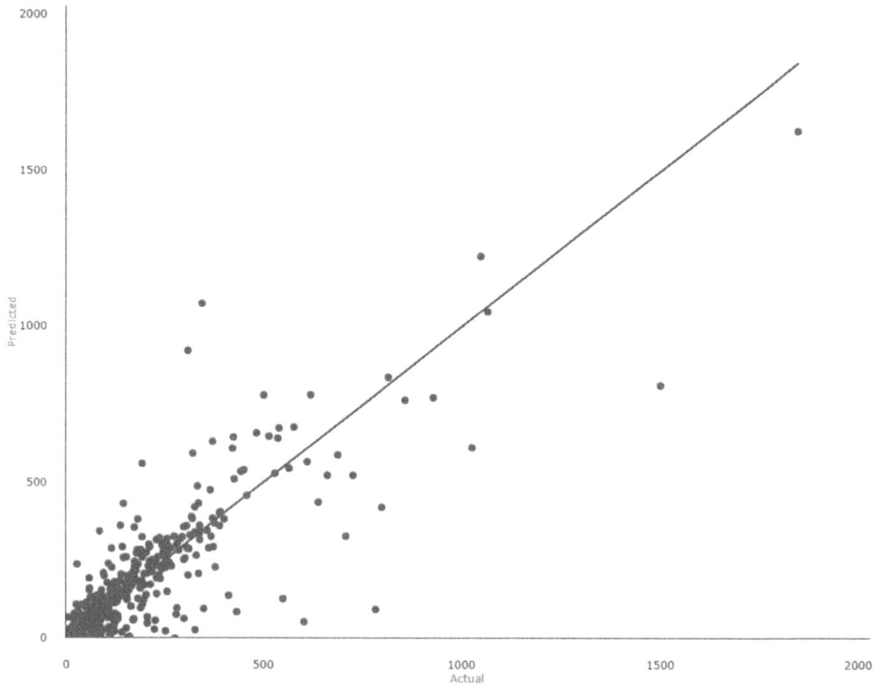

FIGURE 13.8 Training the model over the whole interval of historical data (oil above and gross below).

13.6.4 Step 4: Optimization

The final step is to generate the pareto front. This starts by generating a base case assuming that there will be no changes in the current injection strategy. This base case is then fed to EAs and thousands of variations of it run to find the pareto front. Because the platform allows for MOO, the objectives included maximization of oil production, minimization of water injection capacity, and maximization of NPV. All objectives were optimized over 2 and 15 years to evaluate the short-term and long-term impact of the optimizations. The resulting pareto is shown in Figure 13.9. The black triangle is the base case: average injection of 10,000 m^3/day and an average production of 338 m^3/day for 15 years. The tool was initially set up to include several operational constrains like maximum injection allowed per layer, ranges of liquid production corresponding to the existing pumps, etc.

The initial target scenario selected rendered a potential NPV increase of 8% over the 15 years on an injection redistribution scenario (Figure 13.10; black rhombus). This was selected as it implied no additional or leftover water injection. So, in theory, field implementation would be quite straightforward. This was not quite the case as we will see in the next paragraphs and this target scenario was modified later during the process.

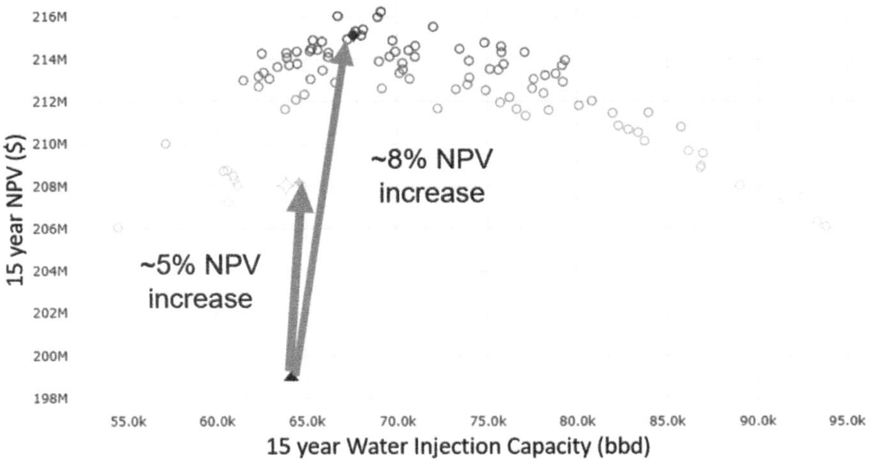

FIGURE 13.9 Pareto front for maximum oil, minimum water and maximum, NPV in 15 years.

The final step, before implementing these changes in the field, was to verify that the selected target scenario met the waterflood conformance criteria and that the prescriptions will be actionable at layer level, i.e., based on current knowledge, ensure a specific layer will admit the flowrate prescribed. It is an important consideration for PAE to ensure that an injector is injecting the amount of

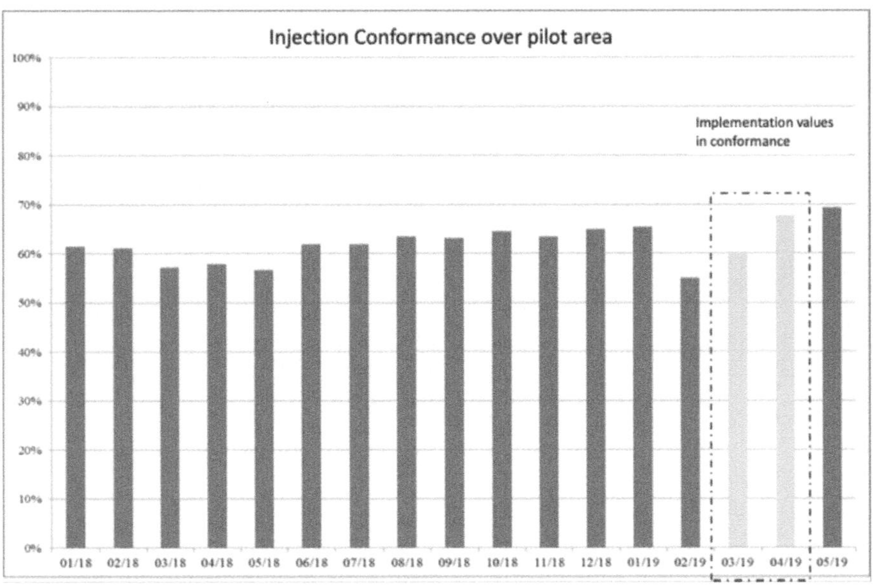

FIGURE 13.10 Historical and recommended injection values in conformance.

water prescribed in each layer. Conformance allows one to measure the injection quality of a waterflooding project. The values obtained are based on averages weighted by real flow vs theoretical flow of the layer. Then it is calculated at well level and project level. Having completed waterflood conformance verification, the target scenario was then finalized and sent to the field for implementation. The entire optimization took about 16 hours to run on 300 workers (4 cores, 4GB VMs) on Google cloud.

13.6.5 Step 5: Field Implementation

The prescribed scenario included changes in all 27 active injectors at each layer as shown in Figure 13.11, where each circle is an injector in a given layer, light gray indicates an increase and black decrease compared with the base case, and the size of the circle indicates the magnitude of the change. Some of the wells in some layers were to be shut-in as shown, e.g., Formation F in well PLMS-820, which went from injecting 59 m^3/day to injecting 0. An interesting observation here is that it is quite complex to comprehend and understand all these changes using basic reservoir engineering, and as such quantitative approaches for optimization are required to identify such optimal scenarios.

One final step before being able to implement in the field was to translate the information provided by the tool to be executable in the field; this means that the prescription of each layer needed to be allocated to the different mandrels in each well. Because the mandrel-layer combination is not identical in every well (i.e., mandrels and layers don't always coincide, sometimes one mandrel may be injecting in two or more layers and other times there may be more than one mandrel

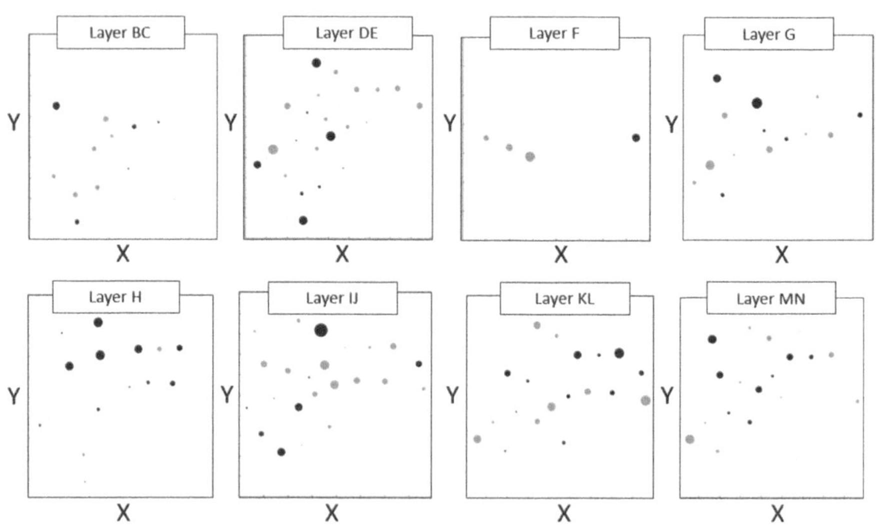

FIGURE 13.11 Layer level injection changes of the target plan.

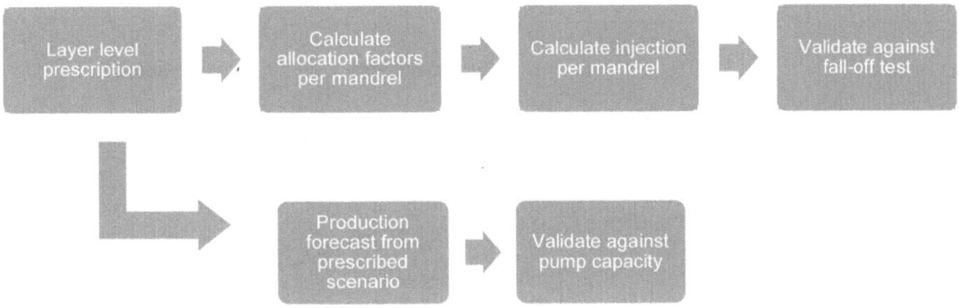

FIGURE 13.12 Pre-implementation of verification process.

injecting in one specific layer), there was a need to translate layer rates to mandrel rates. The resulting water injection prescription was validated against existing fall-off tests to ensure it could be executed in the field. Similarly, at producer level it was necessary to ensure that the resulting liquid production predictions would fall within the ranges of the existing pumps as there was no intention of doing pump changes at this time. The team working in the project designed the workflow shown in Figure 13.12 to execute this task.

This final target scenario differed from the original to some extent as several changes were introduced because of the verifications explained earlier. Field implementation of this final scenario started in February 2019 and finished in April 2019. This final scenario was run through the model, and while it did not fall on the pareto front, it was still expected to increase NPV by 5% compared with the base case. Overall flowrate changes were implemented in all the injectors, representing more than 200 valve operations; approximately 70 radioactive tracers jobs were also performed.

13.7 FIELD RESULTS

Figure 13.13 shows the impact of the first and second optimization cycles performed. The green curve is the daily field production. The original field decline rate can be seen from September 2018 to January 2018, before the start of the implementation of the prescriptions. This is extended as the solid line to represent the base case if optimization was not performed.

After the first implementation, the field reaction was quite fast and production decline flattened for about three months. At this point, some injectors were temporarily closed for workovers producing the drop in production observed in July 2019. Once these injectors came back on line production stabilized and declined at a much lower rate than the original declination. A second optimization cycle was performed in early 2020 which helped maintain the decline to be much lower than originally established. Overall accumulated production increased of close to 15% in the first two years of the project was achieved. The field production has generally behaved very close to the model proving the predictive capability of the tool, as demonstrated by additional blind tests to new events such as injector such downs.

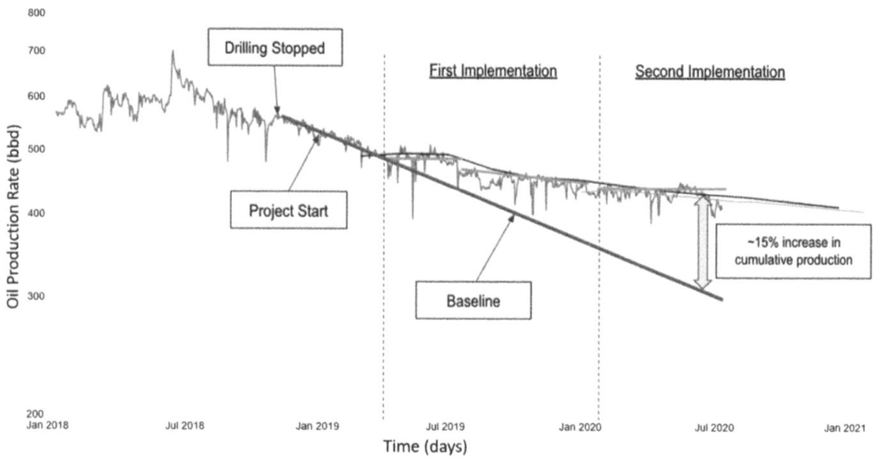

FIGURE 13.13 Daily production chart shows the impact of the optimization.

13.8 CONCLUSIONS

The application of *Data Physics*, a novel modeling and optimization approach, was presented. *Data Physics* Combines physics and machine learning for much faster turnaround times while providing predictive models with long predictive horizons. These models are orders of magnitude faster than conventional simulation models, enabling large-scale closed-loop quantitative optimization. We demonstrated application of this technology to a complex waterflood, where a 15% increase in cumulative production has been achieved since implementation. *Data Physics* models have also been applied to infill drilling optimization, steamflood optimization, fracking optimization, and surface and subsurface back allocation. Finally, like any other model, *Data Physics* models also have limitations, e.g., they can be applied only when significant good quality historical data is available, and when there is a good understanding of the appropriate physics needed for a particular problem. As the oil and gas industry accelerates adoption of digital transformation, technologies like *Data Physics* that enable better use of data in operational decision-making will see increasing demand.

REFERENCES

Alba, Enrique, and José M. Troya. 2001. "Analyzing Synchronous and Asynchronous Parallel Distributed Genetic Algorithms." *Future Generations Computer Systems: FGCS* 17 (4): 451–65.

Asouti, V. G., I. C. Kampolis, and K. C. Giannakoglou. 2009. "A Grid-Enabled Asynchronous Metamodel-Assisted Evolutionary Algorithm for Aerodynamic Optimization." *Genetic Programming and Evolvable Machines* 10 (4): 373–89.

Aziz, K., and A. Settari. 1979. *Petroleum Reservoir Simulation*. Applied Science Publishers, London, 135–39.

Buche, D., Nicol N. Schraudolph, and Petros Koumoutsakos. 2005. "Accelerating Evolutionary Algorithms with Gaussian Process Fitness Function Models." *IEEE Transactions on Systems, Man and Cybernetics. Part C, Applications and Reviews: A Publication of the IEEE Systems, Man, and Cybernetics Society* 35 (2): 183–94.

Emmerich, Michael, Alexios Giotis, Mutlu Özdemir, Thomas Bäck, and Kyriakos Giannakoglou. 2002. "Metamodel—Assisted Evolution Strategies." *Lecture Notes in Computer Science* 2439, 361–70.

Evensen, Geir, 2003. "The Ensemble Kalman Filter: Theoretical Formulation and Practical Implementation." *Ocean Dynamics* 53 (4): 343–67.

Gomez, Javier Eusebio, et al. 2021. "Application of a Combination of Machine Learning and Reservoir Physics to Increase Production in a Mature Waterflood Field in the San Jorge Basin in Argentina." *Abu Dhabi International Petroleum Exhibition & Conference*. OnePetro database, Society of Petroleum Engineers.

Haykin, Simon. 1999. *Neural Networks: A Comprehensive Foundation*. Prentice Hall, Upper Saddle River, NJ.

Herrera, F., M. Lozano, and C. Moraga. 1999. "Hierarchical Distributed Genetic Algorithms." *International Journal of Intelligent Systems* 14 (11): 1099–121.

Jansen, J.-D., R., Brouwer, and S. G. Douma. 2009. "Closed Loop Reservoir Management." *SPE Reservoir Simulation Symposium*. Society of Petroleum Engineers. doi:10.2118/119098-MS

Kampolis, Ioannis C., and Kyriakos C. Giannakoglou. 2008. "A Multilevel Approach to Single- and Multiobjective Aerodynamic Optimization." *Computer Methods in Applied Mechanics and Engineering* 197 (33–40): 2963–75.

Kampolis, Ioannis C., and Kyriakos C. Giannakoglou. 2009. "Distributed Evolutionary Algorithms with Hierarchical Evaluation." *Engineering Optimization* 41 (11): 1037–49.

Kansa, Edward J. 1990. "Multiquadrics—A scattered data approximation scheme with applications to computational fluid-dynamics—I surface approximations and partial derivative estimates." *Computers & Mathematics with Applications* 19 (8–9): 127–45.

Karakasis, Marios K., and Kyriakos C. Giannakoglou. 2006. "On the Use of Metamodel-Assisted, Multi-Objective Evolutionary Algorithms." *Engineering Optimization* 38 (8): 941–57.

Karakasis, Marios K., Alexios P. Giotis, and Kyriakos C. Giannakoglou. 2003. "Inexact Information Aided, Low-Cost, Distributed Genetic Algorithms for Aerodynamic Shape Optimization." *International Journal for Numerical Methods in Fluids* 43 (10–11): 1149–66.

Kyriacou, Stylianos A. 2013. "Evolutionary Algorithm-based Design-Optimization, Methods in Turbomachinery." PhD Thesis, NTUA.

Kyriacou, Stylianos, Pallav Sarma, and Ian Hunt. 2017. "Constrained, multi-objective, steamflood injection redistribution optimization, using a cloud-distributed, metamodel-assisted, memetic optimization algorithm." *SPE Reservoir Characterisation and Simulation Conference and Exhibition*. OnePetro database, Society of Petroleum Engineers.

Lagaris, Isaac E., Aristidis Likas, and Dimitrios I. Fotiadis. 1998. "Artificial Neural Networks for Solving Ordinary and Partial Differential Equations." *IEEE Transactions on Neural Networks* 9 (5): 987–1000.

Lorentzen, Rolf J., Kristin M., Flornes, and Geir Nævdal. 2012. "History Matching Channelized Reservoirs Using the Ensemble Kalman Filter." *SPE Journal* 17 (1): 137–51.

Naevdal, Geir, Liv M. Johnsen, Sigurd I. Aanonsen, and Erlend H. Vefring. 2003. Reservoir Monitoring and Continuous Model Updating Using Ensemble Kalman Filter. *SPE paper 84372 presented at the SPE Annual Technical Conference and Exhibition*, Denver, CO.

Ong, Yew S., and Andy J. Keane. 2004. "Meta-Lamarckian Learning in Memetic Algorithms." *IEEE Transactions on Evolutionary Computation* 8 (2): 99–110.

Ong, Yew S., Prasanth B. Nair, and Andrew J. Keane. 2003. "Evolutionary Optimization of Computationally Expensive Problems via Surrogate Modeling." *AIAA Journal* 41 (4): 687–96.

Raissi, Maziar, Paris Perdikaris, and George E. Karniadakis. 2019. "Physics-Informed Neural Networks: A Deep Learning Framework for Solving Forward and Inverse Problems Involving Nonlinear Partial Differential Equations." *Journal of Computational Physics* 378: 686–707.

Sarma, Pallav, and Wen H. Chen. 2009. "Generalization of the Ensemble Kalman Filter using Kernels for Non-Gaussian Random Field." *The Proceeding of the SPE Reservoir Simulation Symposium*, Houston, TX.

Sarma, Pallav, et al. 2017. "Redistribution of Steam Injection in Heavy Oil Reservoir Management to Improve EOR Economics, Powered by a Unique Integration of Reservoir Physics and Machine Learning." *SPE Latin America and Caribbean Petroleum Engineering Conference*. OnePetro database, Society of Petroleum Engineers.

Smola, Alex J., and Bernhard Schölkopf. 1998. *Learning with Kernels*. Vol. 4. GMD-Forschungszentrum Informationstechnik.

Wen, Xian-Huan, and Wen H. Chen. 2005. "Real-time Reservoir Model Updating Using Ensemble Kalman Filter." *Paper SPE 92991 presented at the SPE Reservoir Simulation Symposium*, Houston, TX.

Zitzler, Eckart, Kalyanmoy Deb, and Lothar Thiele. 2000. "Comparison of Multiobjective Evolutionary Algorithms: Empirical Results." *Evolutionary Computation* 8 (2): 173–95.

14 Deep Neural Network Surrogate Flow Models for History Matching and Uncertainty Quantification

Su Jiang and Louis Durlofsky
Department of Energy Resources Engineering,
Stanford University, Stanford, California

CONTENTS

14.1 INTRODUCTION

The model parameters that characterize subsurface formations are inevitably uncertain, and this uncertainty leads to a wide range of possible outcomes in flow predictions. In the context of oil/gas production or CO_2 storage, for example, the spatial distribution of porosity and permeability, along with rock-fluid properties such as relative permeability and capillary pressure, are at best only known at well locations. Geomodeling software enables the construction of geological realizations that honor these data and are consistent with prior geological concepts, but the range of flow predictions corresponding to these realizations is still usually very large. The application of history matching, which entails the calibration of model

DOI: 10.1201/9781003207009-19

parameters such that the resulting geomodels provide simulation results in essential agreement with field observations, leads to a reduction in prediction uncertainty. The history-matched models are therefore better suited for use in reservoir management decisions. There are, however, many challenges associated with the application of history matching that can limit its effectiveness, as we now discuss.

The computational demands for uncertainty quantification and history matching (also referred to as data assimilation) in subsurface flow settings are very substantial. This is because thousands to millions of forward model evaluations, i.e., flow simulations, may be required for these applications. These evaluations, if performed using high-fidelity simulation models, require the discrete space-time solution of a set of coupled nonlinear partial differential equations (PDEs). Because practical models can contain $O(10^5–10^7)$ grid blocks, and there can be multiple unknowns per block, an individual model can entail hours of computation. Demands can become intractable if a wide range of geological concepts and model uncertainties are considered, or if rigorous data assimilation procedures are applied.

Another complication associated with history matching is the need to maintain geological realism, i.e., consistency with prior models, in the posterior (history-matched) models generated by the data assimilation procedure. This is generally not a problem if the geomodel properties are described by multi-Gaussian spatial distributions, but it can represent a challenge when geometric features such as channels or fractures are present. There has been a considerable amount of recent work on the development of parameterization methods based on deep neural networks (DNNs). These approaches can provide concise representations of geometrically complex systems, with parameters that can be determined independently. One such method will be used in the results presented later in this chapter.

Surrogate or proxy modeling procedures are methods that provide fast but approximate estimates for the key outputs from an underlying numerical model. Such procedures are clearly of interest within the context of history matching and uncertainty quantification, as they can potentially provide the requisite function evaluations with much less computation. Surrogate models can range from simple statistical regressions to reduced-order numerical methods to deep neural network procedures. In some applications, such as production optimization where the goal is to determine the well settings that maximize a scalar objective function (e.g., net present value), methods that predict just the objective function value can be effective. In history matching problems, however, the goal is to construct models that provide detailed well production/injection rates as a function of time, and possibly also spatial distributions of pressure and saturation (if 4D seismic data are available) in agreement with observations. Thus, the requirements for the surrogate model are much more demanding in this case.

In this chapter, we will review a range of surrogate models relevant for history matching and uncertainty quantification in subsurface flow systems. The focus will be on DNN-based surrogate models, though we will discuss a particular family of reduced-order numerical models in the Appendix. Our discussion of DNN models will include methods that contain physics-based components, in the form of numerical residual terms, and approaches based solely on simulation data from a large number of training runs (data-driven methods). Example results with a data-driven DNN surrogate, for flow predictions and history matching of an oil-water problem in a 3D channelized system, will be presented.

14.2 DNN SURROGATE MODELS: LITERATURE REVIEW AND METHODOLOGIES

Our emphasis here will be on surrogate modeling procedures that provide estimates for the spatial distribution and temporal evolution of the global system states. By this we mean that in an oil-water problem, for example, the surrogate model provides estimates of pressure and water saturation in every grid block in the model – not just estimates for well rates, or for a scalar mismatch quantity. Surrogate models of this type may also be applicable for other reservoir management applications such as optimization.

In the past several years, there has been extensive research involving the use of DNNs for surrogate flow modeling. In our discussion here, we will classify these methods into two categories – physics-informed or theory-guided neural networks (referred to as PINNs and TgNNs), and data-driven DNN models. Physics-informed methods seek to incorporate flow physics into the DNN framework by minimizing a physical loss. This loss includes the residual associated with the governing PDEs, as well as error in initial conditions and boundary conditions. In theory, PINNs can be trained without any simulation data (which means no training simulations need to be performed), though they have been shown to benefit when simulation data are also provided. Data-driven DNN methods, by contrast, require large amounts of labeled data (i.e., high-fidelity simulation results) for training.

In this section, we describe the methods in the context of two-phase subsurface flow. We then describe network architectures and loss functions for PINNs/TgNNs and data-driven surrogate models.

14.2.1 Governing Equations for Two-Phase Flow

We consider the flow of two immiscible fluids, which can be viewed as water and oil (in an oil production setting) or water and a nonaqueous phase liquid (in an aquifer remediation setting). The equations for CO_2 and water, in a carbon storage setting, are similar but slightly more complicated. The system is 3D, with gravitational effects included. The governing equations are

$$\nabla\cdot\left(\rho_j\mathbf{u}_j\right)+q_j+\frac{\partial}{\partial t}\left(\phi\rho_j S_j\right)=0,\ j=a,\ na, \tag{14.1}$$

where j denotes phase, a represents the aqueous phase, na represents the non-aqueous or oil phase, ρ_j is the density of phase j, $\mathbf{u}_j$ is the Darcy velocity of phase j, q_j denotes the source/sink term, ϕ is porosity, and S_j is the saturation of phase j. Darcy velocity $\mathbf{u}_j$ is expressed as

$$\mathbf{u}_j=-\frac{\mathbf{k}k_{rj}\left(S_j\right)}{\mu_j\left(p_j\right)}\left(\nabla p_j-\rho_j g\nabla z\right),\ j=a,\ na, \tag{14.2}$$

where **k** denotes the absolute permeability tensor (usually taken to be diagonal), k_{rj} is the relative permeability of phase j, which is generally a nonlinear function of S_j, p_j is the pressure of phase j, z is depth, and μ_j denotes the phase viscosity. In the absence of capillary pressure, $p_a = p_{na} = p$.

Finite volume or finite-element methods are typically applied to discretize Eqs. (14.1) and (14.2). The discrete equations are nonlinear, and Newton's method is used for their solution. The overall simulation procedure can be expressed as

$$\mathbf{x} = [\mathbf{P}, \mathbf{S}] = g(\mathbf{m}), \tag{14.3}$$

where **m** denotes the geomodel that is input to the simulator (e.g., permeability and porosity in every grid block), g indicates the numerical simulation process, and **x** denotes the simulation output, with **P** and **S** the pressure and saturation state variables in every block in the model. With a surrogate model, we represent this process as $\hat{\mathbf{x}} = \tilde{g}(\mathbf{m}, \theta)$, where $\tilde{g}$ denotes the surrogate modeling procedure, $\hat{\mathbf{x}}$ the surrogate model output, and θ the tunable network parameters.

14.2.2 Physics-Informed/Theory-Guided Neural Networks

Raissi et al. (2019) first proposed PINNs for solving nonlinear PDEs. Their method applied fully connected neural networks (FC-NN) to solve various governing equations, including the Navier-Stokes equations for 2D incompressible flow. Zhu et al. (2019) introduced a physics-constrained convolutional encoder-decoder approach. The encoder constructs latent features from the high-dimensional input, while the decoder generates high-dimensional output from the latent variables. A Sobel filter (Gao et al., 2010) was used to estimate the spatial gradients and to calculate residual loss, and the surrogate was trained by minimizing the residual loss without any forward simulations. Zhu et al. (2019) applied this method to predict steady-state single-phase flow responses (the governing flow equation is the single-phase analog of Eqs. 14.1 and 14.2) for new 2D Gaussian and channelized permeability fields. Their procedure is thus applicable for history matching and uncertainty quantification, though temporal evolution was not considered.

Combining data mismatch with physics-based loss in the training objective has been shown to improve prediction accuracy. Wang et al. (2020) introduced a TgNN to predict hydraulic head (essentially pressure) for single-phase compressible flow in 2D Gaussian models. With this approach, forward simulations are performed to provide a labeled training set, and the data mismatch is considered in the training objective. An FC-NN was applied to predict hydraulic head for specified simulation times and locations within the domain. The training objective included data loss, physics loss, and engineering loss (constraints on controls). The original TgNN model was designed to generate the future response of a particular system (fixed geomodel and control settings). TgNN was then shown to provide reasonable temporal extrapolation results, i.e., predictions for future times.

TgNN was further developed by Wang et al. (2021a) to provide flow predictions for new Gaussian geomodel realizations, over a range of log-permeability variance, as well as with different pressure boundary values. Karhunen-Loeve expansion (essentially principal component analysis [PCA]) was applied to parameterize the Gaussian models. A total of 150 simulations were used (along with a physics-based loss) to train the surrogate model. The TgNN model was then applied in a Monte Carlo setting for uncertainty quantification. A theory-guided auto-encoder based on TgNN was developed by Wang et al. (2021c). Permeability fields provided the input, and convolutional neural networks (CNNs) were applied in an encoder-decoder framework to perform image-to-image mapping. CNNs, which are widely used in image processing, entail the application of a series of filtering operations. They are well suited for capturing spatially correlated fields and input-to-output mappings. A range of PINN-type methods, with both residual loss and data mismatch, have also been developed and tested by other investigators. These include He et al. (2020) and Tartakovsky et al. (2020, 2021), who applied PINNs for single-phase steady-state flow and for transport (advection-dispersion) problems.

Two representative PINN architectures are shown in Figure 14.1. Figure 14.1a shows the architecture for an FC-NN, as was applied by Raissi et al. (2019) and Wang et al. (2020). Simulation time and spatial coordinates are inputs to the model. Multiple fully connected layers are used as hidden layers to model the forward process. The output includes the state variables at the specified time and locations. The FC-NN is trained by minimizing the residual loss associated with the governing equations, losses for boundary and initial conditions, etc. (a detailed expression will be given later). The spatial and time derivatives of the governing Eqs. (14.1) and (14.2) are calculated based on auto-differentiation of the network. The FC-NN can be trained with or without any labeled data, i.e., training simulation runs. As noted previously, the FC-NN has been successfully applied for single-phase steady-state and transient flow problems and for advection-dispersion modeling. The residual loss is typically calculated at $O(10^3\text{–}10^5)$ "collocation" points.

Figure 14.1b displays a convolutional encoder-decoder network, of the type developed by Zhu et al. (2019) and Wang et al. (2021c). These networks provide an input image to output image mapping, with the permeability field representing the input. A "time matrix," indicating the times at which simulation output is required, is treated as another input channel. The CNN-based encoder transforms the input image to low-dimensional latent features, whereas the CNN-based decoder constructs the PDE solution from the latent variables. The use of an FC-NN would be impractical with a detailed input map (such as a permeability field) because it would require an extremely large number of parameters to be tuned. A CNN-based encoder-decoder treatment, by contrast, is suitable for such cases and can capture the most important latent features of the spatial distribution. Spatial gradients can be calculated by applying numerical finite difference to the surrogate output or by using a Sobel filter.

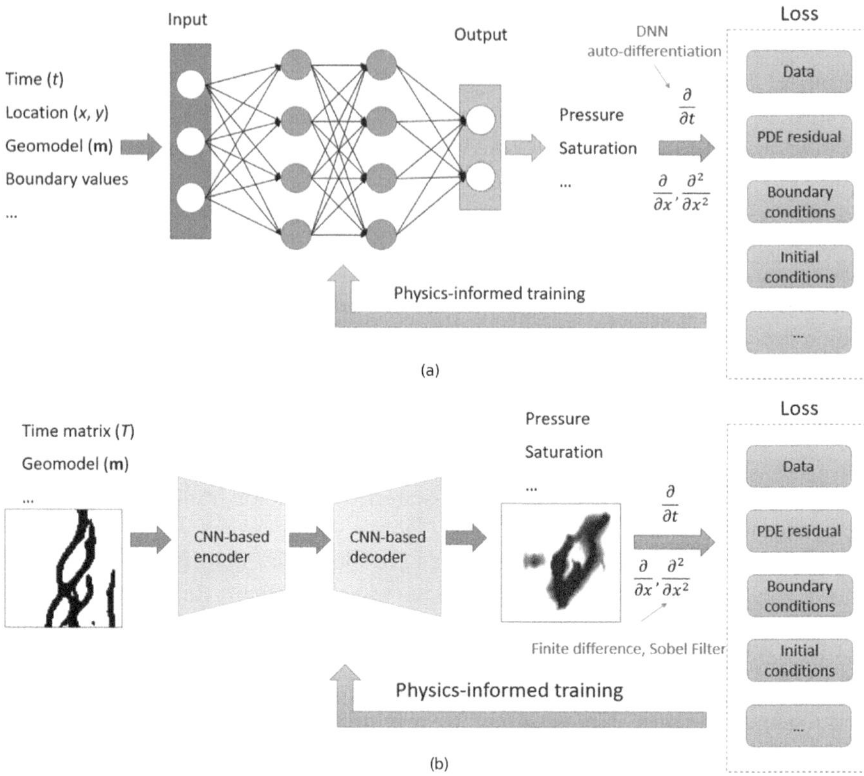

FIGURE 14.1 Neural network architectures for PINN/TgNN procedures (a) surrogate model based on fully connected neural network (after Raissi et al., 2019; Wang et al., 2020) and (b) surrogate model based on convolutional encoder-decoder network (after Zhu et al., 2019; Wang et al., 2021c).

The PINN/TgNN surrogate model is trained by minimizing a loss function that includes violation of the underlying PDEs, initial and boundary conditions, and labeled simulation data (if available). The residual loss L_{res} is expressed as

$$L_{res}(\theta) = \frac{1}{n_c}\frac{1}{n_{ts}}\sum_{i=1}^{n_c}\sum_{t=1}^{n_{ts}}\left\| R\left(\hat{\mathbf{x}}_i^t, \mathbf{x}_i^t\right)\right\|_p^p, \tag{14.4}$$

where n_c denotes the number of collocation points (which are random or prescribed locations at which the physical constraints are satisfied), n_{ts} is the number of simulation time steps, and $R\left(\hat{\mathbf{x}}_i^t, \mathbf{x}_i^t\right)$ is the residual violation, with superscript t indicating time. The losses for boundary and initial conditions are

$$L_{bc}(\theta) = \frac{1}{n_{bc}}\sum_{i=1}^{n_{bc}}\left\|\hat{\mathbf{x}}_i^{bc} - \mathbf{x}_i^{bc}\right)\right\|_2^2, \; L_{ic}(\theta) = \frac{1}{n_{ic}}\sum_{i=1}^{n_{ic}}\left\|\hat{\mathbf{x}}_i^{0} - \mathbf{x}_i^{0}\right)\right\|_2^2, \tag{14.5}$$

where n_{bc} and n_{ic} denote the number of data used to satisfy boundary and initial conditions. The data loss is expressed as

$$L_d(\theta) = \frac{1}{n_{smp}} \sum_{i=1}^{n_{smp}} \left\| \hat{\mathbf{x}}_i - \mathbf{x}_i \right\|_p^p, \tag{14.6}$$

where n_{smp} is the number of training samples. The overall PINN/TgNN loss function L is written as

$$L(\theta) = \lambda_{res} L_{res}(\theta) + \lambda_{bc} L_{bc}(\theta) + \lambda_{ic} L_{ic}(\theta) + \lambda_d L_d(\theta), \tag{14.7}$$

where λ_{res}, λ_{bc}, λ_{ic}, and λ_d denote weightings for each of the loss terms. PINN and TgNN methods include the residual violation in the overall loss L, which means mass balance considerations enter in the training. Because L involves additional terms, and because L_{res} is not driven to zero, these methods will not in general achieve strict mass balance.

Physics-informed methods have not yet been successfully applied for complicated multiphase subsurface flow problems involving heterogeneous 3D models. Recent investigations by Fuks and Tchelepi (2020) and Fraces et al. (2020) suggested that pure PINNs (without training simulation data and using current frameworks) may fail to provide accurate predictions for PDEs with fluctuating or multiscale features, which are common in subsurface flow problems. Fuks and Tchelepi (2020) applied such a method to model the 1D Buckley-Leverett equation. They showed that their PINN provided reasonable predictions for concave fractional flow functions, though it failed when a shock appeared. Fraces et al. (2020) achieved improved results for this problem when data were included, though their results still showed some error. Similar findings were obtained by Wang et al. (2021c). Rodriguez-Torrado et al. (2021) proposed a physics-informed attention-based neural network for this problem and showed it to accurately predict 1D Buckley-Leverett solutions with shocks. In attention-based methods, an additional weight vector is introduced to represent the importance of different inputs. The attention framework improves prediction accuracy in this setting by automatically identifying the shock location. Training time and convergence may, however, represent challenges for attention-based methods in high-dimensional problems. It will be of interest to extend this model to treat realistic 3D multiphase flow systems.

14.2.3 Data-Driven DNN Surrogate Models

The second set of approaches is data-driven DNN surrogate models. Data-driven procedures model the input-output mapping corresponding to the simulation process. All the training samples are labeled with simulation output. Supervised learning is then applied to train the surrogate models. The impact of the spatial distribution of the geomodel properties on the temporal evolution of the flow dynamics can be captured through use of CNNs and recurrent strategies in the

DNN framework. Constraints based on the underlying physics are generally not introduced in these methods, though Jin et al. (2020a) showed that the incorporation of some physics, specifically flux continuity, was effective. Data-driven methods require a large number (e.g., several thousand) of training samples to capture the input-output relationship. Because each sample requires a full-physics high-fidelity flow simulation, the offline (preprocessing) component of these approaches is expensive.

Zhu and Zabaras (2018) developed one of the earliest data-driven procedures. They applied a convolutional encoder-decoder network to predict the velocity and pressure fields in single-phase steady-state flow problems. Mo et al. (2019b) introduced time as another input channel in their convolutional encoder-decoder, which they used to predict pressure and saturation in 2D CO_2 storage problems. An autoregressive method was introduced by Mo et al. (2019a) to capture the temporal evolution of contaminant concentration in 2D transport problems. Mo et al. (2020) further incorporated a "residual-in-residual dense block" to improve prediction accuracy for 2D and 3D solute transport models.

Tang et al. (2020) developed a deep-learning-based surrogate model that used a residual U-Net (R-U-Net) to capture spatial information and a convolutional long short-term memory (convLSTM) recurrent network to capture temporal evolution. LSTMs are widely used to represent temporally correlated features. Tang et al. (2020) first applied their model for 2D oil-water problems with wells under bottom hole pressure (BHP) control. This framework was later extended to treat 3D systems characterized by channelized geomodels (Tang et al., 2021b,c) and 3D CO_2 sequestration problems (Tang et al., 2021a). The latter setup involved coupled flow and quasi-static geomechanics, and the recurrent R-U-Net was shown to provide accurate predictions for several quantities including vertical displacement maps at the Earth's surface. The number of training runs required for the results presented by Tang et al. (2020, 2021a,b,c) ranged from 1500–2500.

Wen et al. (2021) developed a related residual U-Net framework to model CO_2 plume migration in carbon storage problems. They considered an axisymmetric (r-z) system with one vertical injection well at the center of the model. The surrogate model was designed to handle new permeability fields, within the range of the training realizations, and injection variables such as duration and rate. A total of 20,000 simulations was used to train the model. Transfer learning (with hundreds of new samples) was applied to fine-tune the surrogate model when injection duration extrapolation was required. Transfer learning entails the use (after limited retraining) of a neural network for a setup that differs from that considered in the initial training.

Data-driven DNN surrogate models require a large amount of training data, but they are capable of providing accurate solutions for dynamic multiphase flow in complicated settings. CNNs, designed to capture spatial correlations, are commonly used in encoder-decoder networks in data-driven surrogate models. The encoder-decoder CNN architecture facilitates the propagation of spatial information, enabling efficient high-dimensional modeling. The temporal evolution of state variables can be captured by adding time as a continuous input (Mo et al., 2019b), through use of an autoregressive strategy (Mo et al., 2019a), or by using recurrent neural networks (Tang et al., 2020, 2021b).

We now provide an overview of the recurrent residual U-Net (recurrent R-U-Net) introduced by Tang et al. (2020, 2021b). The simulation input $\mathbf{m}$ represents the geomodel, and the simulation output $\mathbf{x}$ includes the pressure $\mathbf{P}$ and saturation $\mathbf{S}$ values in every grid block at a set of n_t time steps (here $n_t < n_{ts}$, where $n_{ts}$is the number of time steps in the simulation). The simulation process is approximated by the recurrent R-U-Net shown in Figure 14.2, which combines the residual U-Net and convolutional LSTM networks. The encoding net in the residual U-Net maps the input permeability field $\mathbf{m}$ to latent features $\mathbf{F}_1, \ldots, \mathbf{F}_5$. Latent feature $\mathbf{F}_5$ is the most compressed feature, and this is fed to the convLSTM cell to construct features $\mathbf{F}_5^1, \mathbf{F}_5^2, \ldots, \mathbf{F}_5^{n_t}$ for all time steps. The decoding net generates the prediction $\hat{\mathbf{x}}^t$ using latent features $\mathbf{F}_1, \ldots, \mathbf{F}_4$ from the encoding net and $\mathbf{F}_5^t$ from the convLSTM cell. See Tang et al. (2020, 2021b) for details on the network architecture.

The minimization applied for training with this procedure can be written as:

$$\theta^* = \underset{\theta}{\operatorname{argmin}} \frac{1}{n_{smp}} \frac{1}{n_t} \sum_{i=1}^{n_{smp}} \sum_{t=1}^{n_t} \left\| \hat{\mathbf{x}}_i^t - \mathbf{x}_i^t \right\|_p^p + \lambda_w \frac{1}{n_{smp}} \frac{1}{n_t} \frac{1}{n_w} \sum_{i=1}^{n_{smp}} \sum_{t=1}^{n_t} \sum_{w=1}^{n_w} \left\| \hat{\mathbf{x}}_i^{t,w} - \mathbf{x}_i^{t,w} \right\|_p^p. \tag{14.8}$$

Here λ_w is an (optional) additional weighting that can be used for the states at the n_w well blocks. Enhanced accuracy at these locations is important for history-matching problems, as it is the well-rate data we typically seek to match. In Tang et al. (2020, 2021b), separate networks were trained for the saturation and pressure fields. It was found that the use of $p = 2$ (L_2 norm) provided better accuracy for saturation, while $p = 1$ gave slightly better accuracy for pressure.

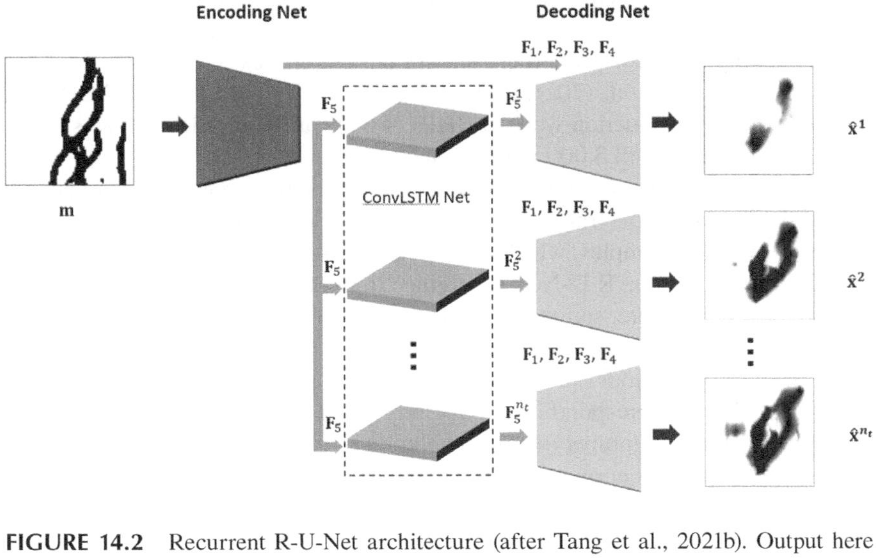

FIGURE 14.2 Recurrent R-U-Net architecture (after Tang et al., 2021b). Output here corresponds to the saturation field at different time steps.

With sufficient training, the data-driven methods discussed earlier appear to be capable of generating accurate predictions for complex dynamic multiphase flow problems in 2D and 3D. A substantial number of training runs are required, however, and the simulation setup is often restricted, e.g., fixed initial and boundary conditions, and has fixed well locations. In addition, data-driven methods do not have explicit mechanisms for satisfying physical constraints such as mass balance. Within the context of the recurrent R-U-Net, mass balance error is difficult to assess because the "effective" time step is generally quite large (recall the solution is estimated at only n_t time steps, with $n_t < n_{ts}$). It is possible that innovative combinations of PINNs/TgNNs and data-driven procedures may circumvent some of the limitations of both families of methods, and work along these lines should be pursued.

14.3 RESULTS USING RECURRENT R-U-NET SURROGATE MODEL

The recurrent R-U-Net surrogate has been successfully applied for oil-water flow in 2D and 3D channelized models (Tang et al., 2020, 2021b,c). In this section, we present results for the 3D channelized system considered by Tang et al. (2021c). Although PINN/TgNN methods have been developed for two-phase flow in 2D Gaussian models (Wang et al., 2021b), their use for 3D channelized systems of the type considered here has not yet been demonstrated. The results presented in this section, therefore, only involve the recurrent R-U-Net method.

The well configuration considered here – three injection wells and five production wells – differs from the setup in Tang et al. (2021c). The 3D channelized models are defined on $80 \times 80 \times 20$ grids. The geomodels are parameterized using the 3D CNN-PCA representation (Liu and Durlofsky, 2021). A total of $N_r = 3{,}000$ realizations are generated using the CNN-PCA model from Tang et al. (2021c). Three of these realizations are shown in Figure 14.3. The setup for the flow problem, including reservoir properties, relative permeabilities, viscosities, etc., is as described by Tang et al. (2021c). The injection wells operate at fixed BHPs of 330 bar, and the production wells at BHPs of 310 bar. High-fidelity flow simulation is performed for all 3,000 realizations.

We use 2,500 of these simulation runs to train the recurrent R-U-Net and 100 for validation within the training procedure. Model performance is assessed with the remaining 400 samples, which are not included in the training and validation sets. Separate recurrent R-U-Nets are trained to predict pressure and saturation at $n_t = 10$ time steps. Water saturation predictions for a typical realization are shown in Figure 14.4 (for clarity, only results for layer 7 are displayed). Panels (a)–(c) show the surrogate model predictions at 50, 400, and 1,000 days, while panels (d)–(f) display the corresponding simulation results. We see that the 3D recurrent R-U-Net accurately captures the time evolution of the saturation front.

Production and injection rates are computed using the pressure and saturation values in well blocks along with a 3D well model. Well-rate predictions for two injectors and two producers, for the geomodel in Figure 14.3b, are presented in

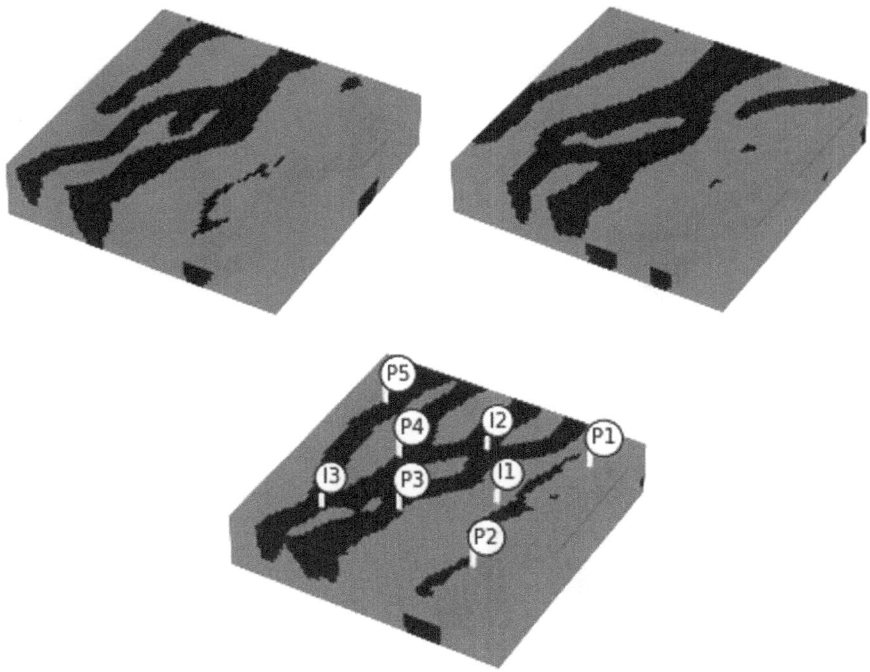

FIGURE 14.3 Three different 3D CNN-PCA realizations honoring rock type at three injectors and five producers shown in (a)–(c), with well locations shown in (c). (a) Model is used as the "true" model in the history-matching example. Cutaway views are shown.

Figure 14.5. The predicted well rates are in generally close agreement with reference high-fidelity simulation results. Predictions for flow statistics over the full set of 400 test-case runs are presented in Figure 14.6. Here we show P_{10}, P_{50}, and P_{90} curves, which correspond to the 10th, 50th, and 90th percentile results over all runs at each of the $n_t = 10$ time steps. Surrogate predictions for these important statistical quantities clearly agree, over a large range of flow responses, with those from high-fidelity simulation.

We now demonstrate the use of the surrogate model for history matching. Rejection sampling (RS) is applied, with the recurrent R-U-Net used to provide flow rate predictions. RS is a rigorous though very expensive data assimilation procedure. It considers a large number of randomly selected geomodels and accepts those that provide close agreement with observed data, while rejecting the samples with large data mismatch. Measurement error is included in the procedure. Please see Tang et al. (2021c) for a detailed description of the RS technique applied here.

The "true" model in this case is the realization shown in Figure 14.3a. Observed data for this assessment are taken to be simulated oil and water rates from the five producers at 100 and 300 days, with random noise, as described by Tang et al. (2021c), added. For RS, we generate 10^6 realizations (using CNN-PCA) and evaluate flow responses for all models using the trained recurrent R-U-Net. A total of

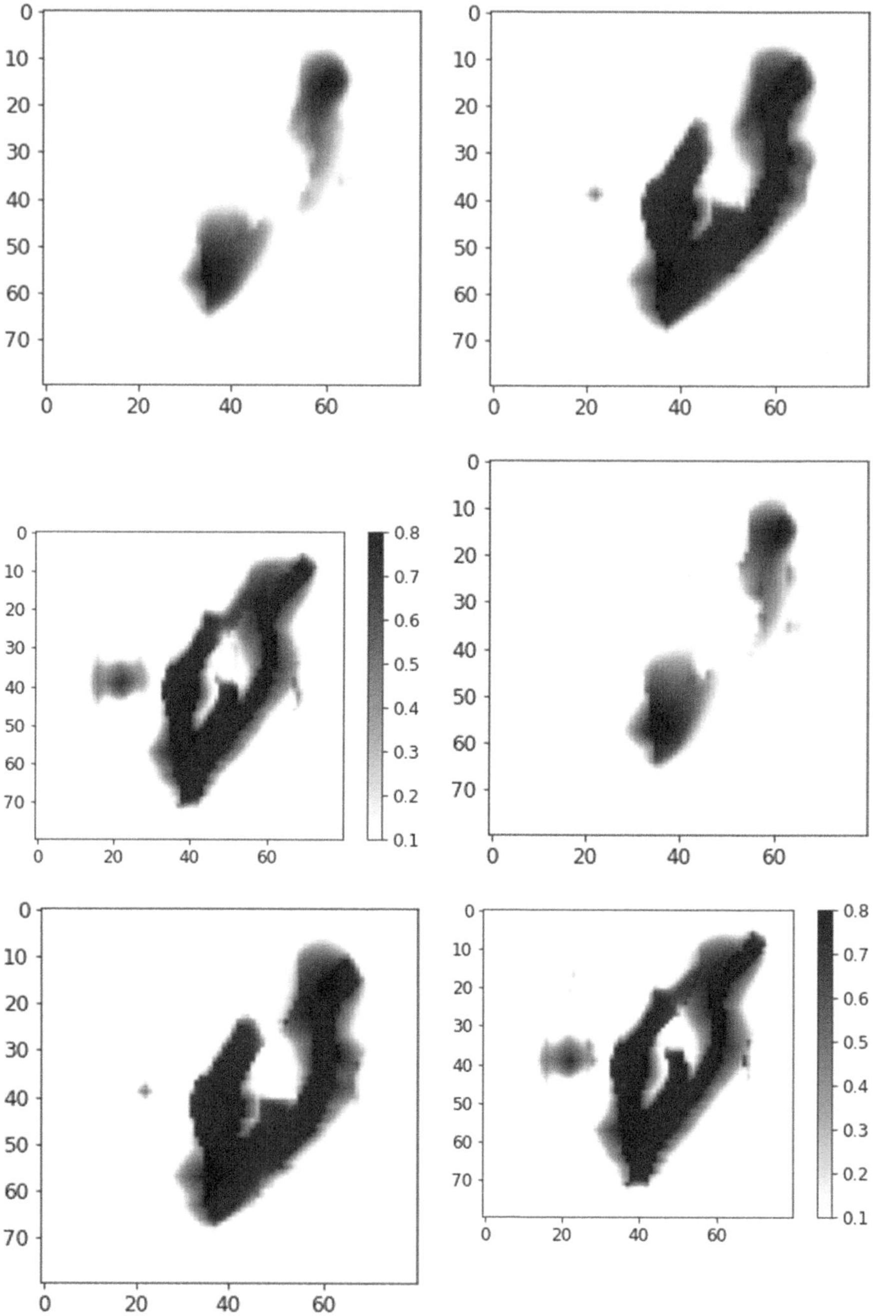

FIGURE 14.4 Water saturation maps for layer 7 of the model shown in Figure 14.3a from 3D recurrent R-U-Net surrogate (a–c) and high-fidelity simulation (d–f) at three different times.

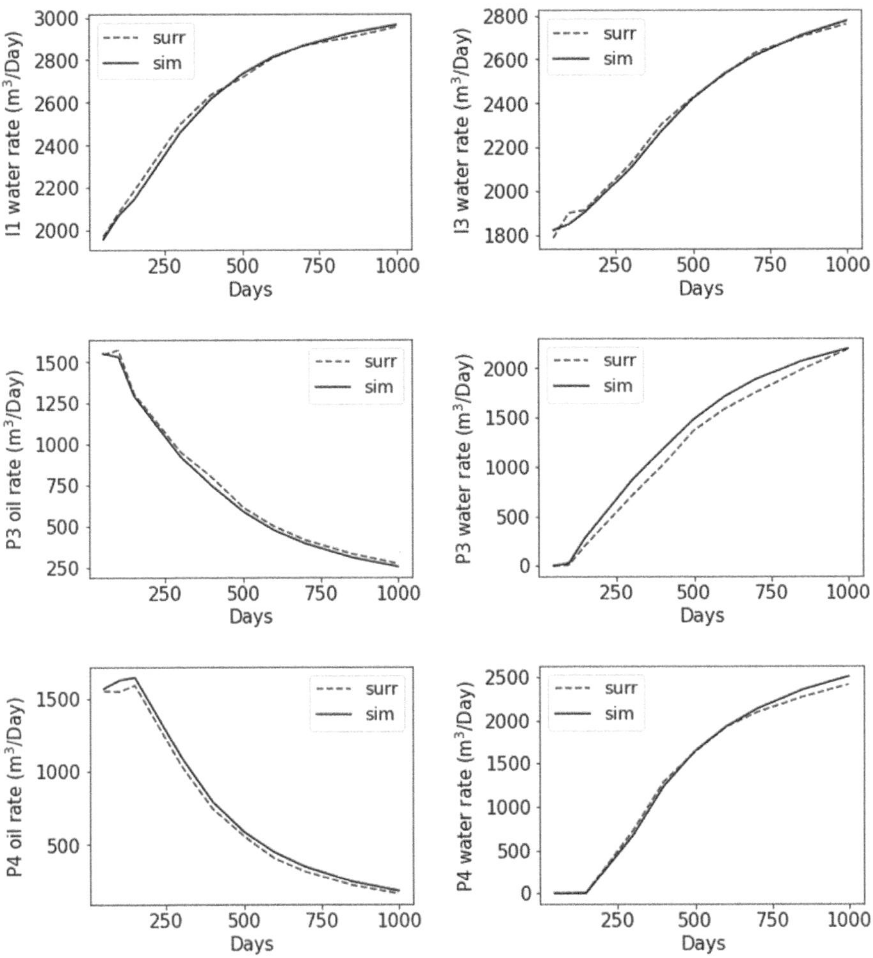

FIGURE 14.5 Comparison of water injection rates for wells I1 and I3 (panels a–b), and oil and water production rates for wells P3 and P4 (panels c–f), for the geomodel shown in Figure 14.3b. Dashed and solid lines represent results from surrogate model and high-fidelity simulation. Selected wells are largest water injectors and oil producers.

70 models are accepted. Prior and posterior (history matched) results are presented in Figure 14.7. The gray region denotes the prior P_{10}–P_{90} range, whereas the dashed curves display the P_{10}, P_{50}, P_{90} surrogate model posterior results. Circles represent observed data and the +'s show the true-model response. It is evident that, even with a small number of observations, RS-based history matching achieves substantial uncertainty reduction for several quantities. The run time (online) speedup achieved using the surrogate model is dramatic – a factor of over 10,000, as reported by Tang et al. (2021b). As discussed earlier, however, the offline computations (training simulations and network training) are quite time-consuming.

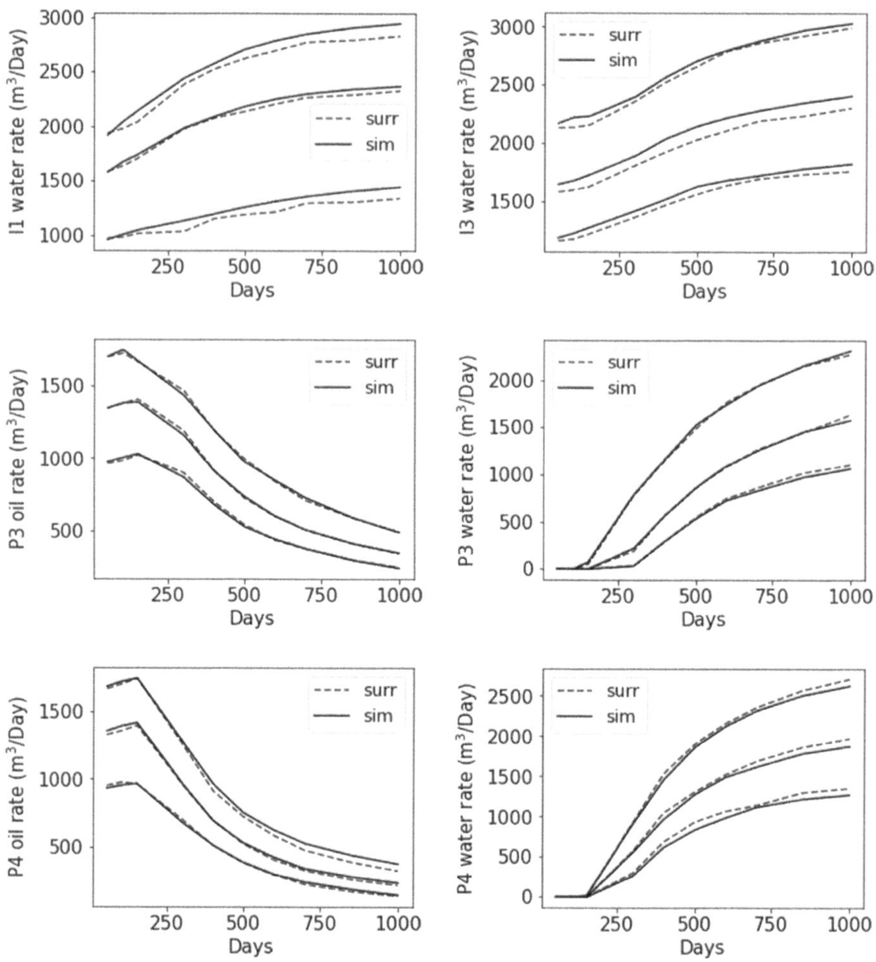

FIGURE 14.6 Comparison of water injection and oil and water production rate statistics for wells I1 andI3 (panels a–b), and P3, and P4 (panels c–f). Lower, middle, and upper curves correspond to P_{10}, P_{50}, and P_{90} results, respectively. Dashed and solid lines represent results from surrogate model and high-fidelity simulation.

Finally, we perform high-fidelity simulations on the posterior models found by RS to evaluate the performance of the surrogate model in this setting. The solid curves show the P_{10}, P_{50}, P_{90} simulation results. We observe reasonably close agreement with surrogate model predictions, though some discrepancies are evident. This may be because the true-model flow response is near the edge of the prior distribution (as is apparent in Figure 14.7), meaning there are relatively few training samples in this region. The fact that we still observe a high degree of surrogate model accuracy for this case is very encouraging.

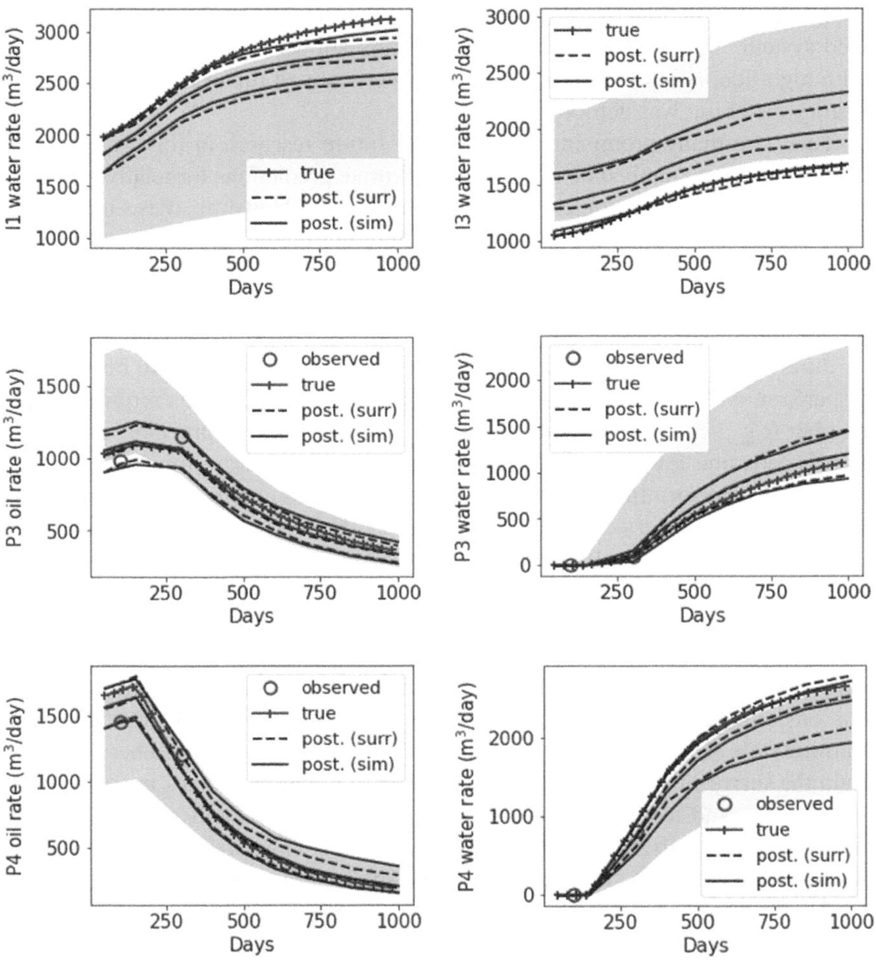

FIGURE 14.7 Prior and posterior injection and production rate results for wells I1 and I3 (panels a–b), and P3 and P4 (panels c–f). Gray regions indicate the prior P_{10}–P_{90} range, dashed and solid curves display P_{10}, P_{50}, P_{90} posterior results for surrogate and high-fidelity simulation models, circles show observed data (which include random noise), and +'s denote "true" (simulated) data. Posterior results are generated using RS, 3D CNN-PCA geomodel generation, and recurrent R-U-Net flow evaluation.

14.4 SUMMARY AND FUTURE OUTLOOK

In this chapter, we described physics-informed and data-driven deep neural network surrogate models (reduced-order numerical models are discussed in the Appendix). We considered two representative frameworks for physics-informed models, involving fully connected and convolutional neural networks. The recurrent R-U-Net data-driven surrogate model was then presented. This model was applied for predictions and history matching for 3D oil-water flow in a channelized system. The surrogate model results were shown to be in close agreement with high-fidelity simulation results, and the efficacy of the overall procedure for history matching was demonstrated.

There are many promising directions for future research in this general area. Physics-informed methods have provided accurate predictions for relatively simple problems with limited (or even zero) training runs. However, these techniques have not yet been demonstrated for realistic 3D cases involving complex multiphase flow physics. The use of other types of networks may enable the extension of these approaches to problems of practical interest. Along these lines, attention-based architectures, as applied by Rodriguez-Torrado et al. (2021), may be beneficial.

Purely data-driven methods such as the recurrent R-U-Net have been shown to perform well for complicated systems, though these methods require a large number (e.g., several thousand) of training simulations. For practical cases with $O(10^6)$ cells and a wide range of uncertainties, even more training samples will be required. Performing these high-fidelity simulations and training the surrogate model will thus be very expensive computationally. Methods to reduce these computational demands should, therefore, be developed. Given that most data-driven methods essentially perform input-to-output image mapping, it may be useful to incorporate some amount of physics-based loss (though perhaps not the full residual as in PINNs) into data-driven surrogate models. The determination of appropriate loss functions could lead to reductions in the number of training simulations required. Another approach is to use multifidelity simulation data in the training step, as introduced by Jiang and Durlofsky (2022). The idea here is to first train the surrogate model with a large number of low-fidelity (e.g., upscaled) flow simulations, and to then apply transfer learning, with a small number of high-fidelity solutions, to fine-tune the surrogate model. This can provide substantial savings in the training simulations required for data-driven surrogate models.

Finally, it will be useful to extend the deep-learning-based methodologies discussed here such that the trained surrogate is applicable for realizations drawn from a range of geological scenarios rather than just for a single scenario. Treatments for varying well controls and locations, and for different reservoir boundary conditions, should also be developed. Computational advances may enable the more robust forward modeling of highly complicated coupled multiphysics problems involving, e.g., combined flow, geomechanical, thermal, and chemical reaction effects (as can arise in CO_2 storage applications). Large-scale high-fidelity simulations of this type may become feasible, but they will likely remain challenging and expensive. Thus, the development of surrogate models for such cases could enable comprehensive history matching, uncertainty quantification, and optimization to be applied.

ACKNOWLEDGMENTS

We are grateful to Meng Tang for providing the recurrent R-U-Net code and for useful discussions on DNN-based surrogate modeling.

REFERENCES

Cardoso, Marco A., and Louis J. Durlofsky. "Linearized reduced-order models for subsurface flow simulation." *Journal of Computational Physics* 229, no. 3 (2010): 681–700.

Cardoso, Marco A., Louis J. Durlofsky, and Pallav Sarma. "Development and application of reduced-order modeling procedures for subsurface flow simulation." *International Journal for Numerical Methods in Engineering* 77, no. 9 (2009): 1322–1350.

Carlberg, Kevin, Charbel Bou-Mosleh, and Charbel Farhat. "Efficient non-linear model reduction via a least-squares Petrov–Galerkin projection and compressive tensor approximations." *International Journal for Numerical Methods in Engineering* 86, no. 2 (2011): 155–181.

Chaturantabut, Saifon, and Danny C. Sorensen. "Nonlinear model reduction via discrete empirical interpolation." *SIAM Journal on Scientific Computing* 32, no. 5 (2010): 2737–2764.

Fraces, Cedric G., Adrien Papaioannou, and Hamdi Tchelepi. "Physics informed deep learning for transport in porous media. Buckley Leverett problem." *arXiv preprint arXiv:2001.05172* (2020).

Fuks, Olga, and Hamdi A. Tchelepi. "Limitations of physics informed machine learning for nonlinear two-phase transport in porous media." *Journal of Machine Learning for Modeling and Computing* 1, no. 1 (2020).

Gao, Wenshuo, Xiaoguang Zhang, Lei Yang, and Huizhong Liu. "An improved Sobel edge detection." In *2010 3rd International Conference on Computer Science and Information Technology*, vol. 5, pp. 67–71. IEEE. (2010).

Ghommem, Mehdi, Eduardo Gildin, and Mohammadreza Ghasemi. "Complexity reduction of multiphase flows in heterogeneous porous media." *SPE Journal* 21, no. 01 (2016): 144–151.

He, Jincong, and Louis J. Durlofsky. "Reduced-order modeling for compositional simulation by use of trajectory piecewise linearization." *SPE Journal* 19, no. 05 (2014): 858–872.

He, Jincong, Pallav Sarma, and Louis J. Durlofsky. "Reduced-order flow modeling and geological parameterization for ensemble-based data assimilation." *Computers & Geosciences* 55 (2013): 54–69.

He, Qizhi, David Barajas-Solano, Guzel Tartakovsky, and Alexandre M. Tartakovsky. "Physics-informed neural networks for multiphysics data assimilation with application to subsurface transport." *Advances in Water Resources* 141 (2020): 103610.

Jiang, Rui, and Louis J. Durlofsky. "Implementation and detailed assessment of a GNAT reduced-order model for subsurface flow simulation." *Journal of Computational Physics* 379 (2019): 192–213.

Jiang, Su, and Louis J. Durlofsky. "Use of multifidelity training data and transfer learning for efficient construction of subsurface flow surrogate models." *arXiv preprint arXiv:2204.11138* (2022).

Jin, Zhaoyang Larry, Timur Garipov, Oleg Volkov, and Louis J. Durlofsky. "Reduced-order modeling of coupled flow and quasistatic geomechanics." *SPE Journal* 25, no. 01 (2020a): 326–346.

Jin, Zhaoyang Larry, Yimin Liu, and Louis J. Durlofsky. "Deep-learning-based surrogate model for reservoir simulation with time-varying well controls." *Journal of Petroleum Science and Engineering* 192 (2020b): 107273.

Liu, Yimin, and Louis J. Durlofsky. "3D CNN-PCA: A deep-learning-based parameterization for complex geomodels." *Computers & Geosciences* 148 (2021): 104676.

Mo, Shaoxing, Nicholas Zabaras, Xiaoqing Shi, and Jichun Wu. "Deep autoregressive neural networks for high-dimensional inverse problems in groundwater contaminant source identification." *Water Resources Research* 55, no. 5 (2019a): 3856–3881.

Mo, Shaoxing, Nicholas Zabaras, Xiaoqing Shi, and Jichun Wu. "Integration of adversarial autoencoders with residual dense convolutional networks for estimation of non-Gaussian hydraulic conductivities." *Water Resources Research* 56, no. 2 (2020): e2019WR026082.

Mo, Shaoxing, Yinhao Zhu, Nicholas Zabaras, Xiaoqing Shi, and Jichun Wu. "Deep convolutional encoder-decoder networks for uncertainty quantification of dynamic multiphase flow in heterogeneous media." *Water Resources Research* 55, no. 1 (2019b): 703–728.

Raissi, Maziar, Paris Perdikaris, and George E. Karniadakis. "Physics-informed neural networks: A deep learning framework for solving forward and inverse problems involving nonlinear partial differential equations." *Journal of Computational Physics* 378 (2019): 686–707.

Rodriguez-Torrado, Ruben, Pablo Ruiz, Luis Cueto-Felgueroso, Michael Cerny Green, Tyler Friesen, Sebastien Matringe, and Julian Togelius. "Physics-informed attention-based neural network for solving non-linear partial differential equations." *arXiv preprint arXiv:2105.07898* (2021).

Tang, Meng, Xin Ju, and Louis J. Durlofsky. "Deep-learning-based coupled flow-geomechanics surrogate model for CO_2 sequestration." *arXiv preprint arXiv: 2105.01334* (2021a).

Tang, Meng, Yimin Liu, and Louis J. Durlofsky. "A deep-learning-based surrogate model for data assimilation in dynamic subsurface flow problems." *Journal of Computational Physics* 413 (2020): 109456.

Tang, Meng, Yimin Liu, and Louis J. Durlofsky. "Deep-learning-based surrogate flow modeling and geological parameterization for data assimilation in 3D subsurface flow." *Computer Methods in Applied Mechanics and Engineering* 376 (2021b): 113636.

Tang, Meng, Yimin Liu, and Louis J. Durlofsky. "History matching complex 3D systems using deep-learning-based surrogate flow modeling and CNN-PCA geological parameterization." In *SPE Reservoir Simulation Conference.* (2021c).

Tartakovsky, Alexandre M., David A. Barajas-Solano, and Qizhi He. "Physics-informed machine learning with conditional Karhunen-Loève expansions." *Journal of Computational Physics* 426 (2021): 109904.

Tartakovsky, Alexandre M., C. Ortiz Marrero, Paris Perdikaris, Guzel D. Tartakovsky, and David Barajas-Solano. "Physics-informed deep neural networks for learning parameters and constitutive relationships in subsurface flow problems." *Water Resources Research* 56, no. 5 (2020): e2019WR026731.

van Doren, Jorn F. M., Renato Markovinović, and Jan-Dirk Jansen. "Reduced-order optimal control of water flooding using proper orthogonal decomposition." *Computational Geosciences* 10, no. 1 (2006): 137–158.

Vermeulen, P. T. M., Arnold W. Heemink, and Chris B. M. Te Stroet. "Reduced models for linear groundwater flow models using empirical orthogonal functions." *Advances in Water Resources* 27, no. 1 (2004): 57–69.

Wang, Nanzhe, Haibin Chang, and Dongxiao Zhang. "Efficient uncertainty quantification for dynamic subsurface flow with surrogate by theory-guided neural network." *Computer Methods in Applied Mechanics and Engineering* 373 (2021a): 113492.

Wang, Nanzhe, Haibin Chang, and Dongxiao Zhang. "Surrogate and inverse modeling for two-phase flow in porous media via theory-guided convolutional neural network." *arXiv preprint arXiv:2110.10080* (2021b).

Wang, Nanzhe, Haibin Chang, and Dongxiao Zhang. "Theory-guided auto-encoder for surrogate construction and inverse modeling." *Computer Methods in Applied Mechanics and Engineering* 385 (2021c): 114037.

Wang, Nanzhe, Dongxiao Zhang, Haibin Chang, and Heng Li. "Deep learning of subsurface flow via theory-guided neural network." *Journal of Hydrology* 584 (2020): 124700.

Wen, Gege, Meng Tang, and Sally M. Benson. "Towards a predictor for CO_2 plume migration using deep neural networks." *International Journal of Greenhouse Gas Control* 105 (2021): 103223.
Xiao, Cong, Olwijn Leeuwenburgh, Hai Xiang Lin, and Arnold Heemink. "Non-intrusive subdomain POD-TPWL for reservoir history matching." *Computational Geosciences* 23, no. 3 (2019): 537–565.
Yang, Yanfang, Mohammadreza Ghasemi, Eduardo Gildin, Yalchin Efendiev, and Victor Calo. "Fast multiscale reservoir simulations with POD-DEIM model reduction." *SPE Journal* 21, no. 06 (2016): 2141–2154.
Yoon, Seonkyoo, Zeid M. Alghareeb, and John R. Williams. "Hyper-reduced-order models for subsurface flow simulation." *SPE Journal* 21, no. 06 (2016): 2128–2140.
Zhu, Yinhao, and Nicholas Zabaras. "Bayesian deep convolutional encoder–decoder networks for surrogate modeling and uncertainty quantification." *Journal of Computational Physics* 366 (2018): 415–447.
Zhu, Yinhao, Nicholas Zabaras, Phaedon-Stelios Koutsourelakis, and Paris Perdikaris. "Physics-constrained deep learning for high-dimensional surrogate modeling and uncertainty quantification without labeled data." *Journal of Computational Physics* 394 (2019): 56–81.

APPENDIX: REDUCED-ORDER MODELS BASED ON PROPER ORTHOGONAL DECOMPOSITION

Reduced-order models (ROMs) represent an important class of numerics-based (rather than data-driven) surrogate modeling procedures. These techniques typically apply proper orthogonal decomposition (POD), which is very similar to PCA, to construct a basis matrix $\Phi \in \mathbb{R}^{n_h \times n_l}$ that relates full-order state variables $\mathbf{x} \in \mathbb{R}^{n_h}$ to reduced-order or latent variables $\xi \in \mathbb{R}^{n_l}$ via the linear relationship $\mathbf{x} = \Phi\ \xi$. Here n_h and n_l represent the number of state variables in the high- and low-dimensional problems. In an oil-water problem, $\mathbf{x}$ denotes the pressure and water saturation in every grid block, and we have $n_h = 2n_b$, where n_b is the number of grid blocks.

The basis matrix is constructed by performing a few full-order training runs and then assembling the state solutions into data matrices, where each column of the data matrix corresponds to the solution at a particular time step. Singular value decomposition of the data matrices, truncated to capture sufficient "energy" while avoiding overfitting, provides the POD basis. POD-based ROMs have mainly been used for production optimization and related problems. In this context, the well settings are varied in the online (run time) computations, relative to those used in training (offline), to maximize an objective function such as net present value or cumulative oil production. These methods have also been used for history matching, as noted in the following.

Techniques that apply POD directly (referred to as POD-only methods) for subsurface flow problems have been developed by a number of researchers, including Vermeulen et al. (2004), van Doren et al. (2006), and Cardoso et al. (2009). In addition to the representation of the solution unknowns ($\mathbf{x} = \Phi\xi$), POD-based methods require a projection treatment. Projection, also referred to as constraint reduction, entails the left-multiplication of the full system of equations to provide n_l equations (corresponding to the n_l unknowns). The two approaches applied for constraint reduction are Galerkin and least-square Petrov Galerkin (LSPG). In a POD-only setting, where the nonlinear governing equations are solved using Newton's method, these treatments correspond to projection using Φ^T or $\Phi^T J^T$, respectively, where $J \in \mathbb{R}^{n_h \times n_h}$ is the full-order Jacobian matrix. The associated matrix multiplications, which must be performed at each Newton iteration, are expensive. Because of this, POD-only methods do not provide substantial (if any) speedup relative to full-order simulation.

To achieve speedup with POD-based methods, two basic approaches have been pursued. These entail the application of local (piecewise) linearization and the use of "gappy" POD. The linearized procedures, referred to as POD-TPWL (trajectory piecewise linearization), involve the construction of low-order ($n_l \times n_l$) matrices in an offline step. These matrices appear in the low-dimensional linear equations that are solved at each time step. Online solutions are extremely fast, with speedups of 100–1,000, because the online solution only involves low-dimensional linear systems.

POD-TPWL methods were first applied for oil-water reservoir simulation problems by Cardoso and Durlofsky (2010). POD-TPWL has subsequently been used, with LSPG, for various reservoir engineering problems, including compositional (He and Durlofsky, 2014) and coupled flow-geomechanical models (Jin et al., 2020a). The use of this approach for history matching has also been explored by, e.g., He et al. (2013) and Xiao et al. (2019). Gappy POD treatments entail a more sophisticated (but more expensive) treatment of nonlinearity. These methods are faster than POD-only approaches, but much slower than POD-TPWL procedures. Their accuracy also lies between these two approaches. Specific implementations include the discrete empirical interpolation method, POD-DEIM (e.g., Chaturantabut and Sorensen, 2010; Ghommem et al., 2016; Yang et al., 2016) and the closely related Gauss-Newton with approximated tensors (GNAT) procedures. GNAT methods, which use LSPG for projection, were originally developed by Carlberg et al. (2011) and have been applied for reservoir simulation by Yoon et al. (2016) and Jiang and Durlofsky (2019).

Many of the DNN methods discussed in Section 14.2 were developed for history matching, whereas most of the ROMs discussed here were used for optimization. In optimization settings, DNN methods may require several hundred training runs, while ROMs typically use just a few (often approximately three to five). Substantial network training (on GPUs) is also commonly required to determine the optimal DNN parameters, and this usually represents much more preprocessing than is entailed with ROMs. The ROMs, however, may only provide accurate predictions within a neighborhood that is "close" to the training runs, whereas DNN surrogates can potentially provide accurate results over a much wider range. In this sense, the ROMs may be preferable with local optimization algorithms, whereas DNN approaches may perform much better with global optimization procedures. See Jin et al. (2020b) for further discussion of the relative advantages and limitations of ROMs and data-driven surrogate models within the context of optimization. Analogous observations would be expected in history-matching settings, though the training demands for both sets of methods are greater in this case.

15 Generalizable Field Development Optimization Using Deep Reinforcement Learning with Field Examples

Jincong He, Yusuf Nasir, and Shusei Tanaka
Chevron Technical Center,
Houston, Texas

CONTENTS

DOI: 10.1201/9781003207009-20

15.1 INTRODUCTION

The economics of oil and gas field development heavily depends on the quality of the field development plan, which includes decisions on the number of wells, well types, well locations, and the drilling sequence. Optimizing these decisions in the development plan can provide a substantial improvement on the net present value (NPV) of the project. Traditionally, such optimization is done in a scenario-specific manner where the solution of an optimization only applies to a specific scenario. In this chapter, we explore generalizable field development optimization (GFDO), where the optimization solution can be generalized to different scenarios. Here, the word scenario refers to the setting of the optimization problem, including the reservoir model description and the set of economical/operational parameters, with or without uncertainty.

In traditional field development optimization, such as those in Yeten et al. (2003), Onwunalu and Durlofsky (2010), Isebor et al. (2014), Kim et al. (2019), and Bouzarkouna et al. (2012), given a scenario, a large number of simulations are performed to achieve an optimal solution. This solution is tied to the scenario being considered (e.g., a particular oil price assumption, a particular reservoir). When the scenario changes (e.g., when a different reservoir or a different oil price assumption is considered), the optimization process will need to start from scratch. Note that even methods that consider optimization under certainty (e.g, Chen et al. 2012;, Kim et al. 2020) are often scenario specific, as the solution is still tied to a particular assumption of the uncertainty distribution. When the assumption on the uncertainty distribution changes, the previous optimization solution is no longer applicable.

For the traditional scenario-specific approaches, the lack of ability to generalize to new scenarios is due to several reasons. Firstly, the traditional approaches only see one scenario at a time and cannot learn from previous optimization of other scenarios. Secondly, most traditional approaches use black box methods such as genetic algorithms (GAs) or particle swarm optimization (PSO). Such optimization methods utilize only the input (decision) parameters and the objective function. They generally do not make use of the state of the reservoir, such as the pressure, saturation, permeability, and porosity maps, which contain important information for development planning.

One attempt to make use of the state of the reservoir and to develop solutions generalizable for different scenarios is the use of reservoir quality maps. Reservoir quality maps are generated by integrating location-based properties (such as remaining oil, pressure, productivity index [PI], etc.) and help identify and highlight favorable locations (sweet spots) for the next drilling well. Works on reservoir quality maps include those that use static properties only (Da Cruz et al. 2004; Maschio et al. 2008) and those that combine static properties with dynamics properties derived from simplified simulations (Guimaraes et al. 2005; Park et al. 2017). For example, Park et al. (2017) proposed to calculate the well potential measurement (WPM) map, which is a product of three heuristic terms: dynamic sweep/drainage capacity, static hydrocarbon richness, and existing well avoidance.

The resulting WPM map can then be used to highlight favorable infill drilling locations for injectors and producers.

The reservoir quality maps are computationally efficient to calculate and thus they can be quickly applied to new scenarios by adjusting the parameters in the formula. Therefore, it can be considered as a generalizable approach. However, the formulas for reservoir quality are usually heuristically constructed and there is no theoretical guarantee of optimality. In addition, reservoir quality maps may be good for finding the location for one additional well. They do not optimize development plans that involve multiple wells that could interact with each other.

Recently, He et al. (2021) formulated the GFDO problem as a Markov decision process (MDP) (Feinberg and Shwartz 2012) and proposed to solve it using deep reinforcement learning (DRL) (Arulkumaran et al. 2017), which is a method for optimizing the MDP by combining deep neural network and reinforcement learning. Similar to the reservoir quality maps approach, in He et al. (2021), a mapping is established between the location-based properties (such as remaining oil, pressure, PI, etc.), which are represented as maps (also called input channels), and action probability map that shows the probability of drilling a well at a certain location. However, instead of using a heuristic formula, He et al. (2021) modeled the mapping (also referred to as the policy) with a deep neural network. This deep neural network is trained in a DRL framework, such that the trained policy (also referred to as the artificial intelligence [AI]) is optimal in maximizing the NPV of the entire field development plan. DRL is a popular framework for training policies to play strategy games, such as the game of Go (Silver et al. 2016, 2017). He et al. (2021) showed that the AI trained by DRL can efficiently provide solutions for new scenarios within a range of applicability. The solutions from AI also statistically outperform an optimized pattern drilling strategy and solutions from human engineers.

The AI trained in He et al. (2021) has limited input channels and uses a prototype simulator for training so it was only applicable to single-phase flow without operational constraints. Nasir et al. (2021) extended the formulation of He et al. (2021) to a two-phase flow problem using the DARTS simulator (Khait 2019) for training. This allows the resulting AI to account for some operational constraints such as the maximum well production rate.

The methodology presented in this chapter is based on He et al. (2021) and Nasir et al. (2021) with some important improvements. In He et al. (2021) and Nasir et al. (2021), the size of the neural network scales directly with the size of the model used in the training (referred to as the common reservoir template) due to the choice of representing possible drilling locations as a vector. In this chapter, a new architecture is presented that eliminates the need for fully connected embedding layers and allows a much larger template model size. In addition, the location mask channel is redesigned to allow for the incorporation of engineering knowledge into the training and the application processes. Furthermore, the methodology in this work is integrated with a commercial simulator to allow for consideration of more practical constraints.

This chapter proceeds as follows. Section 15.2 describes the formulation of the GFDO problem as an MDP, and the DRL framework to solve this MDP.

In Section 15.3, we discuss the specific DRL algorithm and the neural network architecture used in this work. Section 15.4 presents an example in which an AI is trained for field development optimization for general DWGOM fields. It also presents result of the AI when benchmarking with another policy, and when being applied to a real deepwater Gulf of Mexico (DWGOM) reservoir. Summary and further discussion is provided in Section 15.5.

15.2 GENERALIZABLE FIELD DEVELOPMENT OPTIMIZATION

In this section, we briefly describe the GFDO problem for the two-phase flow and its formulation as an MDP. The elements required to solve this MDP in the DRL framework are also introduced. The description here follows that in Nasir et al. (2021).

15.2.1 Problem Description

The governing equations for the subsurface flow are a combination of Darcy's law and the mass conservation law (Aziz and Settari 1986). The solution for the governing equations allows us to predict the liquid flow rate from the wells at any given time.

A typical objective function used in field development optimization is the NPV defined as

$$NPV(\mathbf{u}) = \sum_{n=1}^{N_t} \gamma^{t_n} \left[\Delta t_n \sum_{i=1}^{N_W} \left(\left(p_o - c_{opex} \right) q_{o,n}^i - c_{pw} \; q_{w,n}^i \right) - I_n \, c_w \right], \quad (15.1)$$

where $\mathbf{u}$ is the vector of decision variables that encodes the field development plan such as well count, well location, and drilling sequence; γ^{t_n} is the discounted factor for the period Δt_n; p_o is the oil price, c_{opex} is the operating cost per barrel of oil produced, c_{pw} is the cost per barrel of produced water; $q_{o,n}^i$ and $q_{w,n}^i$ are the oil and water production rate, respectively, for the ith well during the period Δt_n; I_n is the number of well drilled at the time period; and c_w is the cost of the well.

The optimal solution for the optimization $\mathbf{u}^*$ is a function of the parameters in the governing equation, Eq. (15.1), and the constraints, such as the porosity and the oil price p_o. These parameters are referred to as the scenario parameters and denoted as σ (therefore $\mathbf{u}^* = \mathbf{u}^*(\sigma)$). The goal of traditional scenario-specific optimization is to find the value of $\mathbf{u}^*$ given a specific scenario σ. The goal of GFDO is to find the functional form of the function $\mathbf{u}^*(\sigma)$ for a range of σ, such that for a new value of σ, the function can be evaluated to obtain the optimal solution $\mathbf{u}^*$ without optimization. Therefore, GFDO is one level of complexity higher than traditional scenario-specific optimization.

Theoretically, traditional methods such as population-based algorithms (Isebor et al. 2014) can be extended for GFDO by solving the scenario-specific optimization problem repeatedly to generate samples of $(\sigma_1, \mathbf{u}_1^*(\sigma_1)), (\sigma_2, \mathbf{u}_2^*(\sigma_2)), \cdots$.

Regression can then be applied on these samples to obtain the functional form of $\mathbf{u}^*(\sigma)$. In practice, this is only possible when σ is low dimensional. If the dimension of σ is high, the number of samples required for the regression will be high. Because each sample requires an optimization, the computational cost for this approach will become prohibitive.

On the other hand, approaches based on reservoir quality maps (Da Cruz et al. 2004; Maschio et al. 2008) effectively try to build a heuristic functional form for $\mathbf{u}^*(\sigma)$ based on engineering judgments. Such a solution is easy to evaluate in practice but has no theoretical guarantee of optimality nor performance.

In later sections we will explore an approach that efficiently solves GFDO with DRL.

15.2.2 GFDO AS A MARKOV DECISION PROCESS

The GFDO problem can be viewed as an MDP (Feinberg and Shwartz 2012). Figure 15.1 shows the formulation of an MDP. It consists of an environment and an agent. At each decision step t, the agent decides on an action a_t based on its observation o_t, and the environment takes the chosen action a_t as input and advances the system state to the next time step $t+1$ and its state from s_t to s_{t+1}. It also calculates the reward obtained for the current step r_{t+1}. The new observation o_{t+1} that goes into the agent is then a function of the state output s_{t+1} from the environment. If the observation contains all the information in the states (mathematically that means the observation is an invertible function of the state), the MDP is called fully observable. Otherwise, it is called partially observable. The feedback loop in Figure 15.1 terminates at a terminal state, or when the maximum number of time step T is reached. When the system evolves from $t=0$ to termination it is referred to as an episode.

For GFDO, the action a_t we are interested in at time t is whether to drill a well, and if yes, the location and the type of wells (producer or injector) to drill. The decision variable $\mathbf{u}$ is then a collection of a_t at all the time steps, i.e., $\mathbf{u}=[a_1,\cdots,a_T]$. The environment would be the reservoir simulator, which takes in the action and advances the simulation for one decision step. The single-step reward r_t is the NPV that is achieved during decision step t.

In this work, we are interested in model-based optimization, in which a model for the reservoir is available and optimization is defined based on predictions

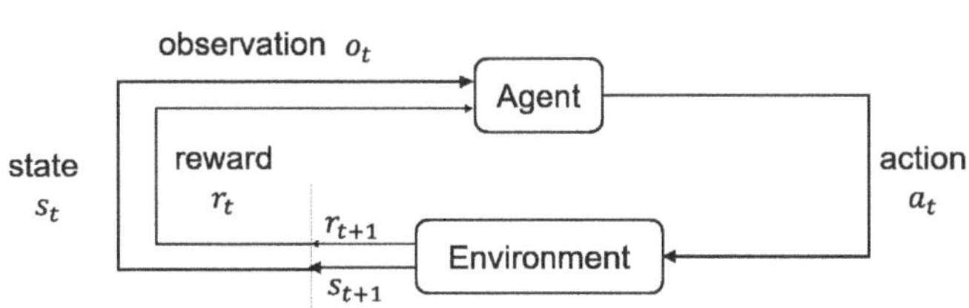

FIGURE 15.1 Markov decision process and the reinforcement learning framework.

from the model. In this case, the state (such as dynamic pressure/saturation distributions) is always fully accessible through the reservoir simulator and the GFDO problem is a fully observable MDP. Therefore, we will not differentiate between observation and state hereafter. The problem would be different if we try to optimize field development action based on limited observation from the wells rather than a reservoir model (as considered in Nasir and Durlofsky (2022) for a subset of the FDO problem). Finally, the agent is a mathematical function that maps the observation o_t to the action a_t, denoted as $a_t = \pi(o_t)$.

15.2.3 Deep Reinforcement Learning for GFDO

DRL is an approach used to obtain the optimal policy for MDPs that has become popular in recent years. The goal of DRL is to obtain the optimal policy $a_t = \pi(o_t)$ that maximizes the expected total reward. It is called "deep" because it models the policy function $\pi(o_t)$ with a deep neural network. "Reinforcement" refers to the fact that the policy function is trained on randomly generated training scenarios, where the update to the policy during the process is based on the rewards r_t provided by the environment. In the rest of this section, we will discuss some common components needed to set up the GFDO problem in the DRL framework.

15.2.4 Generating the Training Scenarios and Range of Applicability

To optimize the policy through reinforcement learning, a large number of training scenarios is needed. Scenarios are described by the list of scenario parameters σ. The training scenarios are generated by sampling from the distribution of these parameters. The distributions of the scenario parameters are referred to as the range of applicability. The optimal policy trained by RL is applicable to new scenarios within this range of applicability.

The definition of the scenario parameters and their ranges are problem specific. Table 15.1 shows an example list of the scenario parameters for primary depletion field development optimization considering oil-water two-phase flow. In general, the more parameters are included as scenario parameters, the more generalizable is the resulting AI. For example, by including oil price in the scenario parameters, the resulting AI would be able to provide optimal development plans for scenarios with different oil prices (here $40–60 per STB). By including the operation constraints on field production rate (usually due to facility capacity), the resulting AI could be applied for new scenarios with different facility capacity designs (e.g., a max. field liquid production rate of 50,000–200,000 STB/day). Also, by considering varying fluid properties such as oil viscosity (here 5–15 cp), the AI can provide optimized field development plan (FDP) for different mobility ratios. In addition, the resulting AI would also have a higher generalization capacity with wider ranges for the scenario parameter. On the other hand, as noted in He et al. (2021), the more scenario parameters considered and the wider the ranges, the more challenging it is to train the AI. The list of most of the scenario parameters and their distributions are provided in Nasir et al. (2021).

TABLE 15.1
Summary of Scenario Parameters

Category	Parameters
Grid	*dx*, *dy*, *dz*
Petrophysical	Permeability, porosity, depth
Fluid: oil	Compressibility, viscosity, density, formation volume factor (FVF)
Fluid: water	Compressibility, viscosity, density, FVF
Fluid: relative permeabilities	End points, critical saturations and Corey exponents
Well	Wellbore radius, skin, productivity index (PI) degradation, drilling time
Operational constraints	Min. bottom hole pressure (BHP), max. liquid production rate for producers, water cut constraints
Economics	Oil price, operating cost, water production cost, drilling cost, discount rate, project life

15.2.5 Definition of Observation

The scenario parameters σ together with the dynamic variables (for two-phase flow they are pressure, saturation, and current time) constitute a complete definition of the states of the system and thus the observation for the agent. However, as argued in He et al. (2021), there is redundancy among this definition of the states as they do not affect the system independently. For example, the skin factor and the wellbore radius always appear together in the well index in the governing equation. Similarly, the permeability and the grid size also always appear together in the variable group of transmissibilities. Using the redundant set of states as input to the agent will increase the complexity and the ill conditioning of the problem, making it more challenging to train the AI. We follow the approach in He et al. (2021) and Nasir et al. (2021) to define the observation as a reduced set of heuristically crafted variables that directly appear in the governing equations, the constraints, or independent scenario parameters. The list of variable groups for observation (also referred to as "input channels") used in this work includes pressure, water saturation, x- and y-directional transmissibilities, oil accumulation, oil and water mobilities, well index, producer drawdown, well location mask, normalized drilling cost, water production cost, discount rate, remaining time, well max. liquid production rate, well min. oil rate under allocation, well max. water-cut, field max. oil rate, field max. liquid rate, PI degradation parameters, and drilling time. Due to limited space, the detailed definitions of the channels are not shown here, but most can be found in Nasir et al. (2021).

Compared with Nasir et al. (2021), the additional channels that are considered in this work include that for the PI degradation model, and that for field oil and liquid production rate constraints. These channels are required for the application to DWFOM, which will be discussed in Section 15.4. They are made possible by the integration of the DRL framework with a commercial simulator.

In addition, the definition of well location mask is also improved over previous works. The well location mask characterizes the locations that are not feasible for drilling. For training, the infeasible region in the location mask is initialized to include all inactive cells, as the episode progresses, the action mask will be dynamically updated to include cells that are close to existing wells. When applying the AI to a new field, the user can customize the initial location mask to exclude any undesirable area, such as those in the aquifer, near a fault, or have low hydrocarbon thickness. The input on the initial location mask during application allows engineering judgments to be incorporated into the AI solutions.

As proposed by He et al. (2021), the concept of a common reservoir template, which is a fixed specification (size, shape, physics considered, input channels) of the reservoir model is used to standardize the definition of different scenarios. This enables the training of an AI that is applicable to different real-life reservoirs of different sizes and shapes.

15.3 DEEP REINFORCEMENT LEARNING ALGORITHMS

DRL combines deep neural networks with the framework of reinforcement learning. The policy function is modeled as a deep neural network with parameter θ, i.e., $\pi(o_t;\theta)$. This network parameter θ is trained using a large number of randomly generated scenarios such that the optimal value $\pi(o_t;\theta^*)$ maximizes the expected total reward over all scenarios.

The architecture of the neural networks depends on the choice of the reinforcement learning algorithms. Similar to He et al. (2021) and Nasir et al. (2021), in this work the method of proximal policy optimization (PPO) is used. We will first briefly describe the mechanism of PPO, and then discuss the architecture of deep neural networks.

15.3.1 Proximal Policy Optimization Method

PPO proposed by Schulman et al. (2017) has been successfully applied in many areas such as robotic control (Bøhn et al. 2019) and supply-chain management (Vanvuchelen et al. 2020). PPO is an "on-policy" algorithm, in which the policy function for the next iteration is obtained based on training samples generated by the policy function at the current iteration. The update to the policy is guided by minimizing a loss function based on gradient information derived from a method called stochastic gradient descent (SGD). The PPO loss function is a proxy for the expected total return but also takes into account terms that safeguard the optimization process.

For each PPO iteration, a large number of episodes are performed. At each decision step of each episode, the observations o_t, o_{t+1} action a_t, and the reward r_t are recorded and form a batch of the training samples. With the batch of training samples at each PPO iteration, SGD is applied to minimize the following loss function

$$L^{PPO} = L^{\pi} + c_{kl}L^{kl} + c_{vf}L^{vf} + c_{ent}L^{ent}, \tag{15.2}$$

where c_{kl}, c_{vf}, and c_{ent} are the weights for each individual loss component. The detail definition of the terms in Eq. (15.2) can be found in He et al. (2021). On a high level, L^{π} is the policy loss, which is a proxy for the expected total return. L^{kl} is the Kullback-Leibler (KL) divergence loss that characterizes the difference between the new and the current policy. It is designed to avoid drastic updates to the policy. L^{ent} is the entropy loss that characterizes the diversity in the action probability distribution. It is designed to avoid premature convergence during optimization. L^{vf} is the difference between the value function $V(o_t)$, which predicts the remaining reward when the observation is o_t and the current policy is followed, and the actual remaining reward from the training samples.

To utilize the PPO algorithm, the policy function $\pi(o_t)$ and the value function $V(o_t)$ need to be parameterized. While there is more than one way to parameterize high-dimensional nonlinear functions, the deep neural network has become a popular choice due to its ease in calculating gradients and adjusting complexity. In particular, the convolutional neural network (CNN) has shown to be particularly efficient for problems with high-dimensional map-based inputs as we have in this problem (Krizhevsky et al. 2012).

The action probability of the well location was represented as a vector of the size $n_x \times n_y + 1$ in He et al. (2021) or $n_x \times n_y + 2$ in Nasir et al. (2021), in which the $n_x \times n_y$ elements represent the probability to drill at any of the $n_x \times n_y$ grid blocks in the reservoir model and the rest represents the decision to drill or not (in this work, wells are assumed to perforate the entire vertical interval). An embedding layer is used to learn the spatial relationship between the 2D feature map of the CNN and this 1D representation of actions. However, this requires the use of fully connected neural networks where its number of coefficients scale with the size of the common reservoir template.

To address this issue, the CNN architecture shown in Figure 15.2 is proposed. In this new architecture, the action probability for the well location is represented as a 2D map. This eliminates the need for flattening and embedding and thus

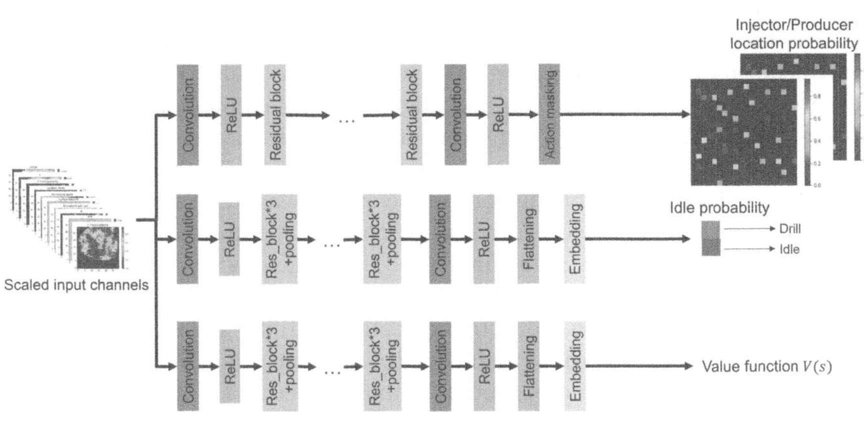

FIGURE 15.2 Neural network architecture.

preserves the spatial relationship of different drilling locations and makes the neural network scalable to larger common reservoir templates. In addition, an action masking layer is added to the policy network. In this layer, the action probabilities for locations that are infeasible for drilling are set to zero according to the location mask input channel. As discussed earlier in Section 15.2, this allows the incorporation of predefined rules and human knowledge into the system. Note that in this work we only considered the placement of producers during primary depletion. However, in cases where there are injectors, a second map for injector location probability can be added to the output with minimal change to the network architecture, as shown in Figure 15.2. The probability of the drill/idle decision is modeled using a separate neural network. The value function network employs the same architecture as the idle action but uses a separate network. The architecture used in the idle/value action network is similar to that used for image classification such as in Krizhevsky et al. (2012). The neural network shown in Figure 15.2 has approximately 800,000 parameters.

15.4 EXAMPLE APPLICATION: BUILDING AN AI FOR DWGOM

15.4.1 Target Reservoir and Common Reservoir Template

The methodology introduced in this work is applied to develop an AI for GFDO for general reservoirs in the DWGOM, which have a unique set of characteristics such as high water depth (up to 10,000 ft), high pressure (over 10,000 psi), and deep reservoirs (over 30,000 ft true vertical depth) in mostly sub-salt environments (Close et al. 2008). It also features a unique set of operational challenges, such as PI degradation (Zhang 2020) in which the PI decreases depending on drawdown and production history, high drilling and facility cost, and extended period of production plateau (constrained by facility capacity).

In this work, the common reservoir template used to characterize the DWGOM reservoirs is a 2D reservoir model of size 80×80, which is more than three times larger than that used in He et al. (2021). This larger template model size is made possible by the improved network architecture in Figure 15.2 and allows for a more detailed description of geological features such as faults. The ranges of the scenario parameters listed in Table 15.1 align with what is commonly observed in general DWGOM fields, rather than from any specific reservoir.

Included in Table 15.1 are four parameters that characterize the behavior of PI degradation. The PI is a cubic function of the drawdown after it reaches a certain threshold. The four parameters represent the coefficient of the cubic function and the threshold value. Also included in Table 15.1 are facility capacity constraints in the form of field max. oil and liquid rate. They are handled using the guide rate balance logic; when the production potential of all the wells exceeds facility capacity, field production will be maintained at a plateau and individual wells will be produced at a target rate proportional to their potential. This complex physics is made possible by using a commercial simulator to simulate the template model. AI is trained to take 20 decision steps (the length of the interval is controlled by the "drilling time per well" parameter).

With the earlier setup, the resulting AI is expected to be generally applicable for reservoirs in DWGOM under primary depletion. However, it may not be applicable for a different reservoir type (such as heavy oil reservoir, channelized reservoir, or fractured reservoir) or under a different recovery mechanism (such as waterflood or gas injection). For different reservoir types and different recovery mechanisms, Table 15.1 needs to be redesigned and a different AI needs to be trained.

15.4.2 Algorithm and Computational Setup

In this work, the PPO implementation in the open-source library Ray RLlib (Liang et al. 2018) is used. The value for the weights for the loss components in Eq. (15.2) are set at $c_{kl} = 0.2$, $c_{vf} = 0.1$, and $c_{ent} = 10^{-3}$. The learning rate in SGD is set at 10^{-5}. The optimal values for these hyperparameters are problem dependent. The values used here are obtained through limited numerical experimentation.

The deep neural network is set up according to the architecture in Figure 15.2. For the location network, seven residual blocks are used. The convolution layers in the residual blocks use 48 channels, kernel size of 3×3, and stride size of 1. Three residual-pooling blocks are used in the idle probability network. Each residual-pooling block consists of two residual blocks followed by a max.-pooling layer. The pooling size used in the pooling layer is 2×2. After three residual-pooling blocks, the size of the feature maps is reduced from 80×80 to 10×10. The convolution layer after the residual-pooling blocks contains only one channel, effectively combining all feature maps into one before it is flattened into a 100×1 vector. The embedding layers in the idle probability network and the value function network have 20 neurons and therefore reduce the dimension of the feature further to 20×1 before it is finally reduced to a 2D output (drill/do not drill) for the idle network and single scalar output for the value network.

The training of the neural network employs a head node with 24 CPU cores and 4 GPU cores, in addition to three worker nodes with 72 CPU cores each. Of the 240 available CPU cores, three are reserved for the main thread, whereas the remaining 237 cores (referred to as workers) are used to run the reservoir simulations. The 4 GPU cores are used to train the neural network in Figure 15.2 by minimizing the loss function in Eq. (15.2) using SGD.

15.4.3 Training Result

Figure 15.3 shows two of the most important metrics for the progress of the training. Figure 15.3a shows the average NPV (normalized by a constant of $6 billion) achieved by current AI policy over the 237 randomly generated training episodes in the current PPO iteration vs the number of training simulations. The average NPV sharply increases for the first 100,000 episodes (simulation runs) before the improvement slows down. The fluctuation is due to statistical noise from the limited sample size (237). Figure 15.3b shows the entropy of the action probability through the training process. As stated earlier, this entropy characterizes how concentrated the action probability distribution is. Decreasing entropy as shown in Figure 15.3b indicates the convergence of the AI policy, or, more plainly speaking, that the AI

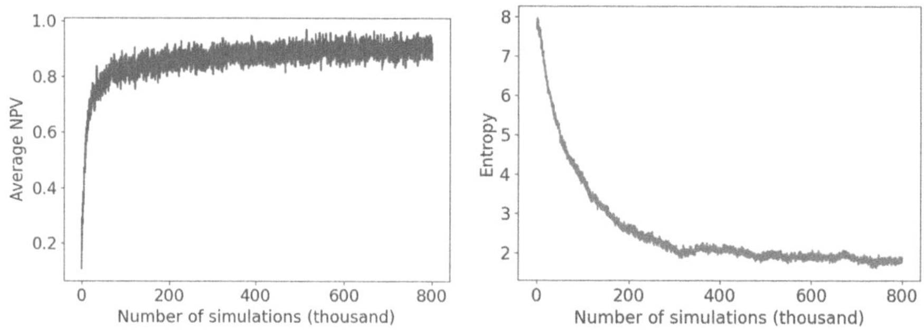

FIGURE 15.3 (a) Progress of the DRL AI during training for (a) NPV and (b) entropy.

is more and more sure about taking certain actions under certain scenarios. In Figure 15.3b, entropy sharply decreases over the first 200,000 episodes before slowing down, likely due to the entropy loss term in Eq. (15.2).

15.4.4 Benchmark with the Max-Spot Agent

Once the DRL AI is trained, it can be applied to obtain development plans for new scenarios in just one simulation. To statistically investigate how good those development plans are, they are benchmarked with the Max-Spot agent introduced in He et al. (2021). The Max-Spot agent performs primitive optimization on well spacing for simple pattern drilling. For any given scenario, the Max-Spot agent evaluates four potential drilling patterns: 4-spot, 5-spot, 9-spot, and 16-spot (wells falling into an inactive cell will be automatically skipped) and returns the pattern with the highest NPV. The DRL AI requires only one simulation to obtain the development plan for a given scenario, whereas the Max-Spot agent requires four. Black box algorithms like GA or PSO can also be viewed as a policy, though it would take thousands of simulations to obtain a development plan and their performances are highly problem/setup dependent. Therefore, from a computational and practical standpoint, it is more meaningful to benchmark the DRL AI with a simple agent such as Max-Spot than with GA or PSO.

Figure 15.4 shows the benchmark result for 100 randomly generated scenarios that are not part of the DRL AI training. The x-axis shows the performance of the DRL AI in terms of NPV. The y-axis shows the NPV of the Max-Spot agent. Each black dot represents one of the 100 scenarios. The dashed line shows the 45° line. It can be seen that all black dots are below the 45° line, meaning the development plan generated by the DRL AI beats the Max-Spot agent 100% of the time.

15.4.5 Blind Test on a Real Field

15.4.5.1 Field X

After the AI is trained, it is expected to be applicable for any primary depletion reservoir in DWGOM within its range of applicability (Table 15.1). In this study,

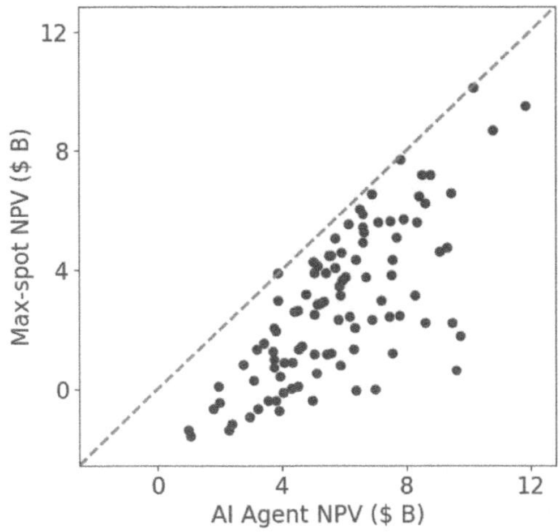

FIGURE 15.4 Comparison of the AI with Max-Spot agent.

Field X from DWGOM is picked to demonstrate the performance of the AI on a real field. It should be noted the AI is trained only with general information about DWGOM without any specific knowledge of Field X.

The size of the original reservoir model for Field X is 160 × 136 × 845. To apply the trained AI to Field X, aggressive upscaling is applied to upscale the reservoir model to the size of the common reservoir template. In addition, in the original model, more than 10 pressure-volume-temperature regions were used, each of which has its own relative permeability tables. Figure 15.5a and 15.5b shows the permeability and the initial water saturation for Field X after upscaling and simplification. There is a weak aquifer on the right and permeability increases moving from the right (down-dip) to the left (up-dip). There are two major faults in the reservoir. Figure 15.5c shows the initial location mask, where the dark color indicates locations that are ineligible to drill. These locations are the inactive cells, the aquifer, and the cells next to the faults. The information in Figure 15.5 will be parts of the input to the DRL AI.

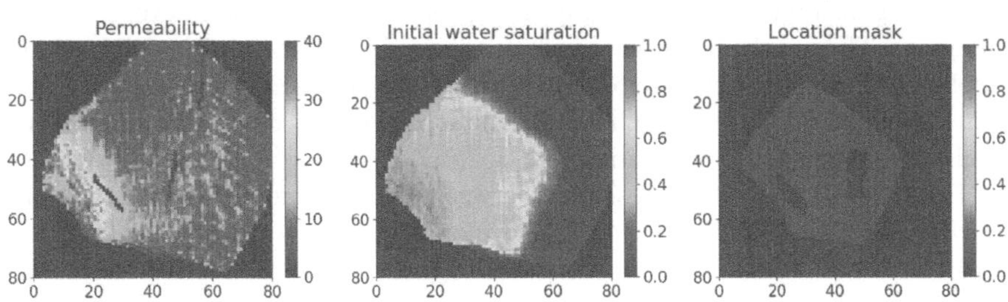

FIGURE 15.5 Selected properties for Field X after rescaling: (a) permeability, (b) initial water saturation, and (c) location mask.

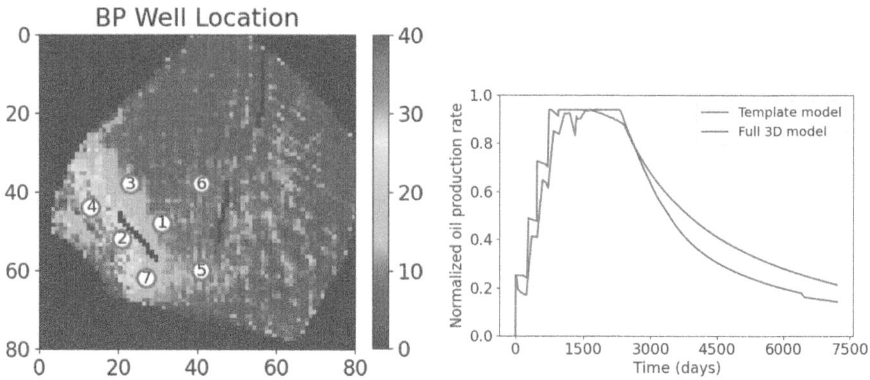

FIGURE 15.6 Field X base development plan: (a) well location and (b) production rate.

Figure 15.6a shows the base plan designed by human engineers. It consists of seven wells, with six evenly distributed on the two sides of the faults on the left, and one stepping out closer to the water oil contact (WOC). It should be noted that this plan was not designed to only maximize project economics. It also considers many other factors such as the lease retention requirement and the value of gathering information and further delineates the reservoir. In this work, it is used to establish a baseline for the DRL AI performance.

Figure 15.6b shows the field oil production rate (normalized by an unspecified amount) for the base plan using the full 3D model and the upscaled template model. While inaccuracy can be observed due to upscaling and simplification, the template model still captures the general trend of the production profile.

15.4.5.2 Selecting an AI

Field X as a real field does not exactly follow the statistics of the training scenarios. In addition, there is also a discrepancy between the full 3D model and the 2D reservoir template as shown in Figure 15.6. Therefore, the best AI snapshot in terms of a blind test with the training scenarios (i.e., the peak of the curve in Figure 15.3) may not be the best for Field X. Following the practice in He et al. (2021) and Nasir et al. (2021), we save snapshots of AI periodically and test all of them against Field X. Specifically, a snapshot of the AI is saved every 50 PPO iterations. This results in 67 AI snapshots. These 67 snapshots of the DRL AI are applied to Field X to generate 67 different field development plans and NPV values (evaluated on the template model). The top 20 plans are then projected back to the full 3D model for validation.

Figure 15.7 shows the cross-plot between the estimated ultimate recovery (EUR) and the NPV for the top 20 development plans from the DRL AI for Field X. The EUR and NPV numbers are determined from simulating the full 3D model and they are all normalized by the numbers of the base plan. The square indicates the base plan, and the circles indicate the 20 AI development plans. The numbers in the shapes indicate the number of producers drilled in each plan.

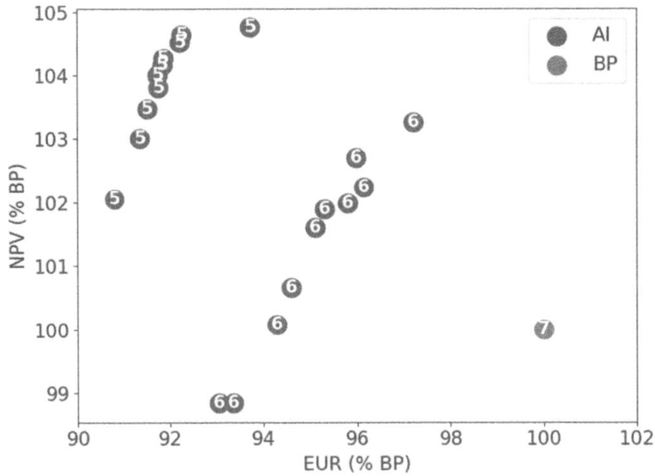

FIGURE 15.7 Comparison of the EUR and NPV of the top 20 AI solutions against the base plan.

It can be seen that the solutions from the DRL AI generally recommend development plans with five to six wells, which have comparable or higher NPV than the base plan with seven wells, although with lower EUR numbers due to having fewer wells. This result suggests that the AI can identify a reasonable well count (comparable or higher NPV than the base plan, which took months to develop) consistently (similar well counts for the top 20 AI solutions). This is impressive considering that Field X was not part of the AI training.

Figure 15.8a shows the best field development plan from the DRL AI in terms of NPV. The plan consists of five wells with four up-dip of the fault in the high permeability region and one down-dip of fault. Figure 15.8b shows the production

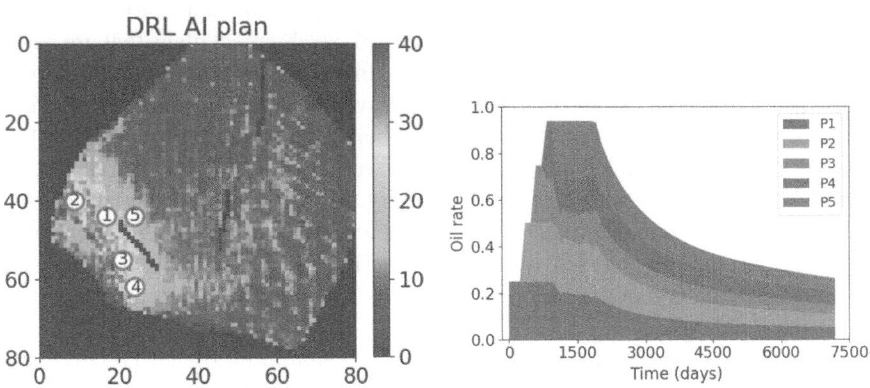

FIGURE 15.8 Best field development plan from the AI for Field X: (a) well location and (b) oil rate.

rate from the DRL AI development plan. It can be seen that all five wells from this optimized plan are all productive and able to sustain a production plateau for a prolonged period.

The solutions from the current AI need to be further fine-tuned to satisfy practical considerations (such as leasing requirement) that informed the base plan. However, the fact the DRL AI can generate development plans comparable to human performance for any given reservoir at a very low cost (each AI solution takes one simulation on the template model to generate and one simulation on the 3D full model to validate) still shows the potential for the method. It is especially useful to get a first-pass development plan for new scenarios, such as when moving to a new reservoir, updating the oil price assumption, or updating the surface facility design.

15.5 CONCLUSION AND FUTURE OUTLOOK

In this chapter, a DRL-based approach is introduced for GFDO. Different from traditional scenario-based approaches in which solutions are tied to a specific scenario (reservoir model, oil price assumption, etc.), the goal in GFDO is to develop a generalizable policy that can generate different optimized field development plans for different scenarios. The GFDO problem can be solved with a DRL framework, in which the problem is formulated as an MDP, and deep neural networks are designed to model the policy function and are trained using the PPO method with a large number of randomly generated training scenarios. Once it is trained, the resulting policy function (referred to as the DRL AI) can be applied to new scenarios within the range of applicability with minimal computational cost.

The methodology presented in this chapter is built upon the previous works of He et al. (2021) and Nasir et al. (2021) with an improved architecture for the neural network, an improved location mask is introduced that combines dynamic masking, and it is trained on a commercial simulator. Benchmarking study and real-field application have demonstrated the potential of the DRL framework for the GFDO problem. It requires on the order of millions of simplified simulations to train. However, once it is trained, the DRL AI is generally applicable to a range of different scenarios and is able to provide field development plans that are superior to the base plan designed by human engineers at a minimal computational cost. While the plans from the AI may not have taken into account all practical considerations (e.g., lease retention, data gathering, surface facility design), they provide useful indications on well count and well placement, and a good starting point for further fine-tuning. In addition, the DRL AI is also particularly useful when the scenario changes, such as when a new reservoir is considered, when the oil price assumption changes, or when the facility capacity assumption changes. It could provide updated development plans for the new scenario with minimal computational cost.

The DRL framework has shown substantial potential for the GFDO problem. There are many interesting directions to further advance the methodology and its practical application. Advances in the methodology could come in the areas of neural network design and the DRL algorithms. In terms of the neural network architecture, 2D CNNs have been used in most of the existing works on this topic, which ignores the vertical heterogeneity of the reservoirs. A 3D CNN (Xiong et al. 2018) could potentially be employed in the DRL framework to enable optimization of the well perforation and trajectory. The increase in computational cost may be alleviated by transfer learning (Shin et al. 2016), where the CNN trained on a smaller-sized template can be used as parts of the CNN for a larger-sized template. In addition, there has been a rapidly growing body of DRL algorithms, many of which, such as IMPALA (Importance Weighted Actor-Learner Architectures) (Espeholt et al. 2018), has shown superior performance over PPO on certain problems. Systematic exploration of such alternatives could improve the performance of the DRL algorithms.

Advances in practical application could come in the areas of AI specialization and optimization under uncertainty. Existing works have only considered primary depletion in oil reservoirs. With the integration with commercial simulator accomplished in this work, the methodology can be extended to other reservoir types such as gas reservoir, channelized reservoir, or fractured reservoir, and to other recovery mechanisms such as water flooding or gas injection. Different reservoir types and recovery mechanisms would have different sets of scenario parameters. Therefore, specialized AI can be separately developed for different problem types. For example, AI for waterflood planning and AI for gas injection planning may need to be separately developed. In addition, existing works have focused on the optimal plan for a deterministic scenario while in practice the subsurface geology is uncertain. Uncertainty can be considered by expanding the input channels to include input from multiple realizations, though that will substantially increase the size of the neural network, and thus the computational cost. The high computational cost could potentially be alleviated by using transfer learning (Shin et al. 2016), where parts of the neural network from deterministic GFDO problems is reused to reduce the complexity of the neural network for the GFDO problem under uncertainty.

ACKNOWLEDGMENTS

We thank Chevron Technical Center for permission to publish this work; Mr. Yakov Dyadkin, Dr. Kainan Wang, and Dr. Xian-Huan Wen for their technical support; and the Chevron asset team on Field X for sharing the simulation model data and providing technical guidance.

REFERENCES

Arulkumaran K, Deisenroth MP, Brundage M, Bharath AA (2017) Deep reinforcement learning: A brief survey. IEEE Signal Processing Magazine 34(6):26–38.

Aziz K, Settari A (1986) Fundamentals of Reservoir Simulation. Elsevier Applied Science Publishers.

Bøhn E, Coates EM, Moe S, Johansen TA (2019) Deep reinforcement learning attitude control of fixed-wing UAVs using proximal policy optimization. In: 2019 International Conference on Unmanned Aircraft Systems (ICUAS), IEEE, pp. 523–533.

Bouzarkouna Z, Ding DY, Auger A (2012) Well placement optimization with the covariance matrix adaptation evolution strategy and meta-models. Computational Geosciences 16(1):75–92.

Chen C, Li G, Reynolds A, et al. (2012) Robust constrained optimization of short- and long-term net present value for closed-loop reservoir management. SPE Journal 17(03):849–864.

Close F, McCavitt RD, Smith B (2008) Deepwater Gulf of Mexico development challenges overview. In: SPE North Africa Technical Conference & Exhibition, OnePetro.

Da Cruz PS, Horne RN, Deutsch CV (2004) The quality map: a tool for reservoir uncertainty quantification and decision making. SPE Reservoir Evaluation & Engineering 7(01):6–14.

Espeholt L, Soyer H, Munos R, Simonyan K, Mnih V, Ward T, Doron Y, Firoiu V, Harley T, Dunning I, Legg S, Kavukcuoglu K (2018) Impala: Scalable distributed deep-RL with importance weighted actor-learner architectures. 1802.01561.

Feinberg EA, Shwartz A (2012) Handbook of Markov Decision Processes: Methods and Applications, vol 40. Springer Science & Business Media.

Guimaraes MdS, Schiozer DJ, Maschio C (2005) Use of streamlines and quality map in the optimization of production strategy of mature oil fields. In: SPE Latin American and Caribbean Petroleum Engineering Conference, OnePetro.

He J, Tang M, Hu C, Tanaka S, Wang K, Wen XH, Nasir Y (2021) Deep reinforcement learning for generalizable field development optimization. SPE Journal 1–20.

Isebor OJ, Echeverría Ciaurri D, Durlofsky LJ, et al. (2014) Generalized field-development optimization with derivative-free procedures. SPE Journal 19(05):891–908.

Khait M (2019) Delft advanced research terra simulator: General purpose reservoir simulator with operator-based linearization. PhD thesis, Delft University of Technology.

Kim J, Kang B, Jeong H, Choe J (2019) Field development optimization using a cooperative micro-particle swarm optimization with parameter integration schemes. Journal of Petroleum Science and Engineering 183:106416.

Kim J, Yang H, Choe J (2020) Robust optimization of the locations and types of multiple wells using CNN based proxy models. Journal of Petroleum Science and Engineering 193:107424.

Krizhevsky A, Sutskever I, Hinton GE (2012) Imagenet classification with deep convolutional neural networks. In: Advances in Neural Information Processing Systems, 1097–1105.

Liang E, Liaw R, Nishihara R, Moritz P, Fox R, Goldberg K, Gonzalez J, Jordan M, Stoica I (2018) RLlib: Abstractions for distributed reinforcement learning. In: International Conference on Machine Learning, pp. 3053–3062.

Maschio C, Nakajima L, Schiozer DJ (2008) Production strategy optimization using genetic algorithm and quality map. In: EUROPEC/EAGE Conference and Exhibition, OnePetro.

Nasir Y, Durlofsky LJ (2022) Deep reinforcement learning for optimal well control in subsurface systems with uncertain geology. arXiv preprint arXiv:220313375.

Nasir Y, He J, Hu C, Tanaka S, Wang K, Wen X (2021) Deep reinforcement learning for constrained field development optimization in subsurface two-phase flow. Frontiers in Applied Mathematics and Statistics 7:54.

Onwunalu JE, Durlofsky LJ (2010) Application of a particle swarm optimization algorithm for determining optimum well location and type. Computational Geosciences 14(1):183–198.

Park HY, Yang C, Al-Aruri AD, Fjerstad PA (2017) Improved decision making with new efficient workflows for well placement optimization. Journal of Petroleum Science and Engineering 152. DOI 10.1016/j.petrol.2017.02.011

Schulman J, Wolski F, Dhariwal P, Radford A, Klimov O (2017) Proximal policy optimization algorithms. arXiv preprint arXiv:170706347.

Shin HC, Roth HR, Gao M, Lu L, Xu Z, Nogues I, Yao J, Mollura D, Summers RM (2016) Deep convolutional neural networks for computer-aided detection: CNN architectures, dataset characteristics and transfer learning. IEEE Transactions on Medical Imaging 35(5):1285–1298.

Silver D, Huang A, Maddison CJ, Guez A, Sifre L, Van Den Driessche G, Schrittwieser J, Antonoglou I, Panneershelvam V, Lanctot M, et al. (2016) Mastering the game of go with deep neural networks and tree search. Nature 529(7587):484–489.

Silver D, Schrittwieser J, Simonyan K, Antonoglou I, Huang A, Guez A, Hubert T, Baker L, Lai M, Bolton A, et al. (2017) Mastering the game of go without human knowledge. Nature 550(7676):354–359.

Vanvuchelen N, Gijsbrechts J, Boute R (2020) Use of proximal policy optimization for the joint replenishment problem. Computers in Industry 119:103239.

Xiong W, Ji X, Ma Y, Wang Y, AlBinHassan NM, Ali MN, Luo Y (2018) Seismic fault detection with convolutional neural network. Geophysics 83(5):O97–O103.

Yeten B, Durlofsky LJ, Aziz K, et al. (2003) Optimization of nonconventional well type, location, and trajectory. SPE Journal 8(03):200–210.

Zhang Y (2020) Integrated PI degradation modeling IPDM. In: SPE Annual Technical Conference and Exhibition, OnePetro.

Section VI

Predictive Maintenance Applications

16 Case Studies Involving Machine Learning for Predictive Maintenance in Oil and Gas Production Operations

Luigi Saputelli[1], *Carlos Palacios*[2], *and Cesar Bravo*[3]
[1]Frontender,
Abu Dhabi, United Arab Emirates
[2]Frontender,
Houston, Texas
[3]Frontender,
Houston, Texas

CONTENTS

DOI: 10.1201/9781003207009-22

16.1 INTRODUCTION

Throughout history, maintenance has moved from reactive maintenance of fixing it when it breaks toward more systematic analysis techniques in terms of root-cause analysis (Rødseth, Fordal and Schjølberg 2018). With the evolution of digitalization and the breakthrough technologies from Industry 4.0, advanced analytics techniques are becoming available.

In the past, many companies considered maintenance as an inevitable source of cost. For these companies, maintenance operations have a corrective function (fix it on failure) and are only executed in emergency conditions. Today, this form of intervention is not acceptable because of certain critical elements such as product quality, plant safety, and increased costs of maintenance departments (Bevilacqua and Braglia 2000).

Predictive maintenance (PdM) allows us to choose optimum intervention frequency by collecting insights from current equipment health condition and inferring equipment failure. Instead of being reactive after failure and/or rule-based preventive maintenance, PdM significantly depends on engineering models and statistical tools to process the data and derive policies to intervene with the equipment at the right time.

PdM is a trend-oriented policy that begins with identifying the states of each component within the equipment and estimates future values of some key indicators that characterize the health condition of a system (Figure 16.1) (Paolanti,

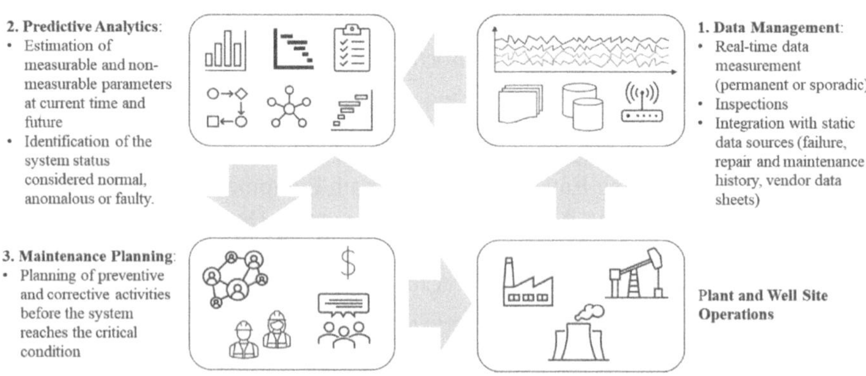

FIGURE 16.1 Predictive maintenance process in oil and gas production operations.

et al. 2018). PdM relies on engineering techniques and statistical tools to process the data and analyze the health condition to predict possible equipment failure (Lee, et al. 2005).

PdM encompasses a variety of topics, including but not limited to failure prediction (remaining useful life [RUL]), failure diagnosis (root-cause analysis), failure detection, failure type classification, and recommendation of mitigation or maintenance actions after failure (Microsoft Azure Team 2015).

Examples of PdM include the following: (1) instructing a machine inspection driven by high vibrations signaling bearing deterioration, (2) degrading motor capacity because of the temperature of drown current, and (3) detecting degradation of rubbing contact parts from the composition measurement of particles in a lubricating oil (Paolanti, et al. 2018).

16.2 PREDICTIVE MAINTENANCE IN OIL AND GAS OPERATIONS – ESP CASE

More than 90% of producing oil wells require some form of artificial lift for pumping production fluids to the surface (Bates, et al. 2004a,b), and the electrical submersible pump (ESP) is one of the most popular artificial lift systems in of upstream operations. An ESP system typically comprises an electric motor, a protector, a gas separator, a multi-stage centrifugal pump, a power cable, a motor controller, a transformer, and a power source. Sensors or downhole gauges are installed for monitoring downhole well and pump operating conditions. Signals from the sensors are transmitted through the power cable to surface remote terminal units (RTUs) allowing continuous sampling of pressures, temperature, vibration, current, and voltage (Fonseca, et al. 2021).

Key use cases involving machine learning for PdM in oil and gas equipment, particularly in ESP, are shown in Table 16.1.

16.2.1 Predictive Maintenance Overview and Ongoing Issues

PdM has typically faced challenges in terms of data, prediction confidence, people, and process. The following is a non-extensive list of such issues with an immediate solution.

- Data: It is a challenge for many operators to maintain a minimum acceptable quantity and quality data and information from PdM (Wurzbach 2002). Additionally, most critical historic data (e.g., data sheets, fault, failures, repairs) resides in manually captured records. A sound data management strategy is required to fully address a PdM process.
- Recall: Machine-learning models exhibit a large number of false positives, leading to poor confidence in the PdM recommendations. Access to a larger set of labeled data and a more profound approach may be adopted to reduce the percentage of incorrect predictions.

TABLE 16.1
Use Cases Involving Machine Learning for Predictive Maintenance

Predictive Maintenance Use Case	Data Analytics and Machine Learning Technique
Analyze and anticipate ESP failure modes (Bermudez, et al. 2014; Grassian, et al. 2017)	Fuzzy expert system to predict ESP developing failure modes and determine the causes of a prior failures, built on a set of operator expert rules and used to formulate the conditional statements.
Normal vs anomalous modes operations status classification prediction for ESP (Guo, et al. 2015)	Binary classification model based on generalized-SVM (support vector machines) learning approach, using a linear kernel model and selected hyperparameters via cross-validation.
Detect and diagnose anomalous modes operations, and prevent impending abnormal events in ESP (Gupta, Saputelli and Nikolaou et al. 2016)	22nd degree time-series dimensionality reduction with Robust PCA (principal component analysis), projected onto a single plane using Hotelling's T^2 statistics to diagnose and prescribe actions to avoid abnormal events.
Predicting anomalous modes operations for improved production performance (Abdelaziz, Lastra and Xiao 2017)	PCA as an unsupervised machine learning (ML) technique to identify hidden patterns, and to detect developing ESP failures and predict remaining operating time before failure
ESP well and component probability of failure prediction 30 days in Advance (Marin, et al. 2019)	Random forest (RF) to achieve objective functions by predicting well failure and failed ESP pump component based on the identified parameters.
Proactive identification of anomaly detection and failure prediction with real-time alarming system (Bhardwaj, et al. 2019)	Two predictive ML models deployed, PCA-based and nonlinear residual model built for the motor vibration, with alarming process for failure identification by constructing dynamic operating envelopes for each ESP.
Prediction of ESPs time to failure 1–3 days before its actual stop (Khabibullin, et al. 2020)	Ensembles of decision trees based on extreme gradient boosting (XGBoost), CatBoost and LightGBM, were used as the main model after binary classification of equipment state immediately before and a long time before a failure.
Automation of ESP failure root cause analyses from both structured and unstructured documents (Karnik, et al. 2021)	Convert textual sentences to equivalent numeric vectors using long short-term memories (LSTMs) and autoencoders, and an ensemble ML model for the classification objective.
Automated esp failure database and reliability analysis platform (Robyn and Chacin 2021)	Integrated system built around an equipment failure database and visualization platform that continually gathers relevant data through the ESP cycle with basic statistics and analytic techniques.

(Continued)

TABLE 16.1 *(Continued)*
Use Cases Involving Machine Learning for Predictive Maintenance

Predictive Maintenance Use Case	Data Analytics and Machine Learning Technique
Predictive maintenance optimum scheduling reduces the wastage of part life, number of events, number of outages, and maintenance cost (Tapan, Sarika and Bansal 2013)	Binary base integer programming model used for preventive maintenance scheduling for large number of complex rotary equipment.
Forecast gas turbines problems at a future date so that the problem can be dealt with in the present (Bangert 2020)	First, turbine state forecasted into the future using an LSTM network. Second, a neural network recognizes normal healthy behavior to identify problem. Third, problem fingerprint diagnosed for similar failure mode, i.e., vibration.
Predicting fouling in heat exchanging equipment for scheduling predictive maintenance (Cadei, et al. 2019)	Ensemble of two different forecast models: RIDGE regression for a long-term degradation speed following operating conditions, and a short-term ARIMA model. Lastly, a one-class SVM to define the normality and abnormality boundary.

- People: Applications of PdM involve intersection of knowledge among various specialties including mechanical and electrical engineering, instrumentation and process control engineers, and data science disciplines. The latter has become a scarce resource to liaise with traditional engineers. Cross-disciplinary knowledge interchange shall close this gap in the future. Traditionally, there has been a culture clash between maintainers and data scientists particularly due to the lack of confidence in PdM models, i.e., poor recall score as presented earlier.
- Process: PdM suggests actions to intervene with equipment based on a forecast of degraded expected performance. Maintenance personnel may lack a proper business process to execute such policies, *e.g., approving to shut down equipment based on vibration when there is a production commitment, or avoiding preventive maintenance because a predictive model does not exhibit any performance loss.* A management system with proper governance may be needed to adopt new PdM policies.
- Deployment: For the most part, many PdM techniques have been tested across industry; however, they remain stuck in research departments and they are not fully deployed in the industrial environment. As it will be shown in the examples in this chapter, the case studies presented are mostly research cases, not commercial ones.
- Economic benefits measurement: The lack of a standard method for quantifying the economic value of its implementation poses a challenges for scaling up PdM practices.

16.2.2 Data Analytics Role in Predictive Maintenance

PdM is about taking actions when the unknown becomes apparent, yet it still is an inexact science and a debatable act for certain audiences (Table 16.1). Some key questions that machine learning shall try to address include the following:

- What is the probability of failure in a given period (next week, next month, etc.)?
- What is the Remaining Useful Life (RUL)?
- How much has the equipment (or certain part) degraded?
- When would an anomaly occur in equipment behavior?
- Does better analysis provide better failure prediction?
- How best to change equipment settings to delay failure? Or to extend its life?

Sound data management is one key enabler for PdM of many oil and gas operations (Figure 16.2), including online equipment condition monitoring (Gupta, et al. 2016). From addressing availability and operability concerns of safety-related equipment to increasing profitability through support system reliability and reduced maintenance costs (Wurzbach 2002) data management plays a critical role. Failure documentation standards have been presented (Alhanati, Solanki and Zahacy 2001) to add continuous value to operations and maintenance. Repurposing legacy pressure data to employ them in a predictive monitoring application allows modern machine learning methodology tracking of pump health (Giunta, et al. 2020).

This chapter provides an overview of PdM use cases in oil and gas operations, identifies the suggested techniques for detecting equipment anomalies, and outlines the challenges in applying such techniques in real-life applications. A general idea of PdM for well integrity management and asset life extension, an outline of the future of PdM, and some ideas for closing existing gaps are also presented

Machine, Equipment or Asset Conditions	Machine, Equipment or Asset Fault History	Maintenance and Repair History
• Date Stamp (past or future) • Equipment or Asset ID • Status (Healthy, Faulty, etc) • Operating Conditions Data (current, voltage, pressure, temperature, vibration, flow rate, fluid conditions) • Control Status (Automatic, Manual) • Overload Status • Hours since comissioning • Hours since last maintenance • Hours since last trip or fault	• Date of Trip, Failure or Fault • Date Service Reestablished • Equipment or Asset ID • Failure mode (flow, electrical) • Failed Items (pump, motor) • Failure Descriptor (leak, burn) • Primary Cause of Failure • Secondary Cause of Failure • Profit Units Lost (Uptime Hours, US$, barrels of oil per day, kWh, etc.)	• Date of Intervention (past or future) • Duration • Equipment or Asset ID • Type of Intervention (Corrective, Preventive, Predictive, Condition-based) • Intervention Category • Job or Tasks Performed • Replaced Components • Profit Units Lost (Uptime Hours, US$, barrels of oil per day, kWh, etc.)

FIGURE 16.2 Predictive maintenance sound data management practice.

Data-driven models are less complex than first-principles models and can precisely capture complex underlying relationships among several variables to gain actionable insight into equipment performance in real time. For example, an ESP health monitoring analytics engine, when integrated within the real-time monitoring workflow, would create an opportunity to advance ESP management from a reactive to a more proactive approach (Gupta, Saputelli and Nikolaou 2016). Such systems shall alert and guide on-site personnel in a timely manner with recommendation on when to intervene with the operation of equipment.

16.3 IDENTIFYING OPTIMUM MAINTENANCE FREQUENCY FROM DATA ANALYTICS

Energy transition to a lower carbon emission industry is driving significant changes in the way operators should be reducing costs. These adjustments affect maintenance plans and make its role even more crucial to stay ahead of the curve (Tajadod, Mohammadali and Rategari 2016).

Best maintenance policies are specific to each equipment running condition. An analytical hierarchical process has been proposed for complex downstream plants where maintenance strategies are arranged in a hierarchic structure and evaluated using only a series of pairwise judgments (Bevilacqua and Braglia 2000). PdM can assist operators to dictate minimum maintenance requirements, which can add immediate value by cutting costs from unnecessary maintenance activities.

16.3.1 Condition-Based Monitoring (CBM)

Condition-based maintenance (CBM) involves monitoring the condition of machine and equipment to predict potential failures in time, assess consequences with probability of occurrence, and take action before the actual failure happens. It is a proactive maintenance strategy based on equipment reliability characteristics where it is imperative to understand the mechanics behind failures (Ferket 2019). CBM relies on the availability of large numbers of sensors in complex equipment such as sub-sea equipment and large processing plants. It involves not only predictive analytics to forecast anomalies and failures, but also rule-based systems to describe the cause-effect interaction among components.

CBM checks several conditions, and if one of them indicates that there is a degradation of the equipment, CBM is directly executed. PdM applies predictive analytics and tools to determine the required actions. As in the case of CBM, PdM also performs the maintenance only when it is necessary. In principle, CBM and PdM are actually addressing the same maintenance policy (Kang, Catal and Tekinerdogan 2021).

16.3.2 Economic Implications

Two issues drive the overall operating cost of ESPs. First, the low system run life results in high numbers of trips or failures, resulting in multiple interruptions to

TABLE 16.2
Annual Cost of Lost Production or Intervention due to ESP Failure

Lost Production Cost		Intervention Cost	
Price of oil barrel	\$50	Onshore conventional	\$5–25K
Typical oil production	150 bopd	Onshore unconventional	\$150–250K
Estimated downtime	2 days	Offshore well	Up to \$1MM
Estimated incidents/year	10		
Estimated lost	$150 \times 2 \times 10 \times \$50 = \$150K$		

Source: Carrillo (2013).

production. Second, each trip or failure is itself very expensive. The annual cost of lost production or intervention due to ESP failures could reach \$150 thousand or \$1 million per well, respectively (Carrillo 2013). Table 16.2 highlights the assumptions and results for a typical economic study. A profit loss indicator (PLI) has shown promising results in measuring, e.g., time losses in production in a monetary term (Rødseth, Fordal and Schjølberg 2018).

16.4 FROM IDENTIFYING ABNORMAL EVENTS TO PREDICTING IMMINENT AND UNPLANNED FAILURES

Data analytics techniques have proven to be very useful in real-time monitoring, enabling production engineers with predictions and pump diagnostics about unexpected problems, such as pump wear, gas interference, tubing leaks, viscous forces, and solids-plugged intakes (Fonseca, et al. 2021).

Data-driven analytical solutions have been proposed to proactively monitor and transform vital statistics related to ESP performance into actionable information that extends equipment life and reduces downtime (Gupta, Saputelli and Nikolaou 2016).

16.4.1 Operating Envelopes for Optimum Equipment Health

One key step in PdM is to be able to determine the condition of equipment and the deviation from the normal operating envelope. Manufacturers and engineering designers determine the conditions for which an equipment shall operate. These conditions may change through the life of the asset because of material degradation and variations in the assumed operating conditions (Bates, et al. 2004).

Ammeter charts have been the primary rotating equipment monitoring method for many decades to correct minor operational issues in ESP performance. This method measures only the electric current drawn by the motor and records it as a function of time on a circular chart (Takács 2009). Changes in input amperage to the motor can be linked only to voltage fluctuations, fluid density, or flow rate; hence, these charts are limited in the extent to which they can define ESP performance. This monitoring method is still very common but requires extensive workforce and regular wellsite visits to collect data and make necessary adjustments to pump and well operations.

Machine learning has been used for decades to learn nonlinear mapping functions for the prediction of rotating equipment parameters using the motion current signature while avoiding the need for costly measurement of system parameters (Bansal, Evans and Jones 2004).

16.4.2 Identifying Abnormal Events (Anomaly Detection) in ESP

Unpredictable and repeated ESP shutdown often occurs due to changes in well conditions (e.g., increased gas oil ratio), improper system design, and operating practices (Takács 2009). Problematic production environments such as high gas volume wells, produced solids, high temperature, and corrosive environments often lead to multiple incidents of trips and failures. A trip happens when key parameters start operating beyond safe operating thresholds and a failure is imminent. It results in both downtime and additional ESP restarts, leading to deferred production losses and high operating costs.

Equipment failure occurs due to changes in its performance over time. Failure of various parts of the ESP unit (pump, engine, cable, etc.) can lead to failure and subsequent repair of a well equipped with an ESP. The operating conditions affect each of these parts and they are an important factor that that should be considered (Khabibullin, et al. 2020).

Fuzzy logic solutions are built on a set of production operation expert rules and used to formulate the conditional statements, e.g., if the motor temperature exceeded a certain limit, the pump can burn out (Bermudez, et al. 2014). The tool determines trends by calculating a slope for the different sensor signals. The closeness of the measured trends to the models is expressed as "conditioners," which represent the likelihood of each trend occurring over a period of time (Grassian, et al. 2017).

16.4.3 Case Study: Robust PCA for Predicting Impending Failures in ESP

Multivariate statistical techniques for dimensionality reduction, such as principal component analysis (PCA), offer tremendous opportunities for detecting impending problems long before they occur, diagnose potential causes, and prescribe preventive action (Gupta, Saputelli and Nikolaou 2016). PCA is a statistical method to explain covariance structure of data by means of a small number of components (Hubert, Rousseeuw and Verdonck 2009). The overall workflow involves building a robust PCA model that uses selected decision variables and a target variable (Figure 16.3).

To build a data-driven model, key parameters that capture ESP performance are selected as input decision variables including intake pressure, discharge pressure, and pump head as well as motor diagnostic parameters such as vibration,

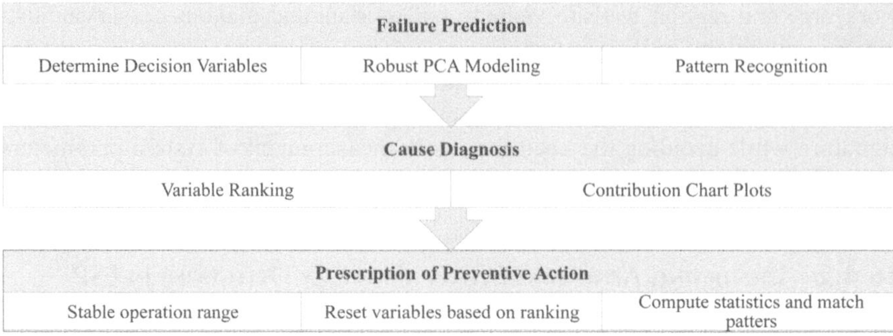

FIGURE 16.3 ESP analytics workflow (Gupta, Saputelli and Nikolaou 2016).

leakage current, and motor temperature. Other computed variables such as delta pressure, liquid rates, and pump efficiency can be inferred from simple engineering expressions or manufacturer curves and used as decision variables.

Data Collection. Data was obtained at a frequency of 1 minute for a 2-year time period for multiple wells, which saw several failure and trip events during this time period. An event log file containing information during a trip or failure was also obtained. This record was used to study the behavior of the patterns based on the selected parameters obtained from the historian long before, immediately before, and exactly during the trip or failure event.

In PCA, the classical variance (which is being maximized) and the classical covariance matrix (which is being decomposed) are very sensitive to anomalous observations in classical PCA (Hubert, Rousseeuw and Verdonck 2009). Therefore, as a consequence, the first components are often attracted toward outlying points and may not capture the variation of the regular observations. This can lead to the conclusion that data reduction based on classical PCA becomes unreliable if outliers are present in the data. In ESP data, there is a very high probability of encountering outliers or erroneous data from sensors, so the robust PCA technique was adopted to eliminate bad sensor data from the input feed to the model

Model Building. A data-driven model using PCA was constructed to reduce the dimensionality of a historical dataset containing 22 parameters and to extract important features or dimensions that capture most of the variability in the historical data.

PCA provides a simple yet powerful way of predicting abnormal behavior in ESP. Normal operating condition forms a cluster around the origin, whereas unstable behavior leads to a trip formed away from that for the dataset representing normal operations (Figure 16.4).

Predictive Model. Historical data during a known unstable time period was selected as new input to the model; such new input belonged to the time window when deviation from normal operating behavior started until a trip, or a failure, happened.

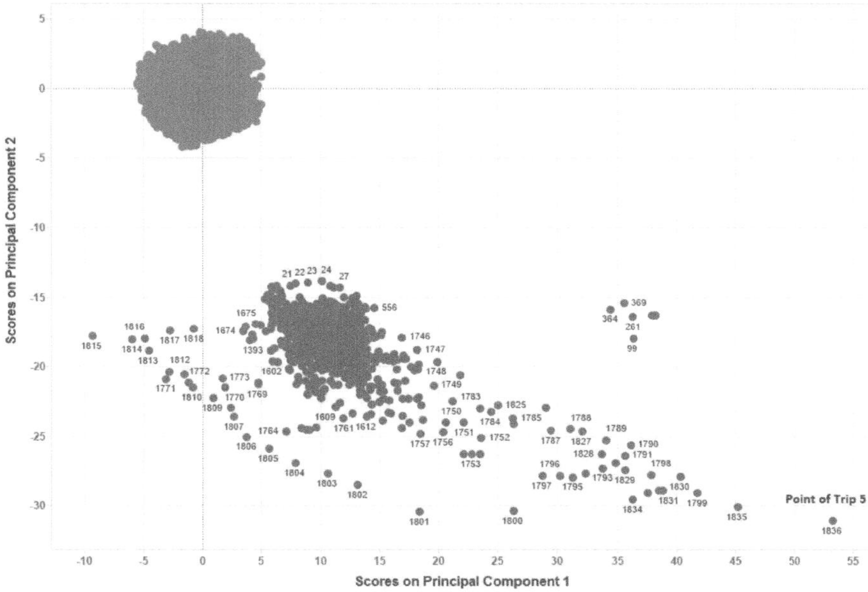

FIGURE 16.4 Predicting ESP abnormal behavior from scores in principal component 1 and 2.

Diagnostic Model. A diagnostic model to identify cause of a trip was also developed by ranking the contribution of each of the decision variables toward a trip. In this manner, the issue can be diagnosed, and investigative actions can be prioritized based on the possible anomalies that can happen due to the highest and possibly correlated higher ranking variables (Figure 16.5). In this case, motor temperature had the highest contribution for the first half of the unstable time period.

Prescriptive Model. A prescriptive model to proactively suggest corrective action to address the issue was also developed.

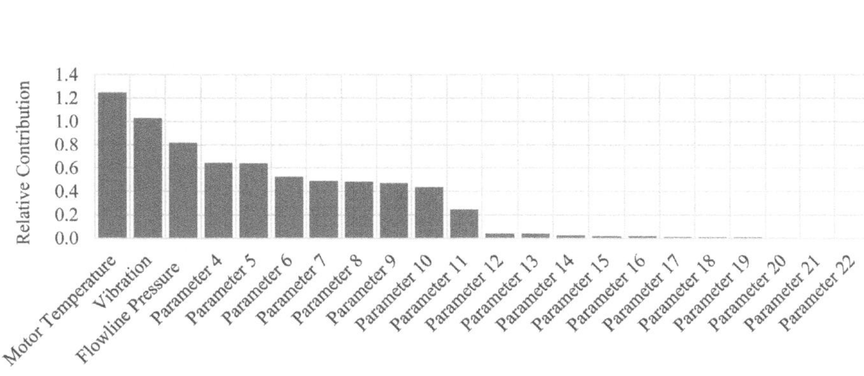

FIGURE 16.5 Diagnosing ESP failure by ranking contributions.

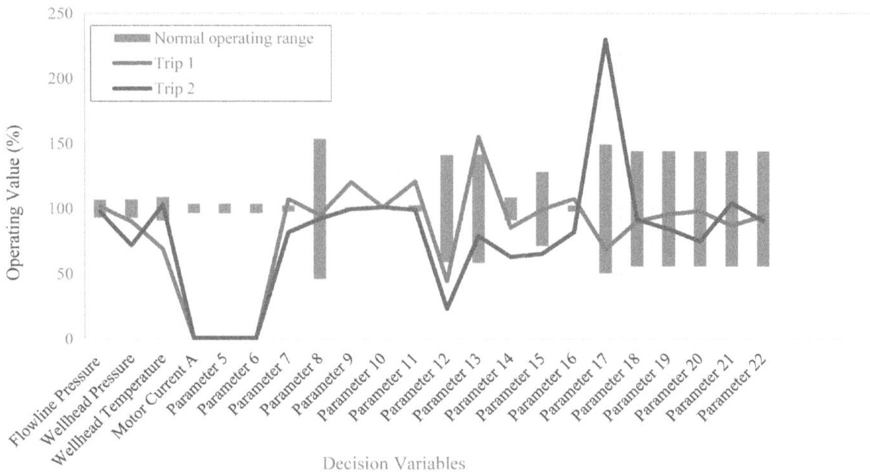

FIGURE 16.6 Prescriptive indicator suggestion to correct ESP trip.

The normal operating range for each of the parameters was determined from the two solutions obtained for each parameter using Eq. (16.1). These are represented as vertical bars (Figure 16.6). The line graphs represent the reading of the various parameters for two separate trip events, Trip 1 and Trip 2.

To attain a normal operating state, all parameters should lie within their stable bands. At that point, the scores of the two principal components would lie in the top cluster (Figure 16.4).

The prescriptive plot can help determine which of the decision variables are operating outside their safe limits in real time and take corrective measures to bring them back to their stable operating bands. However, not all parameters can be controlled directly, e.g., vibration. At the same time, change in one will have an impact on the others, because parameters are interrelated. Priority is assigned to parameters having higher ranking, which is determined using the diagnostic model.

Therefore, by modifying parameters, which can be adjusted by human action (e.g., choke, motor frequency), other parameters are indirectly adjusted to operate inside their respective stable bands. This behavior is also replicated in the predictive plot for each new time step.

16.4.4 Failure Classification

Fault classification and understanding the context of the root cause of failure is also a key enabler of PdM. Problems with system design, equipment specification, manufacturing, installation, and day-to-day operation can be identified and corrected, contributing to increased run lives, lower operating costs, and increased profits. An approach to establish consistent practices for collecting, tracking, and sharing ESP run life and failure information is based on two key elements: a general and common dataset and a standard nomenclature for coding ESP failure information (Alhanati, Solanki and Zahacy 2001). Structured and unstructured data across equipment, production, operations, and failure reports to automate root-cause identification and analysis post breakdown (Karnik, et al. 2021).

Another enabler for fault classification is the method described by failure mode and effect analysis (FMEA), which is a structured approach to discover ways in which a process can fail (failures) and ways those failures lead to dangerous outcomes (effects) (AIAG 1993).

16.4.5 Remaining Useful Life Prediction

RUL is the length of time that an item, component, or system is likely to be able to function in accordance with its intended purpose before it requires repair or replacement (Okoh, et al. 2014). By taking RUL into account, engineers can schedule maintenance, optimize operating efficiency, and avoid unplanned downtime. Multiple RUL models are available based on the application and availability of data (Table 16.3).

Empirical models in predicting remaining life phenomenon are simple statistical models based on the actual operating time of the equipment. The Kaplan-Meier estimator is used to predict the probability of survival of equipment after a certain period of life or event (Kaplan and Meier 1958) when there is some information about an item survival time but precise survival time is unknown, because such item is no longer under observation and the event was not directly observed.

For the curious reader, some practical examples of RUL calculations using open-source machine learning techniques are publicly available:

1. Predicting remaining useful life of a machine using LSTM (Saraswat 2021).
2. Recurrent Neural Networks for Predictive Maintenance (Griffo 2017).
3. Remaining Useful Life prediction for Wind Turbines (Mammadov 2021).

16.4.6 The Rise of Digital Twins in Predictive Maintenance

Often it is too difficult to create the fault conditions necessary for training a PdM algorithm on the actual machine. An "idealistic" solution to this challenge would be to build a digital twin that would be able to replicate real faults including corrosion, abrasion, material breakages, and fatigue. Today, multiple fundamental models support such phenomena, but they lack sufficient field data to provide a high confidence level. With more available asset data, meaningful features could be extracted and then used to develop a suitable machine learning model that represents the physical asset. Simulated failure data could be then used to design a fault-detection algorithm. If such model structure can be

TABLE 16.3
Remaining Useful Life Prediction Models

Method	Available Data (Application)	Model	Method
Similarity Models (run-to-failure history) estimate RUL using a residual comparison-based similarity model	Large dataset of degradation path histories for an ensemble of similar manufactured machines to the same specifications	Hash Similarity	Compares similarities in mean, power, minimum, or maximum values for the data
	Match signal shapes from multiple machines manufactured to the same specifications	Pairwise Similarity	Compares the degradation profile of two similar components
	Known degradation dynamics for an ensemble of similar components	Residual Similarity	Residual comparison-based similarity model
Degradation Models (known failure threshold) extrapolate past behavior to predict the future condition	Log-scale signal or non-cumulative damage (useful when your system does not experience cumulative degradation)	Linear Degradation	Describes degradation behavior as a linear stochastic process with an offset term
	Cumulative damage (useful when the test component experiences cumulative degradation)	Exponential Degradation	Describes degradation behavior as an exponential stochastic process with an offset term
Survival Models (lifetime data with or without covariates) estimate RUL using a proportional hazard survival model	Lifetime data only (have limited failure times for an ensemble of similar components)	Reliability survival	Estimates the probability distribution coefficients from a collection of failure-time data
	Lifetime data and covariate (use historical information about the life span of components and associated covariates)	Covariate survival	Fit model coefficients using a collection of failure-time data and associated covariates

Source: Mathworks (2021).

found, then a self-tuning digital twin process can be automated, enabling quick adjustment to varying process conditions (Rao 2020). Digital twins could also gain popularity in PdM because of the advances in high-performing computing and Internet of Things (IoT) technologies.

Real-time analysis of gas turbine air inlet systems enables continuous performance improvement through the use of digital twins, which are used to verify if the existing system is performing as expected. Digital twin solutions can measure filter system deterioration over time and detect anomalies, which can then be used for making operational decisions, e.g., replacing the evaporator media with more efficient media to get a higher increase in power output (Gautam and Kohn 2019).

Digital twins combine advanced real-time analytics tools with traditional engineering groups (computer-aided engineering [CAD]) enabling the project's front-end engineering and design (FEED) and commissioning stages far more closely, allowing them to perform some of the latter virtually using the digital twin of a physical asset on remote locations, such as rotating equipment (Settemsdal 2021).

16.5 PREDICTIVE MAINTENANCE IN CORROSION AND ASSET INTEGRITY MANAGEMENT

16.5.1 Corrosion Predictive Models

Corrosion is a common issue faced by aging facilities associated with costly repairs (Koch, et al. 2002). Corrosion is increasingly a significant challenge to the offshore industry and attributed to an aging worldwide offshore fleet; assets being kept in operation for prolonged periods of time; units operating beyond their original "design basis"; and newer larger vessels in deeper, harsher environments with less opportunity for shipyard repair (Koch, Ripley and McKay 2013).

Corrosion management starts from the correct material design suitable for the predominant environment exposed, followed by periodic data collection through sensors and visual inspections to monitor material thickness and environment conditions including pressure, temperature, oxygen, carbon dioxide, and hydrogen sulfide (Kermani 2012). Corrosion rate prediction models have been proposed to reduce the impact of unplanned downtime caused by corrosion (Table 16.4).

Corrosion reaction involves giving off and receiving electrons via a conductive metallic path as well as exchange of ions through a conductive medium. The loss of metal per unit of surface and time defines the metallic corrosion rate (Andrade and Alonso 1996). It can be estimated from the weight loss of an "intrusive" corrosion coupon after exposure to a corrosive environment, expressed as mils (thousandths of an inch) per year penetration (mpy). Corrosion rate (R) is calculated assuming uniform corrosion over the entire surface of the coupon and it is measured in mpy.

TABLE 16.4
Machine Learning Use Cases for Corrosion

Use Case for Corrosion	Data Analytics and Machine Learning Technique
Forecasting corrosion rate of cooling water (Shanrang, et al. 2010)	Least squares support vector machine (LS-SVM): water quality factors related with corrosion were selected as input variables and the corrosion rate were selected as output variables.
Predicting CO_2 internal corrosion in oil and gas pipelines (Hatami, et al. 2016; Zubir, Aziz and Jaafar 2018; Zukhrufany 2018; Palacios 2019)	Neural network, SVM, and random forest (RF) with synthetic data generated via random Gaussian function and de Waard-Milliams model, for CO_2 internal corrosion. The input parameters of the model are temperature, CO_2 partial pressure, flow velocity, H_2S, CO_2, T, P, pH, flow regime, material chemical composition, microstructure, oil API, water-cut, water composition, etc.
Compute corrosion risks using the failure probabilities by estimating the corrosion defect depth growth of aged pipelines (Ossai 2019)	PCA, particle swarm optimization (PSO), neural networks, gradient boosting machine (GBM), and RF to estimate the corrosion defect depth growth of aged pipelines.
Determine the likelihood of stress corrosion cracking (SCC) behavior in boiling water reactors (Sturrock and Bogaerts 1996)	Classical and non-parametric statistics including decision trees and fuzzy logic to predict the likelihood of SCC for varying conditions of temperature, chloride level, oxygen content, and metallurgical condition.
Quantifying a prediction for the future evolution of damage to assess integrity of pipelines transporting hydrocarbons over long distances (Masi, et al. 2015)	Ensemble of ANNs is generated for quantifying internal corrosion prediction.
Corrosion detection via automated visual inspection (Nash, Drummond and Birbilis 2019)	Fully convolutional network applied to a large dataset of 250 images with semantic segmentations labeled by non-experts and a dataset of 10 images labeled by subject-matter experts.
Automated detection of pipe corrosion of the water supply and waste disposal systems (Hoang and Tran 2019)	Image texture including statistical measurement of image colors, gray-level co-occurrence matrix, and gray-level run length employed to extract features of the pipe surface. Support vector machine optimized by differential flower pollination is then used to construct a decision boundary that can recognize corroded and intact pipe surfaces (dataset consisting of 2,000 image samples).

$$R = \frac{K \cdot m}{A \cdot \rho \cdot t} \tag{16.1}$$

where K is a constant for the process (e.g., 22,300 for an iron coupon), m is the material loss in grs, ρ is the coupon metal density of the coupon (g/cm^3), A is the area of coupon (sq. in.), and t is the time of exposure in corrosive environment (days).

Nondestructive examination (NDE) data analysis can be used to determine percentage material lost, and it is effective for detecting and quantifying corrosion damage. Yet, NDE methods are most effective when integrated with other equipment data (Figure 16.2). The NDE methods include, among others, computed tomography, thermography, and radiography (Rutherford and Luzar 1999).

16.5.2 Well Integrity Management

Well integrity management includes the technical, operational, and organizational practices to oversee well mechanical structure from design to abandonment, reducing the risk of uncontrolled release of formation fluids. Well integrity forms a fundamental part of the licensing requirements.

A well integrity management system (WIMS) is a stepping stone that allows data from various sources of well life-cycle information to be integrated seamlessly including well register, well construction envelope and operating limits, well status (operational, integrity, barriers, annuli and valve status), barrier management, well failure matrix, and leak rate criteria (Table 16.5). As in any other PdM program, well integrity management relies on (1) well design and operating envelope, (2) inspection and verification, and (3) failure management (Sparke, Conway and Copping 2011).

TABLE 16.5
Machine Learning Use Cases for Well Integrity Management

Use Case for Well Integrity Management	Data Analytics and Machine Learning Technique
Automated detection of clinging P arrivals in the flexural wave data for oil and gas wells' cement evaluation (Gkortsas, Bose and Zeroug 2019).	ML highlights axial and azimuthal regions, where the annular fill behind casing is confirmed to be a solid with additionally a tight range for its compressional wave velocity, rendering a more quantitative and more autonomous cement diagnosis.
Predict and diagnose risk status of gas/lift wells, failure, and operating envelope compliance (Yakoot, et al. 2011).	Logistic regression, decision tree, random forest, Gaussian, k-nearest neighbor, and support vector machine classifiers on data from valves' status, conductor movement, ship status, tower staffing, rates, and pressures.

If determined by standard risk assessment practices, well capacities can be downgraded where some of its operating limits are revised to accommodate certain integrity issues and extend the operability of the well. A reasonable grace period is given while developing an action plan to fix the integrity issue.

16.5.3 Role of Predictive Maintenance in Asset Life Extension

Structural assessment, real-time monitoring, and PdM are the three main aspects of the life extension process for the offshore structure (Bhowmik 2019). Some corresponding use cases are given in Table 16.6.

A fatigue monitoring system is used to improve the monitor in-place fatigue response in a top tensioned production riser (TTR). Fatigue performance is a key input for any life extension requests, and long-term integrity and its performance is dependent on the operating condition during the life of the field including fluid contents and top tension (Mercan, Campbell and Clay 2021). Frequency-based transfer functions are pre-determined using a riser finite-element analysis (FEA) model that correlates acceleration at the measurement location with bending moment at the location of interest for fatigue. The PdM schedule is then developed based on the remaining fatigue life (Bhowmik 2019).

TABLE 16.6
Machine Learning Use Cases for Asset Life Extension

Use Case for Asset Life Extension	Data Analytics and Machine Learning Technique
Gas pipelines for structural health monitoring (defect location prediction) integrated with intelligent PIG devices and pipe crawling robots (Zajam, Joshi and Bhattacharya 2019).	SVM classifier (separates the segment of data corresponding to defect region) and wavelet analysis on vibration (spectral components of frequency) response in the detection of various kinds of defects present in gas pipelines.
Risk-based inspection screening assessment for oil and gas production and processing units (Rachman and Chandima Ratnayake 2019).	Machine learning model (logistic regression, SVM, *k*-NN, gradient boosting decision trees, AdaBoost, random forests); the best performing model achieves accuracy and precision of 92.33% and 84.58%, respectively.
Monitoring assessment enables real-time load and flexible riser fatigue evaluation and validation of design predictions (Viana, Khalid and Shankar 2020).	Nonlinear autoregressive exogeneous neural network (NARXnet) and convolutional neural network (CNN) evaluated to predict accurately flexible riser fatigue damage based on actual metocean conditions.
Generate the predictive maintenance schedule from the fatigue prediction model (Bhowmik 2019).	ML using a recurrent neural network (RNN) was used to generate the PdM schedule based on the data-driven fatigue prediction model created from the measurement data (Bhowmik 2019).

A risk-based corrosion management process for offshore structures provides confidence to key stakeholders accounting for life extension, reducing the risk of corrosion failure, and lowering the cost of inspection and maintenance (Koch, Ripley and McKay 2013).

16.6 FUTURE OUTLOOK

State-of-the-art innovative workflows for detection, diagnosis, and prevention of potential equipment problems in real time have been discussed using machine learning techniques and multivariate statistics. Such workflows rely on data being collected by downhole and surface sensors. Key challenges discussed include the lack of enough sensor data, the slow pace of adoption of data-driven applications in the commercial arena, the inability to quantify economic benefits of PdM, and the cultural clash between maintenance personnel and data scientists.

Upstream industry must catch-up instrumenting wells, fields, and plants with adequate measurement support to fulfill PdM goals. The evolution of key digital technologies will enable a better road for machine learning use cases in oil and gas PdM, including cost-effective wellsite sensors and massive data capture through wearable devices. With lower cost sensors coming to market in the next decades, the PdM future looks for a new beginning.

Machine learning applications are still in the infancy stage in PdM in the oil and gas industry. Although data-driven models claim to increase compressor and pump uptime, minimize maintenance efforts, reduce intervention costs, and safeguard operations, their application has not been fully deployed across organizations. Many opportunities to apply predictive data analytics and extend the approaches described here include automating alarm settings through reinforced learning, extending machine learning models to classify all categories of failure or abnormal behavior, and creating a health and condition management to all forms of artificial lift and other mechanical systems. Major effort is required to define use cases and embed the change into existing processes.

Numerous opportunities exist for young professionals to perform and find ways to grow their career in PdM powered by machine learning technologies. By closing the gaps in terms of knowledge skills, technical challenges including operating standards and organization alignment will be addressed. These challenges can be met through the implementation of quality training to PdM personnel and regular updating and re-certification of key technology holders. Better communication and engagement between the maintenance personnel and data scientist is also a must for these applications to succeed.

Machine learning core technologies including deep learning and explainable artificial intelligence (AI) continuous progress shall be a key factor for enhancing accuracy in many PdM applications, which in turn will enable more transformative and value-added processes. Better model confidence toward data-driven PdM in addition to increased adoption through large organizations will enhance the use of machine learning; therefore, it will provide oxygen to the oil and gas industry for navigating through the next decades (Pandey, et al. 2020).

REFERENCES

Abdelaziz, M., R. Lastra, and J. J. Xiao. 2017. "ESP Data Analytics: Predicting Failures for Improved Production Performance." *Abu Dhabi International Petroleum Exhibition & Conference*. Abu Dhabi, UAE. doi:10.2118/188513-MS.

AIAG. 1993. *Potential Failure Mode and Effect Analysis*. Automotive Industry Action Group. Accessed December 24, 2021.

Alhanati, F. J. S., S. C. Solanki, and T. A. Zahacy. 2001. "ESP Failures: Can We Talk the Same Language?" *SPE Gulf Coast Section Electric Submersible Pump Workshop*. Houston, TX. doi:10.2118/148333-MS.

Andrade, Carmen, and C. Alonso. 1996. "Corrosion Rate Monitoring in the Laboratory and On-Site." *Construction and Building Materials* 10 (5): 315–328. Accessed December 29, 2021. https://sciencedirect.com/science/article/pii/0950061895000445.

Bangert, P. 2020. "Predictive Maintenance for Gas Turbines." *International Petroleum Technology Conference*. Dhahran, Kingdom of Saudi Arabia. doi:10.2523/IPTC-19864-Abstract.

Bansal, Dheeraj, David Evans, and Barrie Jones. 2004. "A Real-Time Predictive Maintenance System for Machine Systems." *International Journal of Machine Tools & Manufacture* 44 (7): 759–766. Accessed December 21, 2021. https://sciencedirect.com/science/article/abs/pii/s0890695504000392.

Bates, R., C. Cosad, L. Fielder, A. Kosmala, S. Hudson, G. Romero, and V. Shanmugan. 2004a. "Taking the Pulse of Producing Wells—ESP Surveillance." *Oilfield Review* Summer: 16–25.

Bates, R., L. Cosad, A. Fielder, A. Kosmala, and S. Hu. 2004b. *Taking the Pulse of Producing Wells—ESP Surveillance*.

Bermudez, F., G. Carvajal, G. Moricca, Joydeep Dhar, F. Adam, A. K. Al-Jasmi, H. K. Goel, and H. Nasr. 2014. A Fuzzy Logic Application to Monitor and Predict Unexpected Behavior in Electric Submersible Pumps (Part of KwIDF Project). doi:10.2118/167820-MS.

Bevilacqua, M., and M. Braglia. 2000. "The Analytic Hierarchy Process Applied to Maintenance Strategy Selection." *Reliability Engineering & System Safety* 70 (1): 71–83. Accessed December 24, 2021. doi:10.1016/S0951-8320(00)00047-8.

Bhardwaj, A. S., R. Saraf, G. G. Nair, and V. Sridharam. 2019. "Real-Time Monitoring and Predictive Failure Identification for Electrical Submersible Pumps." *Abu Dhabi International Petroleum Exhibition & Conference*. Abu Dhabi, UAE,. doi:10.2118/197911-MS.

Bhowmik, S. 2019. "Life Extension of Offshore Structure Using Machine Learning." *Offshore Technology Conference Brasil*. Rio de Janeiro. doi:10.4043/29759-MS.

Cadei, Luca, Corneo, Andrea, Milana, Diletta, Loffreno, Danilo, Lancia, Lorenzo, Montini, Marco, Rossi, Gianmarco, Purlalli, Elisabetta, Fier, Piero, and Francesco Carducci 2019. "Advanced Analytics for Predictive Maintenance with Limited Data: Exploring the Fouling Problem in Heat Exchanging Equipment." *Abu Dhabi International Petroleum Exhibition & Conference*. Abu Dhabi, UAE. doi:10.2118/197355-MS.

Carrillo, W. 2013. "Prognostics for Oil & Gas Artificial Lift applications." *PHM Conference*. New Orleans: GE Oil and Gas.

Ferket, J. 2019. "Condition Monitoring, Continuous Condition Monitoring or Predictive Maintenance? That's the Question!" *Abu Dhabi International Petroleum Exhibition & Conference*. Abu Dhabi, UAE. doi:10.2118/197170-MS.

Fonseca, D., A. Salazar, E. Villareal, G. Echaiz, and A. Leite. 2021. "Downhole Telemetry Systems to Monitor Electric Submersible Pumps Parameters in Oil Well." *IEEE Access* 99: 1–1. 10.1109/ACCESS.2020.3049116.

Gautam, M., and J. Kohn. 2019. "Predictive Maintenance of Gas Turbine Air Inlet Systems for Enhanced Profitability as a Function of Environmental Conditions." *Abu Dhabi International Petroleum Exhibition & Conference*. Abu Dhabi, UAE. doi:10.2118/197814-MS.

Giunta, G., G. Bernasconi, R. Giro, and C. Simone. 2020. "Digital Transformation of Historical Data for Advanced Predictive Maintenance." *Abu Dhabi International Petroleum Exhibition & Conference*. Abu Dhabi. doi:10.2118/202906-MS.

Gkortsas, Vasileios-Marios, Sandip Bose, and Smaine Zeroug. 2019. *Machine Learning for the Automated Detection of Diagnosis-Revealing Features on Leaky Flexural Wave Imager Data*. Accessed December 30, 2021. https://aip.scitation.org/doi/abs/10.1063/1.5099774.

Grassian, D., M. Bahatem, T. Scott, and D. Olsen. 2017. "Application of a Fuzzy Expert System to Analyze and Anticipate ESP Failure Modes." *Abu Dhabi International Petroleum Exhibition & Conference*. Abu Dhabi, UAE. doi:10.2118/188305-MS.

Griffo, U. 2017. "Predictive Maintenance using LSTM." *Github*. Accessed December 26, 2021. https://github.com/umbertogriffo/Predictive-Maintenance-using-LSTM.

Guo, D., C. Raghavendra, K. Yao, M. Harding, A. Anvar, and P. Anil. 2015. "Data Driven Approach to Failure Prediction for Electrical Submersible Pump Systems." *SPE Western Regional Meeting*. Garden Grove, California. doi:10.2118/174062-MS.

Gupta, S., M. Nikolaou, L. Saputelli, and C. Bravo. 2016. "ESP Health Monitoring KPI: A Real-Time Predictive Analytics Application." *SPE Intelligent Energy International Conference and Exhibition*. Aberdeen, Scotland, UK. doi:10.2118/181009-MS.

Gupta, Supriya, Luigi Saputelli, and Michael Nikolaou. 2016. *Applying Big Data Analytics to Detect, Diagnose, and Prevent Impending Failures in Electric Submersible Pumps*. Accessed December 24, 2021. https://onepetro.org/conference-paper/spe-181510-ms.

Hatami, Sobhan, Alireza Ghaderi-Ardakani, Mohammad Niknejad-Khomami, Farzan Karimi-Malekabadi, Mohammad Reza Rasaei, Amir H. Mohammadi, and Amir H. Mohammadi. 2016. "On the Prediction of CO2 Corrosion in Petroleum Industry." *Journal of Supercritical Fluids* 117: 108–112.

Hoang, Nhat-Duc, and Van-Duc Tran. 2019. "Image Processing-Based Detection of Pipe Corrosion Using Texture Analysis and Metaheuristic-Optimized Machine Learning Approach." *Computational Intelligence and Neuroscience* 2019: 1–13.

Hubert, Mia, Peter J. Rousseeuw, and Tim Verdonck. 2009. "Robust PCA for Skewed Data and Its Outlier Map." *Computational Statistics & Data Analysis* 53 (6): 2264–2274. Accessed December 24, 2021. https://sciencedirect.com/science/article/pii/s0167947308002879.

Kang, Z., Catal C., and B. Tekinerdogan. 2021. "Remaining Useful Life (RUL) Prediction of Equipment in Production Lines Using Artificial Neural Networks." *Sensors (Basel)* 21 (3): 932. doi:10.3390/s21030932.

Kaplan, E. L., and P. Meier. 1958. "Nonparametric Estimation from Incomplete Observations." *Journal of the American Statistical Association* 53 (282): 457–481.

Karnik, S., N. Yenuganti, B. F. Jusri, S. Gupta, P. Nirgudkar, M. Mohajer, and M. Asim. 2021. "Automated ESP Failure Root Cause Identification and Analyses Using Machine Learning and Natural Language Processing Technologies." *SPE Gulf Coast Section Electric Submersible Pumps Symposium*. The Woodlands, Texas, USA: SPE. doi:10.2118/204519-MS.

Kermani, Bijan. 2012. *Recommended Practice for Corrosion Management of Pipelines in Oil & Gas Production and Transportation*. Accessed December 29, 2021. https://amazon.com/recommended-management-production-transportation-publications/dp/1907975330.

Khabibullin, R., A. Shabonas, N. Gurbatov, N. Sergeevich, and A. Timonov. 2020. "Prediction of ESPs Failure Using ML at Western Siberia Oilfields with Large Number of Wells." *SPE Russian Petroleum Technology Conference*. doi:10.2118/201881-MS.

Koch, G., G. Ripley, and D. McKay. 2013. *The New Risk Perspective in Corrosion Management*. https://asmedigitalcollection.asme.org/mts/proceedings/mts2013/99403/131/271488.

Koch, G. H., M. P. H. Brongers, Neil G. Thompson, Y. P. Virmani, and Joe H. Payer. 2002. *Corrosion Cost and Preventive Strategies in the US*. https://rosap.ntl.bts.gov/view/dot/40697.

Lee, J., H. D. Ardakani, S. Yang, and B. Bagheri. 2005. "Industrial Big Data Analytics and Cyber-Physical Systems for Future Maintenance & Service Innovation." *Procedia CIRP* 38: 3–7. doi:10.1016/j.procir.2015.08.026.

Mammadov, I. 2021. "Predictive Maintenance for Wind Turbines Dashboard." *Github*. Accessed December 26, 2021. https://github.com/iameminmammadov/dash-predictive-maintenance.

Marin, A., S. Busaidy, M. Murad, I. Balushi, A. Riyami, S. Jahwari, A. Ghadani, et al. 2019. "ESP Well and Component Failure Prediction in Advance using Engineered Analytics - A Breakthrough in Minimizing Unscheduled Subsurface Deferments." ADIPEC. Abu Dhabi, UAE. doi:10.2118/197806-MS.

Masi, Giulia De, Manuela Gentile, Roberta Vichi, Roberto Bruschi, and Giovanna Gabetta. 2015. *Machine Learning Approach to Corrosion Assessment in Subsea Pipelines*. Accessed December 30, 2021. https://ieeexplore.ieee.org/document/7271592.

Mathworks. 2021. *Three Ways to Estimate Remaining Useful Life*. Mathworks. Accessed December 26, 2021. www.mathworks.com.

Mercan, B., M. Campbell, and T. Clay. 2021. "Fatigue Monitoring and Life Extension for Top Tensioned Production Riser Systems." *Offshore Technology Conference*. Houston, Texas. doi:10.4043/31259-MS.

Microsoft Azure Team. 2015. *Predictive Maintenance: Step 2A of 3, Train and Evaluate Regression Models*. Accessed December 26, 2021. https://https://gallery.azure.ai/Experiment/Predictive-Maintenance-Step-2A-of-3-train-and-evaluate-regression-models-2.

Nash, Will, Tom Drummond, and Nick Birbilis. 2019. *Deep Learning AI for Corrosion Detection*. Accessed December 30, 2021. https://research.monash.edu/en/publications/deep-learning-ai-for-corrosion-detection.

Okoh, C., R. Roy, J. Mehnen, and L. Redding. 2014. "Overview of Remaining Useful Life Prediction Techniques in Through-life Engineering Services." *Procedia CIRP* 16: 158–163. doi:10.1016/j.procir.2014.02.006.

Ossai, Chinedu I. 2019. *A Data-Driven Machine Learning Approach for Corrosion Risk Assessment—A Comparative Study*. Accessed December 29, 2021. https://mdpi.com/2504-2289/3/2/28/pdf.

Palacios, C. A. 2019. *Corrosion and Asset Integrity Management for Upstream Installations in the Oil/Gas Industry: The Journey of a Corrosion Integrity Engineer - Real Life Experiences*. Houston: Basenji Studios, LLC.

Pandey, Y. N., A. Rastogi, S. Kainkaryam, S. Bhattacharya, and L. Saputelli. 2020. *Machine Learning in the Oil and Gas Industry: Including Geosciences, Reservoir Engineering, and Production Engineering with Python*. Berkeley, CA: Apress. doi:10.1007/978-1-4842-6094-4.

Paolanti, M., L. Romero, A. Felicetti, A. Mancini, E. Frontoni, and J. Loncarski. 2018. "Machine Learning approach for Predictive Maintenance in Industry 4.0." 14th IEEE/ASME International Conference on Mechatronic and Embedded Systems and Applications (MESA). Oulu, Finalnd.

Rachman, Andika, and R. M. Chandima Ratnayake. 2019. "Machine Learning Approach for Risk-Based Inspection Screening Assessment." *Reliability Engineering & System Safety* 185: 518–532. Accessed December 30, 2021. https://sciencedirect.com/science/article/pii/s0951832018301662.

Rao, S. V. 2020. "Using a Digital Twin in Predictive Maintenance." *Journal of Petroleum Technology* 72: 42–44. doi:10.2118/0820-0042-JPT.

Robyn, R., and J. Chacin. 2021. "Development and Implementation of an Automated ESP Failure Database and Reliability Analysis Platform." *SPE Gulf Coast Section Electric Submersible Pumps Symposium*. doi:10.2118/204516-MS.

Rødseth, Harald, Jon Martin Fordal, and Per Schjølberg. 2018. *The Journey Towards World Class Maintenance with Profit Loss Indicator.* Accessed December 24, 2021. https://link.springer.com/chapter/10.1007/978-981-13-2375-1_25.

Rutherford, Paul S., and Joe J. Luzar. 1999. *Corrosion Quantification by Different Nondestructive Inspection Methods.* Accessed December 29, 2021. https://spiedigitallibrary.org/conference-proceedings-of-spie/3586/1/corrosion-quantification-by-different-nondestructive-inspection-methods/10.1117/12.339881.full.

Saraswat, A. 2021. *Predicting Remaining Useful Life of a Machine.* March 7. Accessed December 26, 2021. https://cloudxlab.com/blog/predicting-remaining-useful-life-of-a-machine/.

Settemsdal, S. 2021. "Machine Learning and Artificial Intelligence as a Complement to Condition Monitoring in a Predictive Maintenance Setting." *SPE Oil and Gas India Conference and Exhibition*. Mumbai, India. doi:10.2118/194590-MS.

Shanrang, Yang, Liu Xiuwei, Cao Shengxian, Zhao Bo, Hu Yanping, Liu Fan, Men Hong, and Xu Zhiming. 2010. *Forecasting Corrosion Rate of Coolingwater Based on Least Squares Support Vector Machine.* Accessed December 29, 2021. https://dl.acm.org/citation.cfm?id=1955655.

Sparke, S. J., R. Conway, and S. Copping. 2011. "The Seven Pillars of Well Integrity Management": The Design and Implementation of a Well Integrity Management System." *SPE/ICoTA Coiled Tubing & Well Intervention Conference and Exhibition*. doi:10.2118/142449-MS.

Sturrock, Charles P., and W. F. Bogaerts. 1996. *Computer Learning Systems in Corrosion.* Accessed December 30, 2021. http://osti.gov/scitech/biblio/379721-computer-learning-systems-corrosion.

Tajadod, M., A. Mohammadali, and A. Rategari. 2016. "A Comparison of Multi-Criteria Decision Making Approaches for Maintenance Strategy Selection A Case Study." *International Journal of Strategic Decision Sciences*. 7(3): 51–69.

Takács, G. 2009. Electrical Submersible Pumps Manual: Design, Operations, and Maintenance. Oxford, UK: Gulf Professional Publishing, Elsevier.

Tapan, D., C. Sarika, and S. Bansal. 2013. "Predictive Maintenance and Deterministic Maintenance Scheduling for Continuous Process Industry Rotary Equipments." *SPE Middle East Intelligent Energy Conference and Exhibition*. Manama, Bahrain. doi:10.2118/167470-MS.

Viana, P., A. Khalid, and S. Shankar. 2020. "Flexible Riser Life Extension Associating Machine Learning and Innovative Monitoring." *The 30th International Ocean and Polar Engineering Conference*.

Wurzbach, Richard Norman. 2002. *Developing a Comprehensive Training Curriculum for Integrated Predictive Maintenance.* Accessed December 24, 2021. https://spiedigitallibrary.org/conference-proceedings-of-spie/4710/1/developing-a-comprehensive-training-curriculum-for-integrated-predictive-maintenance/10.1117/12.459583.full.

Yakoot, M., S. Ragab, A. Salem, and M. Omar. 2011. "Yakoot, Mostafa Sa'eed, Ragab, Adel Mohamed Salem, and Omar Mahmoud. "Machine Learning Application for Gas Lift Performance and Well Integrity." *SPE Europec featured at 82nd EAGE Conference and Exhibition*. Amsterdam, The Netherlands. doi:10.2118/205134-MS.

Zajam, Saurabh, Tushar Joshi, and Bishakh Bhattacharya. 2019. "Application of Wavelet Analysis and Machine Learning on Vibration Data from Gas Pipelines for Structural Health Monitoring." *Procedia Structural Integrity* 14: 712–719. Accessed December 30, 2021.

Zubir, Wan Mohammad Aflah Mohammad, Izzatdin Abdul Aziz, and Jafreezal Jaafar. 2018. *Evaluation of Machine Learning Algorithms in Predicting CO_2 Internal Corrosion in Oil and Gas Pipelines*. Accessed December 29, 2021. https://link.springer.com/chapter/10.1007/978-3-030-00211-4_22.

Zukhrufany, Stiffi. 2018. *The Utilization of Supervised Machine Learning in Predicting Corrosion to Support Preventing Pipelines Leakage in Oil and Gas Industry*. Accessed December 30, 2021. https://uis.brage.unit.no/uis-xmlui/handle/11250/2565865.

17 Machine Learning for Multiphase Flow Metering

Patrick Bangert[1,2]
[1]Samsung SDSA, San Jose, California
[2]Algorithmica Technologies Inc., Cupertino, California

CONTENTS

17.1 MULTIPHASE FLOW AND ITS METERS

The substance that emerges from an oil well is a mixture of crude oil, gas, water, and particulate matter such as sand and bits of rock. This mixture is known as ***multiphase flow***. Usually, the particulate component is relatively small because if it were not, pumping this flow would be too damaging on the equipment. In fact, when bursts of particulates flow into the well, this usually leads to an unplanned outage of the pump. Two important figures are generally used to get a handle on this flow. The ***water-cut*** (WC) is the relative amount of water in the total liquid flow and the ***gas-oil ratio*** (GOR) is the ratio of the gas flow rate to the oil flow rate. Some wells approach a WC above 90% and GORs range across the full range of possible values from 0.1% to 99%.

We need to know how much oil, water, and gas are in the mixture to correctly attribute the total production volume to the wellsites and, most importantly, to their owners. As wells, fields, and pipelines are often co-owned or co-used by multiple parties, it is important to keep track of how much any one party transports into a common system. In practice, large sums of money rely on relatively inaccurate production estimates. The essential desire in multiphase flow metering is the accurate measurement of the flow rates in each of the four possible

DOI: 10.1201/9781003207009-23

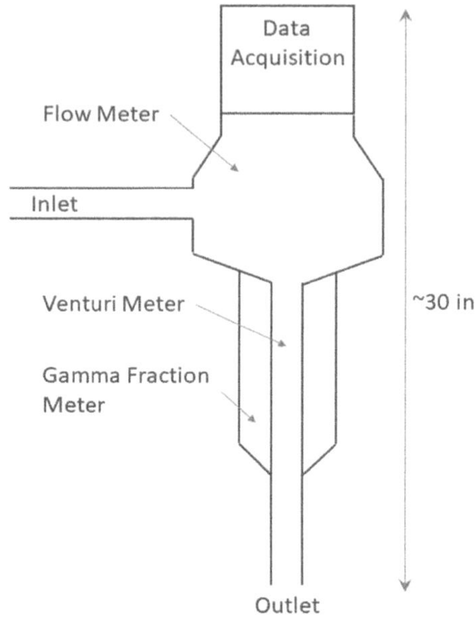

FIGURE 17.1 Diagram of a physical multiphase flow meter in cross-section.

components of the flow versus the total flow rate. That measurement is the topic of this chapter.

The accurate measurement of other interesting quantities such as temperature, pressure, total flow rate, and even valve opening positions at various points of an oil well is both easy and cheap. The relevant sensors are readily available and are tough enough to survive the harsh conditions often present at wellsites. Unfortunately, measuring how much gas, oil, and water is in the flow is much harder and therefore more expensive. Multiphase flow meters exist as physical devices on the commercial market typically as gamma-ray emitters and therefore radioactive devices (Figure 17.1). The accuracy of these meters is about ±7% in each phase's percentage relative to the total flow rate (Islami rad and Gholipour Peyvandi, 2019) and highly dependent on WC and GOR. Furthermore, these meters are fragile and fail more often than other field instruments (Al-Kadem et al., 2014) such that dedicated failure mode analyses are being researched (Barbariol, Feltresi and Susto, 2020a, 2020b).

In an effort to overcome the drawbacks and the financial expense of installing a multiphase flow meter at every wellsite, the industry performs well tests. These are events that last between a few hours to a few days in which a well is attached to a separator that separates the phases each of which is then flow measured on its own. This process sometimes requires the transport of equipment and manual hooking up of the well to the separator (Figure 17.2). These well tests provide accurate relative flow rates that are then usually assumed to be constant until the next well test for this well. Depending on the restrictions of

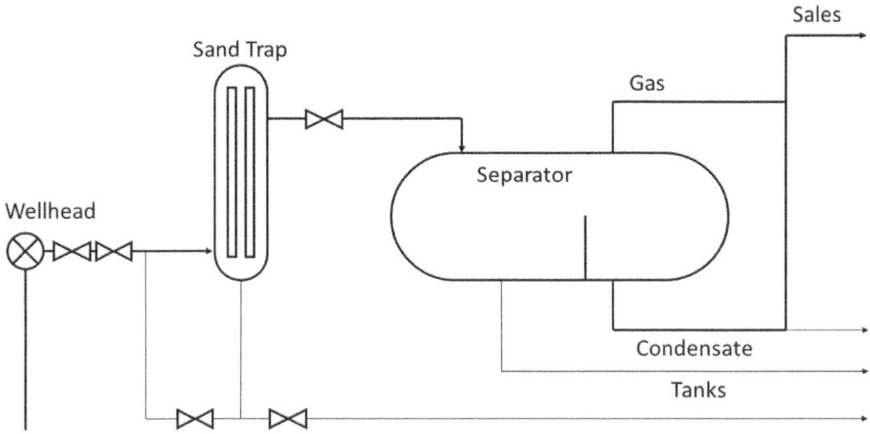

FIGURE 17.2 Well test setup.

the field, it may be a few to many weeks between well tests of any one well. The assumption that things stay constant is known to be false, but it is commonly the only available choice.

Due to well tests, we have data where we simultaneously measure the easy wellsite data (temperatures, pressures, valve data, etc.) and the multiphase flows. The question arises: Can we use this data to establish a formula that would allow us to *calculate* the multiphase flow data from the wellsite data *without* measuring it – with either a well test or a multiphase flow meter?

This question has been approached (for many decades (Kabir and Hasan, 1986)) in three major ways (Bikmukhametov and Jäschke, 2020a). First, it gave rise to a significant effort to model the system using the laws of physics – an effort that we shall refer to as the first-principles approach (FP) (Correa et al., 2008; Backi, Grimes and Skogestad, 2018). Second, some researchers have attempted this using the techniques of machine learning (ML) and/or artificial intelligence (Bishop, 1995; Stief et al., 2019; Gryzlov et al., 2021). Third, hybrid approaches where both methods are combined have also been introduced (Hotvedt, Grimstad and Imsland, 2020; Bikmukhametov and Jäschke, 2022; Hotvedt 2022). In brief, all three approaches have met with success and have, at this time, been capable of delivering results comparable in accuracy to the physical multiphase flow meter.

The benefits of FP are that the techniques are understandable to engineers, so this approach engenders more trust. However, FP requires fine-tuning to every wellsite and therefore does not scale in the sense that doubling the sites will also double the cost. ML, on the other hand, is a black box method that is poorly understood by engineers but scales extremely well and lends itself to monitoring a fleet of wells from the cloud. In addition, the accuracy and versatility of ML grows with the growth of the data pool fed into it so we can certainly expect ML-based approaches to become scientifically better than FP over time. Given the scientific and economic advantage of ML, it seems to be a matter of time before this approach overtakes FP in industrial applications. A more detailed comparison of FP and ML approaches is available in Bikmukhametov and Jäschke (2020a, 2020b).

17.2 SOFT SENSING

A ***soft sensor*** is a formula that allows us to calculate something instead of measuring it. An example is the fitness tracker on your wrist. By measuring the number of steps you take and your pulse, it calculates how many calories you have burned that day – without measuring them. The point of this is that steps and pulse are easy and cheap to measure while calorie consumption would be difficult and expensive to measure. Ultimately, the fitness tracker *becomes* the calorimeter!

It is the purpose of ML to obtain a formula – also called a model – that correctly represents the relationship between two sets of variables. The inputs are the variables that will be known in the field application and the outputs are the variables to be calculated by the model. ML expects to receive a table containing both sets of variables and will then attempt to find a good model to connect them.

First, the human operator decides on the type of formula to be attempted. This can be as simple as a linear function or one of the more complex ML models like a support-vector machine, random forest, neural network, and so on. All of these model types have a number of numerical parameters.

Second, the parameters of the selected model type are now adjusted until the model fits the data in the best possible way. The goodness of the fit is usually measured in the least-squares sense – the differences between the model outputs and the real known outputs are computed, squared (to prevent underestimates and overestimates from canceling each other out), and then summed up. The recipe that does this is called an ***algorithm***. The algorithm is model type specific and itself has parameters that can (and usually should) be tuned in a process called ***hyperparameter tuning***.

The outcome is a model type with specific numerical values for its parameters. This full model can now be used just like any other formula by putting the inputs in, calculating the output, and returning it to the user in lieu of a real measurement. If the model is accurate over all possible input values, the soft sensor is attractive because it can be copied and scaled at nearly no cost. Maintenance is cheap as well because only the central model execution engine ever needs to be maintained, as opposed to a geographically diverse fleet of devices.

17.3 METER AND WELL TEST DATA

In the same way, we want to have a ***virtual multiphase flow meter*** (VMFM). This is a formula like any other from mathematics: $y = \text{VMFM}(x)$. What are x and y? The quantities x are usually called the inputs as they are the quantities we measure at the wellsite. These would be (1) downhole pressure, (2) downhole temperature, (3) wellhead pressure, (4) wellhead temperature, (5) production choke, (6) pressure downstream choke, (7) gas lift rate, (8) gas lift pressure, (9) gas lift choke, and (10) flowline pressure. The two downhole measurements are often absent as they are more costly to install. The quantities y are called the outputs and are the three flows: (1) oil, (2) gas, (3) water flow rates.

It is very important to note that all empirical measurements are uncertain. Any particular value for, say, the downhole temperature is not exactly equal to the true temperature of the reservoir. Sensors have an internal tolerance. Sensors drift. The local temperature where the sensor is may be somewhat different from the average temperature of the reservoir as a whole. The presence of the well and its equipment may be distorting the local temperature profile.

The sensor has an inherent uncertainty, the measurement chain from sensor to data historian adds a readout uncertainty, and the locally changing conditions at the precise location of the sensor will add variability that is not characteristic of the average situation that one is usually concerned about. The importance of this lies in the fact that if we use uncertain data in a computation, then the computed result inherits uncertainty from each of its data sources. We can determine the uncertainty in the computed value if we know the measurement uncertainty in the original data. In mathematical modeling, it is as important to have a good model as it is to know how accurate that model is.

A good review of soft sensing methods applied to the VMFM problem is in Yan et al. (2018). In this chapter, we will not review these methods but refer the reader to the paper. The reason is that the ML algorithm chosen to do the modeling is fairly unimportant – it does not feature in the top five items needed to make a good VMFM. This may be surprising to many as the literature on ML tends to focus on this part of the workflow. The fact is that the correct and concrete definition of the problem, gathering of data, cleaning of the data, feature engineering, and finally deployment (including business transformation) are the important aspects that influence model accuracy and final outcome orders of magnitude more than the choice between a neural network and a decision tree.

17.4 FEATURE ENGINEERING

Often, the raw input data from sensors is not informative enough for ML to accurately adjust the model parameters. To bring out the information content of the data in a better way, it is best to transform the raw data first. For instance, domain experts might know that the difference between two raw data points is particularly interesting for the problem at hand. This difference then becomes a new column in the data table and a new input into the model. We may even choose to remove the two original columns in favor of just their difference thereby reducing the dimensionality of the input data. In this way, ***feature engineering*** consists of both designing new input data columns – ***features*** – and selecting the right features to present to the algorithm. Another approach to feature engineering for VMFM is described in Shaban and Tavoularis (2014).

For VMFM, we have found that information from the last few minutes is important as the whole well system is slow to react with many instantaneous ups and downs that have little consequence and must therefore be "averaged out." The data in these time windows can be informatively summarized by four standard statistical quantities: (1) average, (2) standard deviation, (3) skewness, and (4) kurtosis. Furthermore, there are macroscopic delays between cause and effect in the system. For example, if the downhole pressure has a window of 20 minutes and a delay of 40 minutes, we will then summarize the downhole pressure by the four previously mentioned statistical quantities derived from all the downhole pressures measured between 40 and 60 minutes before the present moment. The dimensionality of the input data is thus multiplied by four.

The dimensionality of the input data is important as the model becomes more complex with more inputs. A more complex model has more parameters. More parameters are harder to tune and thus require more experimental data for their tuning. In industrial applications, we usually have to proceed in reverse; however, the total amount of available data is fixed. Based on a fixed amount of data and thus information, we can only afford a certain complexity of model, i.e., a certain number of model parameters. This limitation in turn limits the number of input variables and the model that we can attempt to use. These restrictions are a very complex matter in ML and are known as the ***bias-variance-complexity trade-off***. Oversimplified, it is in our interest to minimize the number of input dimensions.

The most popular and robust way to achieve this dimensionality reduction is called ***principal component analysis (PCA)*** (Bangert, 2021). The procedure rotates the dataset so that the directions of greatest variance are aligned with the coordinate axes and then removes the dimensions that see very little variance. Depending on where the cutoff for little variance is set, this method usually removes 30%–60% of the dimensions in a standard oil and gas dataset because the information content is duplicated across several sensor channels; for example, the downhole and wellhead temperatures are not independent, so they can be compacted to less than two dimensions. The same is true of the pressures and the choke settings.

We will now start looking at one particular dataset of 158 well tests taken from 23 wells in one reservoir. There are 67,000 measurements separated by 1 minute from each other in time. Some well tests lasted longer than others, but this indicates an average duration of 424 minutes or 7 hours per well test. Each measurement consists of the aforementioned 13 numbers. There are 10 inputs: (1) downhole pressure, (2) downhole temperature, (3) wellhead pressure, (4) wellhead temperature, (5) production choke, (6) pressure downstream choke, (7) gas lift rate, (8) gas lift pressure, (9) gas lift choke, and (10) flowline pressure. There are three outputs: (1) oil, (2) gas, and (3) water flow rates.

Figure 17.3 displays a sample well test output in terms of oil, gas, and water rates for one well test lasting 422 minutes. We note that the gas rate is approximately randomly distributed about a mean value and could be reasonably represented by its average. However, we note that the oil and water rates are structured.

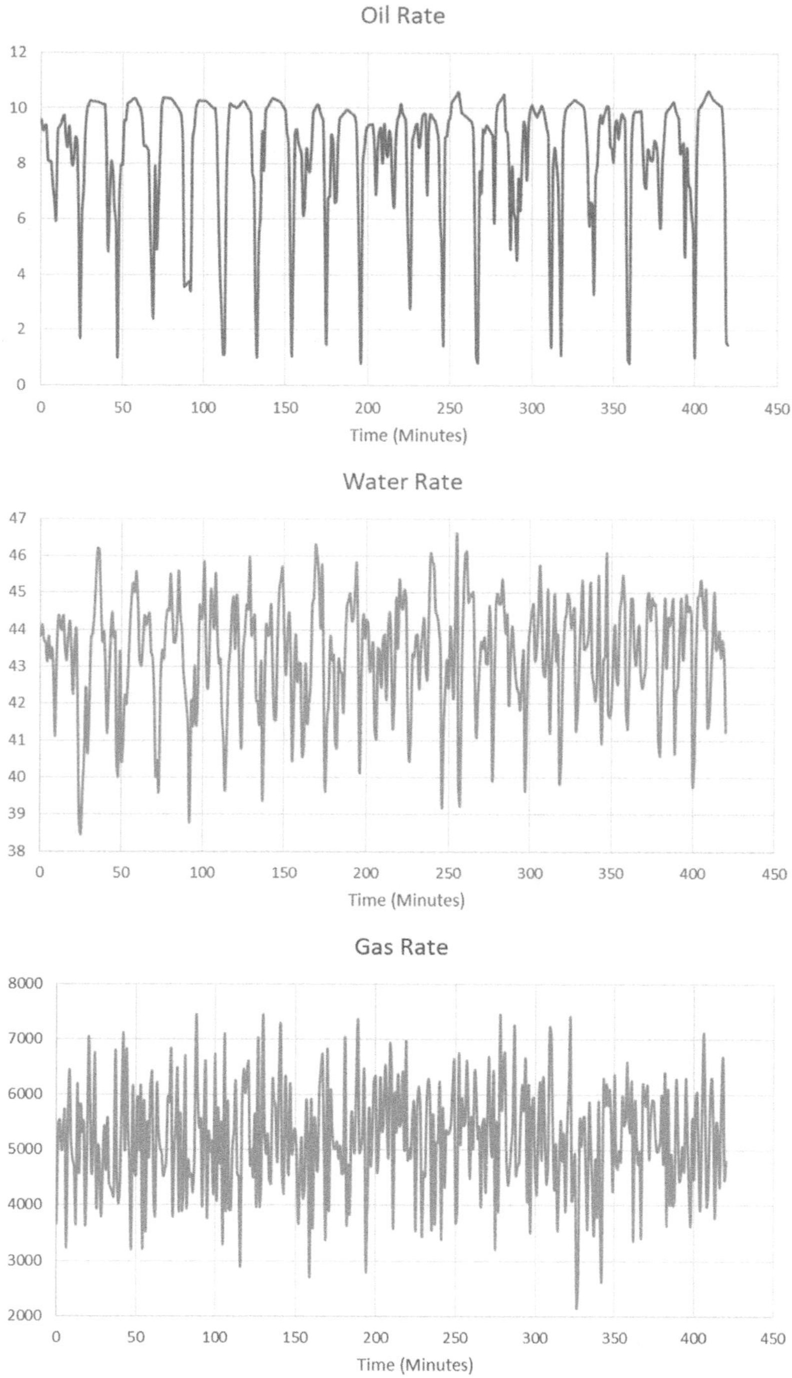

FIGURE 17.3 Oil, water, and gas rates for one sample well test over 422 minutes.

Particularly the oil rate regularly goes up and down between 2 and 10 barrels per hour. This is a significant variation known as slugging. Depending on the conditions of the well, slugging has a different period and behavior and significantly influences the flow rate. We cannot assume that the average flow rate is a reasonable representation of the long-term flow from this well, i.e., the total flow in between this well test and the next well test (weeks or months later) may differ significantly from the average rate over this well test.

For each of the inputs, and each well test, we now construct a moving window of a certain length in time and compute the moving (1) average, (2) standard deviation, (3) skewness, and (4) kurtosis in that time window. This exercise turns the 10 raw input variables into 40 inputs. We then apply PCA to lower the dimensionality to 20, which we found experimentally by demanding that 99% of the dataset variance be retained. The question of how long the time window should be must be experimentally determined. This is an example of a hyperparameter of the whole methodology that must be tuned to the problem to reveal the most information possible. For our dataset, we determined that 40 minutes is the most optimal time-window size by choosing various sizes, performing the full modeling, and choosing the one that achieves the best performance without changing any of the other hyperparameters. Such experiments – known as hyperparameter tuning – must be done for each case and represents a major effort and time commitment for the data scientist.

After all this pre-processing, the data is divided into three parts. A randomly chosen 80% of the well tests are used to train the model (training dataset), a further 10% are used to determine whether the model training has converged to a meaningful answer (test dataset), and the last 10% are used to gauge the accuracy of the model (validation data). This third dataset is therefore not involved in any way with the process of making the model and thus serves as the true baseline by which we determine whether we have succeeded.

A neural network model of the long short-term memory (LSTM) type was then trained on the data and its hyperparameters tuned; the results are discussed next. Similar and other ML approaches have been reported in the literature such as LSTM (Andrianov, 2018a, 2018b), neural networks (Hasanvand and Berneti, 2015; Al-Qutami et al., 2018; Omrani et al., 2018), and a comparison of models (Khan et al., 2019).

17.5 VIRTUAL METERING

For the present dataset, we find that the oil flow rate can be reproduced accurately to 10%, gas flow to 9%, and water flow to 8% for all the well tests examined. These percentages are calculated by taking the difference between model output and measurement and then dividing by the true measurement. In the statement that something can be reproduced accurately to 10%, the 10% is not an average but an extremum. This is very important. All the well tests must be reproduced with a percentage difference of *at most* 10% for this claim to hold. Thus, many well tests are reproduced more accurately. The well tests that are inaccurate are those for which the relevant phase had nearly no flow, so there was too little representative data for these edge cases. These are the wells that have high WC (almost all the flow is water and little oil), low WC (almost all the flow is oil and little water), or high GOR (almost all the flow is gas and little oil). The case that the gas flow is low does not apply to our case as all wells are gas injection wells, so due to the injection there is always a significant gas outflow.

The accuracy of a model can be judged in several ways. The simplest is to put the known output of the validation data next to the model output and look at the difference. That is how the earlier percentage figures were arrived at. A more sophisticated way is to graph the distribution of these differences, also called ***residuals***. A residual is merely the difference between model output and true measurement. This value may be negative or positive. The best value of a residual is zero because this is where the model agrees perfectly with reality. Looking over the list of all residual values, we divide the range from least to greatest residual into several bins and then count the number of residuals for each bin. The outcome of this histogramming process, once we divide the counts by the number of residuals in the entire exercise, is known as a distribution.

Figure 17.4 is an example of such a residual distribution. On the horizontal axis is the size of the residual, which is negative if the measurement is greater than the model output and positive otherwise. On the vertical axis is the probability of getting this residual value over all the data points in the validation dataset. The probability is merely the number of such points divided by the total number of data points.

In Figure 17.4, the black line is the distribution itself for the training dataset. The green shaded area is the 90% confidence interval, i.e., the interval resulting from selecting 90% of the area under the curve such that the area is centered about the mean of the distribution. In our case, this is from –30 to 30 barrels per hour. The solid line is the same distribution for the validation dataset. As the distributions for the training and validation datasets are more or less identical, we can rest assured that the model generalizes quite well to cases that were unseen by the process that made the model.

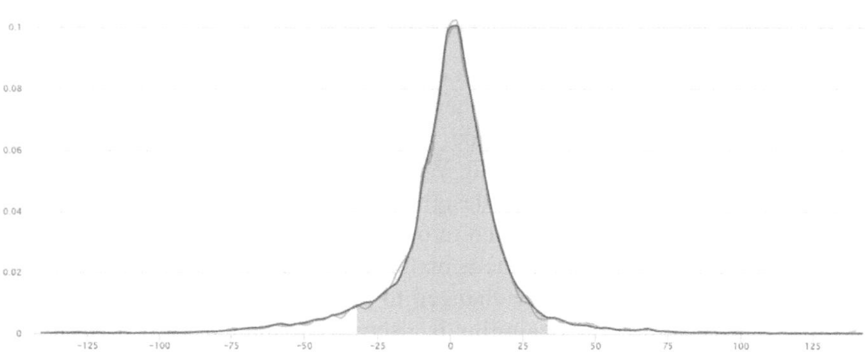

FIGURE 17.4 The residual distribution plot for the oil flow rate.

The residual distribution immediately communicates several crucial attributes of a model. First, we want the distribution to be centered on zero – the most typical residual value should be zero indicating that most of the time the model agrees with the measurement completely or at least within the agreed margin of error. If the most common residual is not zero, there is a systematic error in the model, so the architecture of the model must be reexamined, or there is a fundamental problem with the training dataset.

Second, we want the distribution to be symmetric about its mean so that neither negative nor positive residuals are overrepresented. This indicates that any errors are random and not systematic. This would then seem to be either inherent noise of the data, missing variables that were not measured, or something similar as opposed to a fundamental modeling mistake. Failing symmetry, or worse more than one peak, would indicate an error in modeling.

Third, we want the tails on either end of the distribution to fall off exponentially. This is hard to achieve in practice because a real dataset is typically not large enough to really see an exponential drop-off. The real drop-off should, however, be much faster than linear so that the residuals are by and large clustered tightly around the mean.

All of these features can be seen at a glance from Figure 17.4 indicating that the model is basically sound. From the horizontal axis we may read off that the residuals are mostly in the range between –30 and 30 barrels per hour. This means that we sometimes underestimate and sometimes overestimate the oil flow rate by an amount that is almost always less than 30 barrels per hour away from the true result. The symmetry indicates that every overestimate is balanced by another underestimate.

We remind ourselves at this point of the original problem to be solved. We want to be able to attribute oil, water, and gas flows correctly in between well tests. We are not concerned with estimating them accurately and correctly at one moment in time but only in aggregate over a macroscopic period of time lasting many days or weeks. The over- und underestimations don't affect this as they cancel out over time. *On average*, the soft sensor is therefore correct.

17.6 MACHINE LEARNING WORKFLOW

As illustrated in this chapter, there is a natural flow of steps in getting to the final model. We briefly review these steps here.

First, the use case must be made very clear. From the point of view of ML, the use case must be numerical. Whatever precedes making the situation numerical must be included in the design. Whatever follows the numerical output of the model must also be taken into account. Typically, the model output becomes part of a larger IT system or flows into a human decision-making process. These systems must be changed to include this new element. Failure to think of this in the very beginning is a serious mistake as it then often gets ignored. This is the principle cause for failure of ML projects in the industry. Based on these elements, the project must be funded and staffed correctly, which is also crucial. This initial process includes defining success numerically. Criteria must be found by which the model is "good enough" for the purpose and "practically acceptable" for the people who have to use it to make decisions. That is not at all easy.

Second, the data needs to be collected. This process generally starts with a careful consultation between domain experts and data scientists in which criteria are defined that the data must satisfy. On this basis, data is either found, if it already exists, or methods are designed to acquire it. If the data does not yet exist but must be freshly acquired, then this generally means (in the oil and gas industry) that physical devices must be purchased and installed. Such a digitization process has the potential to be expensive and time-consuming and distract from the actual task of making an ML model.

Third, the data must be cleaned. This means removing data points from the dataset that are either measurement errors, outliers that we do not want to consider (as opposed to outliers that are important to the problem), or not representative of the problem we want to solve. It also means augmenting the data with human domain expert input and feedback that may be needed to correctly identify scenarios. This injection of domain knowledge is often called annotation or labeling in the ML literature.

Fourth, feature engineering is performed by which we first add new features by some computational process. In this study, we added moving windows and computed statistics over them. We could also take differences or ratios or other more complex combinations of existing data columns and make new columns. This is feature generation. Having enlarged the number of features, we must now select the features that are going to be informative in solving our problem. We typically remove features if they correlate poorly with the desired output or if they correlate highly with another feature. What we want is a small number of independent and highly informative features. This is feature selection. Finally, it is usually expeditious to add a feature transformation method like PCA to reach the final feature list.

Fifth, the right model architecture is chosen. Whether this is a neural network, a recurrent neural network (such as the LSTM used here), a support vector machine, a decision tree, or one of the many other kinds of available model types must be determined. Within each of these, there are architectural parameters to be selected such as the number of layers and number of neurons per layer in a neural network. Currently, there is not good a priori way of selecting the model. We are compelled to do this experimentally by trial and error.

Sixth, the algorithm that trains the selected model architecture has hyperparameters that are now tuned. This process can be automated by adaptive search, which is a multidimensional search process that selects the new point to be tested as a function of the points so far tested and usually converges in a realistic number of trials. It is, however, still a trial and error process. We note that the combination of model selection and hyperparameter tuning requires a large number of training runs to be executed – typically many thousands and, if we want high performance, perhaps orders of magnitude more – and this is time-consuming as the humans must wait for the computers to finish. Parallel processing on several computers in a data center is expedient.

Seventh, the model must be embedded in the IT systems that feed it the input data and receive the output. The model must be observed from time to time to determine whether it has drifted. A model itself does not change over time, but the source data usually does as reality usually changes over time. If and when this happens, the model must be adjusted. We must never assume that a model once made will persist in its accuracy and applicability for the long term!

Eighth, all involved stakeholders and users must now agree to trust the model and use it in their human decision-making or action taking jobs. This is the most important step because, if they do not, all the other steps are useless. It is only the human decision-making at this point that generates whatever value the ML model was supposed to deliver. This is why it is so important to design for this at the very start and – crucially – to reserve funding and staff to make this process happen. It is not easy and typically requires much deliberating.

All in all, this is the full workflow of ML presented in a nutshell. Each step is important. If we are in a hurry and must do things quickly or cheaply, the only steps that may be omitted or saved for later are steps 4 (feature engineering) and 6 (hyperparameter tuning), and step 5 (model selection) can be abbreviated by selecting what an ML expert believes to be a reasonable first guess. All the other steps cannot be skipped. If they are, the project will likely fail.

17.7 CONCLUSION

At this point, we must carefully think about the industrial application to gauge whether the model is fit for purpose. The real purpose of multiphase flow metering is the correct attribution of production volumes to wells and owners. The total flow rates are known and so the (uncertain) partial flow rates can be correctly normalized to the total so that the distribution into the phases is always sensible. If the model outputs every hour or every day are all subject to the random errors as displayed in Figure 17.4, these errors cancel out at the end of the month or year. This means that the financial accounting of who produced how much oil and gas in total over the long term is accurate.

If the use case demanded near-perfect accuracy at any one time, the physical meter might be better, but this is not a certainty. Three-phase separation and lease automatic custody transfer (LACT) units are the most accurate physical sensing methods available, but these are not necessarily significantly more accurate at any one time. Alongside the financial investment and logistical effort, the use case must be considered carefully. However, this use case does not demand this, and the VMFM will get more accurate with more data.

Apart from mathematical steps of feature engineering, selecting the best model architecture, and carefully tuning hyperparameters, it is the data itself that has, by far, the largest influence on the quality of the final model. Modeling challenges are most readily and effectively solved by increasing the dataset. We cannot, however, simply add more lines. For example, duplicating an existing data point would increase the size but adds no information. Inserting well tests from another reservoir would increase the dataset but this may add wrong information if the second reservoir has different behavior than the reservoir in question.

Using language from statistics, the data must be ***significant*** and ***representative***. Data is significant if it adds information about the problem at hand. Duplication, or near duplication, does not add information. Data is representative if it contains the information about the problem we want to address. Data from another reservoir may or may not represent this reservoir. Data from one type of operation may or may not represent another type of operation, e.g., if we only included well tests of the high WC and ignored those of low WC. The low WC events are real and cannot be ignored.

Contained in both concepts is the question of ***imbalance***. Most datasets contain many points of the typical or normal case and few points for rare cases. However, in most problems, the rare cases are the important ones because the model often has to pay particular attention to them as they typically pose the most opportunity or danger. We may therefore be faced with having to over-represent or under-represent certain cases to highlight to the ML algorithm what we are concerned about. This means the data must be representative of the problem we are solving and not necessarily of the overall situation!

Getting more good data is both the secret and the challenge of ML methods in oil and gas. Every operator has a little data, like the dataset mentioned earlier. If several operators could collect their data into a coherent larger pool, a much better model could result. An independent company could curate this dataset and provide only the model and conclusions back to the operators instead of providing the raw data to everyone. This would protect data privacy but still enable the true strength of ML – learning from large datasets. Not only do large datasets contain more information, they contain more rare edge cases and these will make the model more robust in all cases.

Deploying the model relies on the model having access to the input data. For this, the model will either need to be executed on the wellsite or the measurement data be continuously transferred to some form of data center or cloud processing facility. The IT infrastructure to support this is usually referred to as ***digitization*** or the ***digital transformation*** and is part of an effort known as the ***digital oil field***.

The oil and gas industry has thus far approached ML methods with great skepticism and small budgets. The result has been a large number of proof-of-concept projects that usually result in proving the concept but not in getting funded into implementation. Poorly defined goals, moving targets, and an unwillingness to change a running system are some common reasons for not going further. It must be recognized that moving from a scientific modeling result to an operational cloud service takes time and investment that must both be budgeted for. The human component in a change management project is absolutely essential and must also be planned from the start for the project to be successful in being adopted by the diverse teams of people who are involved (Bangert, 2021).

Human change management, or business transformation as it has come to be called in recent years, is absolutely essential in deploying ML models. The models themselves calculate quantities and provide them. The decisions made and actions taken on their basis is generally a human matter. Humans must trust the ML model output and they must change their human processes to take this novel aspect into account. The model discussed here would lead to a change in the financial accounting practices in co-owned oil fields and thus gets to the very heart of the industry. For a single-owner field, it would result in the retirement of underperforming wells. Anchoring the model in the organization is crucial to making an impact and that is often more difficult than making the model (Crompton et al. 2020).

In conclusion, this chapter demonstrated the basic ability and advantages of ML to construct a VMFM from experimental data. Three main challenges to adoption were outlined as (1) digitization to provide the necessary data for the model, (2) collaboration between operators to produce a global and large dataset, and (3) implementation of the model in real IT infrastructure with the necessary human change management. The next 5 years will most likely see these challenges resolved not only for VMFM but for many more cases. A more open, collaborative, connected oil and gas industry awaits.

REFERENCES

M. Al-Kadem, F. T. Al-Khelaiwi, A. Al-Mashhad and M. Al-Dabbous, "A Decade of Experience with Multiphase Flow Meters," Paper presented at the International Petroleum Technology Conference, Kuala Lumpur, Malaysia, December 2014. https://doi.org/10.2523/IPTC-18162-MS

T. Al-Qutami, R. Ibrahim, I. Ismail and M. Ishak, "Virtual Multiphase Flow Metering Using Diverse Neural Network Ensemble and Adaptive Simulated Annealing," *Expert Systems with Applications*, vol. 93, pp. 72–85, 2018. https://doi.org/10.1016/j.eswa.2017.10.014.

N. Andrianov, "A Machine Learning Approach for Virtual Flow," in 3rd IFAC Workshop on Automatic Control in Offshore Oil and Gas Production, Esbjerg, Denmark, 2018a.

N. Andrianov, "A Machine Learning Approach for Virtual Flow Metering and Forecasting," *IFAC Papers Online*, vol. 51, pp. 191–196, 2018b. https://doi.org/10.1016/j.ifacol.2018.06.376.

C. J. Backi, B. A. Grimes and S. Skogestad, "A Control- and Estimation-Oriented Gravity Separator Model for Oil and Gas Applications Based upon First-Principles," *Industrial & Engineering Chemistry Research*, 57, no. 21, pp. 7201–7217, 2018.

P. Bangert, *Machine Learning and Data Science in the Oil and Gas Industry*, Elsevier, 2021.

T. Barbariol, E. Feltresi and G. Susto, "Self-Diagnosis of Multiphase Flow Meters through Machine Learning-Based Anomaly Detection," *Energies*, 2020a. https://www.mdpi.com/1996-1073/13/12/3136

T. Barbariol, E. Feltresi and G. Susto, "A Machine Learning-Based System for Self-Diagnosis Multiphase Flow Meters," in *International Petroleum Technology Conference*, Dhahran, Saudi Arabia, 2020b.

T. Bikmukhametov and J. Jäschke, "First Principles and Machine Learning Virtual Flow Metering: A Literature Review," *Journal of Petroleum Science and Engineering*, vol. 184, 2020a. https://doi.org/10.1016/j.petrol.2019.106487.

T. Bikmukhametov and J. Jäschke, "Combining Machine Learning and Process Engineering Physics Towards Enhanced Accuracy and Explainability of Data-Driven Models," *Computers and Chemical Engineering*, vol. 138, 2020b. https://doi.org/10.1016/j.compchemeng.2020.106834.

T. Bikmukhametov and J. Jäschke, "Hybrid Machine Learning Modeling of Engineering Systems – A Probabilistic Perspective Tested on a Multiphase Flow Modeling Case Study," 2022. https://arxiv.org/abs/2205.09196.

C. M. Bishop, "Multiphase Flow Monitoring in Oil Pipelines," In: Murray, A.F. (ed). *Applications of Neural Networks*. Springer, Boston, MA, 1995. https://doi.org/10.1007/978-1-4757-2379-3_6.

A. A. Correa, L. X. Benedict, D. A. Young, E. Schwegler and S. A. Bonev, "First-Principles Multiphase Equation of State of Carbon Under Extreme Conditions," *Physical Review B*, vol. 78, p. 24101, 2008.

J. Crompton, P. Bangert, M. Bladow, T. Edwards, H. Gilman, M. Hauser, M. Mikulis and S. Smart, "Don't Forget People in Your Digital-Transformation Programs: Change Management Is an Essential Competency," *Journal of Petroleum Technology, June,* 2020.

A. Gryzlov, L. Mironova, S. Safonov and M. Arsalan, "Evaluation of Machine Learning Methods for Prediction of Multiphase Production Rates," in *SPE Symposium: Artificial Intelligence - Towards a Resilient and Efficient Energy Industry*, Virtual, 2021.

M. Hasanvand and S. Berneti, "Predicting Oil Flow Rate due to Multiphase Flow Meter by Using an Artificial Neural Network," *Energy Sources, Part A: Recovery, Utilization, and Environmental Effects*, vol. 37, pp. 840–845, 2015. https://doi.org/10.1080/15567036.2011.590865.

M. Hotvedt, "On a Hybrid Approach to Model Learning Applied to Virtual Flow Metering," Norges teknisk-naturvitenskapelige universitet (NTNU), 2022.

M. Hotvedt, B. Grimstad and L. Imsland, "Developing a Hybrid Data-Driven, Mechanistic Virtual Flow Meter - a Case Study," *IFAC-PapersOnLine*, vol. 53, no. 2, pp. 11692–11697, 2020.

S. Islami rad and R. Gholipour Peyvandi, "Precise Volume Fraction Measurement for Three-Phase Flow Meter Using 137Cs Gamma Source and One Detector," *Radiochimica Acta*, 108, pp. 159–164, 2019.

C. Kabir and A. Hasan, "A Study of Multiphase Flow Behavior in Vertical Oil Wells: Part II-Field Application," in *SPE California Regional Meeting*, Oakland, California, 1986.

M. Khan, S. Alnuaim, Z. Tariq and A. Abdulraheem, "Machine Learning Application for Oil Rate Prediction in Artificial Gas Lift Wells," in *SPE Middle East Oil and Gas Show and Conference*, Manama, Bahrain, 2019.

P. Omrani, I. Dobrovolschi, S. Belfroid, P. Kronberger and E. Munoz, "Improving the Accuracy of Virtual Flow Metering and Back-Allocation through Machine Learning," in *Abu Dhabi Internation Petroleum Exhibition & Conference*, 2018.

H. Shaban and S. Tavoularis, "Measurement of Gas and Liquid Flow Rates in Two-Phase Pipe Flows by the Application of Machine Learning Techniques to Differential Pressure Signals," *International Journal of Multiphase Flow*, 2014. https://doi.org/10.1016/j.ijmultiphaseflow.2014.08.012.

A. Stief, R. Tan, Y. Cao, J. R. Ottewill, N. F. Thornhill and J. Baranowski, "A Heterogeneous Benchmark Dataset for Data Analytics: Multiphase Flow Facility Case Study," *Journal of Process Control*, vol. 79, pp. 41–55, 2019.

Y. Yan, L. Wang, T. Wang, X. Wang, Y. Hu and Q. Duan, "Application of Soft Computing Techniques to Multiphase Flow Measurement: A Review," *Flow Measurement and Instrumentation*, vol. 60, pp. 30–43, 2018. https://doi.org/10.1016/j.flowmeasinst.2018.02.017.

Section VII

Summary and Future Outlook

18 Machine Learning Applications in Subsurface Energy Resource Management

Future Prognosis

Srikanta Mishra
Battelle Memorial Institute,
Columbus, Ohio

CONTENTS

18.1 INTRODUCTION

In previous chapters, the authors have presented an assessment of the future outlook for ML applications in each of their topical areas. ML-based data-driven models clearly have great potential to add value across the full spectrum of subsurface operations such as reservoir characterization, drilling, production data analysis, reservoir modeling, and predictive maintenance. Key challenges to wider adoption of these approaches come from a general lack of data plus improper curation and management, potential lack of foundational understanding about ML among subject matter experts, and tepid acceptance of black box models from stakeholders.

The authors have also discussed the specific challenges and roadblocks associated with ML applications in their area of expertise. A closer look at the points raised therein suggests that they can be synthesized in terms of several overarching and cross-cutting themes. These issues, and their underlying specifics, are discussed next. The intent here is to point the reader toward a forward-looking and prioritized set of issues that should be addressed to help advance the next generation of ML-based models for subsurface applications.

DOI: 10.1201/9781003207009-25

18.2 DATA ACQUISITION, GENERATION, AND FILTERING

- While acquisition of high-frequency data is becoming common place, the need for comprehensive QA/QC checks, and seamless integration of data from disparate sources with the data analytics pipeline to provide a unified view (preferably in a cloud computing environment) remains challenging. Integration of data streams with intelligent hardware also needs to be more widespread for real-time measurement and control applications.
- Often, field data are sparse and must be supplemented with synthetic data generated from full-physics models before data-hungry ML algorithms can be trained, tested, and deployed. There are few protocols that provide guidelines for such activities with geoscience-specific constraints (e.g., data sparsity/gaps, noise, uncertainty in physics and geology). Furthermore, industry standard datasets for benchmarking machine learning (ML) algorithm performance are lacking.
- In many cases, field data are operationally impacted and/or noisy – requiring significant pre-processing to detect and remove outliers and identify natural groups within the dataset. This remains an unappreciated area in the ML landscape and requires greater attention in terms of automated procedures and standardization of workflows.

18.3 ML MODEL BUILDING AND GENERALIZATION

- The standard suite of popular ML algorithms used for subsurface applications mirror those from other disciplines, viz.: supervised learning models (e.g., decision tree, random forest, gradient boosting machine, support vector machine, artificial neural networks), and deep learning models (e.g., convolutional neural networks, long short-term memory). Of particular promise beyond these "usual suspects," especially for model generalization, are (1) transfer learning, i.e., applying models from a data-rich environment to another setting with limited data, and (2) reinforcement learning, i.e., where software agents learn to modify their behavior by performing certain actions and observing the consequent results/rewards.
- Other subsurface-oriented issues to consider include: (1) broader adoption of physics-informed ML (or data physics) formulations; (2) hybrid models that integrate ML-based model elements in a physics-based construct; (3) robust approaches for data-driven discovery of underlying physics rather than simply fitting an input-output model; (4) computationally efficient strategies for spatiotemporal dimension reduction (especially in state variables such as pressure and saturation); (5) the ability to train from, and make predictions across, multiple geologic scenarios in a probabilistic framework (similar to geostatistics); and (6) renewed focus on model interpretability vis-à-vis the underlying geoscientific processes/parameters.

18.4 MODEL DEPLOYMENT AND STAKEHOLDER ENGAGEMENT

- A key challenge for practical situations is life-cycle model management of real-time ML models (MLOps), especially with respect to effective and automatic model (re)training at periodic intervals as needed without human intervention. Another issue is the possibility of using automated machine learning (AutoML) to build/train ML models which would free subject-matter experts (SMEs) from the drudgery of model building to focus on operational optimization. However, the usual caveats of "*understand, then apply*" continue to be relevant.
- One unrealized aspirational goal has been the routine use of ML-assisted models for real-time operational control and optimization in a closed-loop environment. As things stand, most ML models appear to be deployed in a shadow or advisory mode. The frequency with which real-time field observations can be regularly assimilated to update and re-train models (whether purely data-driven or hybrid ML and physics-based), the transferability of learnings from pre-trained models to a different set of operating conditions, the issue of model certification (i.e., is the model "good enough" for deployment without manual intervention), and automated agent-based reinforcement learning from past actions and their consequences are issues that need to be addressed for greater utilization of ML-based closed-loop control.
- Management and decision-maker buy-in will continue to be an important choke point for widespread acceptance and adoption of ML in subsurface operations. Unreasonable expectations about data-driven model accuracy, fear about basing decisions on model predictions that are not transparent, unwillingness to change existing physics-based protocols, and hesitation in moving from a scientific modeling paradigm to a data-driven operational cloud service requiring new funds (without proper quantification of economic benefits of ML applications) are all key barriers. These need to be addressed at the organizational level to bring about a change in culture that is more accepting of ML or ML-assisted physics-based modeling while being cognizant of its limitations.

As we look ahead at the next 3–5 years, one can anticipate multiple advances in all facets of ML-based applied subsurface data analytics, i.e., data management and organization, analytics and knowledge discovery, and visualization and decision support. At the risk of having to articulate some possible outcomes suggested by a cloudy crystal ball, here are some (*fearless*?!) predictions about what we are likely to encounter in the near future:

1. Better real-time data ingestion, intelligent storage and retrieval protocols, and ML-based data cleanup/filtering that is integrated with next-generation hardware and data lakes/warehouses;
2. Greater application of automated model building, testing and re-training in conjunction with closed-loop operational control systems;

3. Additional emphasis on physics-embedded, physics-assisted, or hybrid data-physics model formulations vs using off-the-shelf software routines;
4. Improved focus on model generalizability (especially for heterogeneous systems) using transfer and reinforcement learning;
5. More emphasis on model interpretability, visualization, virtual learning, communication of ML caveats, and articulation of economic/value-added benefits from ML-based tools and workflows to better engage managers and decision-makers; and
6. Broader integration of foundational and applied data science concepts into petroleum engineering and geoscience curricula to produce better-informed SMEs.

Onward and upward then, as we look forward to a new era of ML-aided subsurface energy resource management related problem-solving!

Index

O

P

R

S

T

U

V

W

X